Globalization and the Least Developed Countries

Globalization and the Least Developed Countries

Potentials and Pitfalls

David Bigman

www.cabi.org

CABI is a trading name of CAB International

CABI Head Office
Nosworthy Way
Wallingford
Oxfordshire OX10 8DE
UK

Tel: +44 (0)1491 832111
Fax: +44 (0)1491 833508
E-mail: cabi@cabi.org
Website: www.cabi.org

CABI North American Office
875 Massachusetts Avenue
7th Floor
Cambridge, MA 02139
USA

Tel: +1 617 395 4056
Fax: +1 617 354 6875
E-mail: cabi-nao@cabi.org

A catalogue record for this book is available from the British Library, London, UK.

Library of Congress Cataloging-in-Publication Data
Bigman, David.
Globalization and the least developed countries : potentials and pitfalls / by David Bigman.
p. cm.
Includes bibliographical references and index.
ISBN 978-1-84593-308-1 (alk. paper) -- ISBN 978-1-84593-309-8 (ebook)
1. Rural development--Developing countries. 2. Agriculture--Economic aspects--Developing countries. I. Title.

HN981.C6B55 2007
303.48 ' 21724--dc22

2007016922

ISBN-13: 978 1 84593 308 1

Typeset by SPi, Pondicherry, India.
Printed and bound in the UK by Biddles Ltd., Kings Lynn.

Contents

Acronyms

ACP	African, Caribbean and Pacific countries
ADB	Asian Development Bank
AEC	African Economic Community
AfDB	African Development Bank
AGOA	US African Growth and Opportunity Act
ATC	Agreement on Textiles and Clothing
CIF/FOB	cost, insurance, freight/free on board margin
COMESA	Common Market for Eastern and Southern Africa
CSOs	Civil Society Organizations
EA	East Asia
EAC	East African Community
EBA	Everything but Arms
EPAs	Economic Partnership Agreements
ESAF	Enhanced Structural Adjustment Facility
FDI	foreign direct investment
FTA	Free Trade Area
GATT	General Agreement on Tariffs and Trade
GNI	Gross national income
GNP	Gross national product
GSP	Generalized System of Preferences
HIPC	Highly Indebted Poor Countries Initiative
HYVs	high-yielding varieties (of seeds)
IDA	International Development Association of the World Bank Group
IMF	International Monetary Fund
ISO	International Organization for Standardization
LAC	Latin America and the Caribbean
LDCs	least developed countries
MDB	Multilateral Development Bank

MDGs	Millennium Development Goals
MDRI	Multilateral Debt Relief Initiative
MFA	Multi-Fibre Arrangement
MFN	most-favoured nation
NEPAD	New Partnership for Africa's Development
NTBs	non-tariff barriers
NYT	*New York Times*
OAU	Organization of African Unity
ODA	official development assistance
OECD	Organization for Economic Cooperation and Development
PBA	performance-based allocation system
PRSP	Poverty Reduction Strategy Programme
PTAs	Preferential Trade Agreements
RTAs	Regional Trade Agreements
SA	South Asia
SSA	sub-Saharan Africa
TFP	total factor productivity
WTO	World Trade Organization

Preface

Globalization, in the wider sense of the term, ranging from the structure and organization of production and trade to new technological innovations, is perhaps the only subject on which you can read and learn a lot, not only in professional literature, but also in the daily newspaper. So much is happening, at such a rapid pace, in so many parts of the world and with such a profound impact on the lives of nearly every person in nearly all countries, that it has become more than one of the subjects that economists are working on, and a full-time occupation of hordes of writers in all media and all means of communication. I could even claim that I cannot see a subject in today's world as important and as interesting as this, but I am very much aware that all economists who work on all other subjects within the broad area of economics would make the same claim.

I was attracted to this subject during my work on food security and poverty that started at the World Bank quite some time ago. It was therefore almost predictable that my interest in this subject would focus on its impact on the least developed countries (LDCs). This is my ninth book, but the second on globalization. My first book also focused on these countries. At the same time, this focus may seem almost arcane, because these countries were affected the least by globalization. The fact that most of the LDCs and the majority of their population remained on the sidelines of the major developments that took place in the global economy and benefited so little from the huge affluence that globalization has brought to the world's economy, requires the special attention of all of us because of its very profound impact on our civilization.

I came to know and became interested in the developing countries during my work at the IMF, the World Bank and ISNAR-CGIAR. During my work in these organizations I visited many countries and became fascinated at the wide diversity of cultures and civilizations, and came to know many people, many of whom I had contact with for many years and some until today. They helped in the work on this book as much as my colleagues who helped me with professional advice, and commented very methodically on drafts or in seminars. I taught courses on this subject for several years at Wageningen University, Ben-Gurion University and the Open University in Israel. I sponsored the work of students who came from many developing countries and I can say that I have learned from them as much as, I hope, they have learned from me.

I will remain indebted to, and ask forgiveness of, all those whose names I may have omitted while writing this preface and may remember only later. I would also like to thank the Wageningen Universiteit Fonds that supported me during my stay at Wageningen University and the Stanford Center for International Development for a very enjoyable summer I spent there.

I am greatly indebted to my wife Petra who helped throughout my work on this project and, on top of all that, took the time to go through the manuscript and edit it thoroughly. I dedicate this book to Petra in recognition not only of her great help but also of my good fortune in sharing my life with her.

I would also like to thank Meredith Carroll, Rachel Cutts and Claire Parfitt from CAB International for their great assistance during this entire process and during my repeated but inevitable delays in sending the manuscript. I am grateful in particular to Meredith for her good advice and artistic choice in giving the book its final touch.

The biggest problem with the work on this subject is that as soon as I finished the manuscript, the next day's newspaper came, or the weekly magazine or one of the many journals, and then I realized how many subjects I had left out, which made it absolutely necessary to start all over again or perhaps give it another try. I hope that the readers of this book will share the revelations I had and the perspective it gave me when I wrote it.

David Bigman

Introduction and Overview

In our rapidly changing world, globalization has become one of those emotionally loaded words, like 'communism' or 'capitalism', that imply much more than a technical definition of relations between people and commodities; indeed, these terms challenge people to take a position: Is it 'good' or 'bad'? Is it 'desirable' or 'undesirable'? Compare this to concepts such as the 'balance of payments' or the 'gross national product' that clearly represent accounting summations of the total value of products traded or produced in a given period of time – with respect to which the question as to whether one is 'for it' or 'against it' is obviously meaningless.

Some of the emotions associated with the phenomenon of globalization are unloaded at the annual meetings of the International Monetary Fund (IMF) and the World Bank which bring together Ministers of Finance from all countries to discuss world trade and the advancement of globalization. Once upon a time, these meetings passed nearly unnoticed, but today, they bring together large crowds of demonstrators outside the meeting halls to express their antagonism by shouting slogans against globalization and against whatever the dignitaries are discussing inside the halls. There are similar passionate protests at the G8 summits and the meetings of the World Economic Forum in Davos, Switzerland, and often, the rhetoric of these demonstrations echoes rousing revolutionary calls to fight for justice and equality.

However, inside the meeting halls, the delegates are discussing globalization as a process that has allowed millions of people in many countries all over the world to rise up from the abyss of poverty and reach a standard of living that restored their human dignity and gave them hope for an even better future. The opportunities given by globalization to these people would seem ample justification for the hope and belief that, over time, globalization can and will bring similar benefits to many more people everywhere.

The passionate protesters outside the meeting halls have a very different view: for many of them, globalization is almost an incarnation of evil, and the economic principles that they see as the basis for globalization are to them some version of 'satanic verses'. They blame globalization for lost jobs, insecure incomes and reduced freedom for workers and employees, the decaying environment, the exploitation of young children and women in poor countries, the flow of migrants from the poor developing countries and the rising inequalities as transnational corporations and their CEOs amass ever more wealth and unprecedented power.

The protests against globalization and the institutions that promote it tend to see in globalization the source of all malaise and ignore its great potential and the contribution it has already made to raise the standard of living of many millions of people to levels of affluence that humankind has never known before. At the same time, protesters are fully justified to blame globalization for the rise in income inequality between the working people and the affluent elites. The proliferation of outsourcing, the intense competition of cheap imports from the developing countries, and the migrant workers taking their jobs, lowering their wages and reducing the benefits that local workers fought so hard over many years.

Economists tend to have a longer time horizon and focus on the benefits that the economy as well as these working people will have in the long run. The hardships brought by globalization and the adjustments that the economy must make are temporary and the inevitable price of the transition to a 'new equilibrium' in which everybody will be better off. Over time those who lost their jobs will learn new skills, find new jobs and benefit from higher incomes and cheaper consumer products, and eventually the standard of living of everyone will rise.

The bright side of globalization is symbolized most vividly by the spectacular economic development of the East Asian (EA) countries; indeed, the region's countries – particularly China and, more recently, also India – managed to raise their standard of living to levels that, just a generation ago, would have seemed like a dream. The sharp reduction in the incidence of poverty in China from well over 50% of the population in 1980 to well under 10% just two decades later is the most dramatic manifestation of globalization's potential to better the lives of millions. Large numbers of poor people, particularly in Asia, have already benefited from the technology-driven spread of more open and better connected relations between people and countries, and this integration into global markets has lifted millions out of poverty.

Although the positive force of globalization has already been proven by the almost miraculous growth of the EA countries, there is a growing opposition in many countries, both developing and developed, to some of the principles, mechanisms and institutions that are seen as the main pillars of globalization. Trade liberalization and multilateral trade agreements brought huge benefits to consumers all over the world and created millions of new jobs, but they also have a dark side and are blamed for lost jobs and stagnant or even declining wages. In many developing countries, trade agreements, especially the agreements coordinated by the World Trade Organization (WTO), are perceived as compromising their sovereignty to set their own priorities and economic policies. In many developed countries, political parties have taken up the fight of the anti-globalization protesters by advocating stiff restrictions on trade liberalization and foreign migrants. Labour unions and some non-government organizations (NGOs) have become more radical in their protests against globalization due to their growing perception that free trade does not work in their favour, and that both they and the poor in the least developed countries (LDCs) are being taken advantage of by transnational corporations and countries that take away their jobs.

The most passionate protests against globalization are fuelled by the sharp rise in inequality, both within and between countries. The rise in inequality is most visibly manifested by the accumulation of unimaginable wealth in the hands of a narrow corporate elite and the super rich, and by the growing gap between the LDCs and the developed countries.

The leaders of the developing countries and their economic ministers or trade representatives to meetings of the international or the regional development organizations protest the fact that world trade is not really free; some countries tightly restrict their imports and the developed countries promote their exports, particularly of agricultural products, with various subsidies that are damaging for the LDCs. The collaborative framework of the multilateral trade agreements has failed with the collapse of the negotiations at the Doha Round, and the promise for a level playing field in which all countries have free and equal access to all markets is, in practice, grossly violated. Prices are manipulated by monopolies, markets are not competitive and many countries still do not open their economies but impose high tariffs or various administrative restrictions, including food safety standards. The transnational corporations and the rich countries are the 'imperialistic' or 'capitalistic' powers in disguise that take control over their economies under the pretext of free trade, exploit their natural resources and their workers and pay their farmers a pittance for their agricultural products, on which these corporations make millions.

The rising global tide is raising most boats, but yachts are rising much faster and higher than all other boats, and that tide is leaving behind many rafts loaded with poor people who try to find a better life, perhaps in Europe's promised shores, but many of them would not be able to make it.

The nearly one billion people who still live in chronic poverty despite the great affluence that globalization has brought to the majority of the world's population are of great concern to the entire international community. In 2000, the representatives of all countries that are members of the United Nations (UN) declared their commitment to the Millennium Development Goals (MDGs) aimed at reducing the level of global poverty and raising the living standards of the poor. But while poverty is still widespread in a large number of countries in all continents, including some of the more developed countries, the core of the global poverty problem is in the LDCs, particularly in sub-Saharan Africa (SSA). In these countries, the number of poor people even increased since 2000, the economies remained stagnant and their prospects of achieving the MDGs are very dim.

The largest number of poor people still concentrates in the Asian countries, particularly in China and India, but the rapid growth of these countries and the sharp reduction in their poverty in the last two decades make it highly likely that these countries, as well as their poorer neighbours, including Vietnam and Bangladesh, will be able to meet the MDGs and, over time, resolve their poverty problem. In some African countries there are promising signals that they will also manage to end their prolonged stagnation and resume their growth. However, even in these countries poverty remains rampant and the African countries plus the nearly 20 other LDCs will remain the hard core of world poverty.

An analysis of the world poverty problem, the policy options to reduce poverty and the stumbling blocks that prevent countries to achieve this goal must therefore focus first and foremost on the LDCs. This is also why in this book I have elected to narrow down the analysis from a general analysis of global poverty to a more focused analysis of poverty in the LDCs. At the same time, I expanded the scope of the analysis in order to highlight the wider problems that these countries are facing, examined the impact of globalization on their economies and evaluated their policy options and the obstacles that they must overcome in order to reduce their poverty in the coming decade.

The book evaluates the policies of the LDCs and the decisions that they now face against the backdrop of the changes in the structure of the global economy and in the globalization process itself. The analysis in the book pays particular attention to the following trendsetting changes:

- The changes in the structure and organization of world production and trade with the expansion of world trade across all countries and its restructuring as an effect of the regional trade agreements and the incentives they provide for trade diversion.
- The growing dominance and greater monopolistic power of transnational corporations and their impact on the distribution of production across countries with outsourcing and offshoring and the mounting flows of capital between countries.
- The transition from a bipolar world economy that was dominated by the USA and the EU in the second half of the 20th century to a multipolar global economy with the growing economic power and influence of China, India and the large Latin American (LA) countries. In the coming years, Russia is also likely to join this club.
- The declining power and influence of the multilateral trade agreements and the WTO that became obvious with the failure to reach an agreement at the Doha Round, and the decline in the power, influence and resources of the multinational development organizations, the IMF and the World Bank. Although these changes started to unravel in the mid-1990s, they are still ongoing and are likely to have a greater impact in the coming decade and shape the next phase of globalization.

A large number of books and articles have already been written on the globalization process and its impact on the economic, political and social relations between countries and people, as well as on the developing countries and on world poverty. The book is obviously not able to provide a thorough review of all this literature. Instead, its more moderate objective is to review the main writings on the impact of the globalization process on the developing countries and on world poverty and to focus this review on the implications for the LDCs. The book provides a survey of the pros and cons of the globalization process for the developing countries with a specific focus on the large changes that the globalization process has been going through in the recent years.

The book is not aimed, therefore, at providing an evaluation of the past policies or the programmes that were implemented either by the countries themselves or by the international development organizations in order to deal with the poverty problems. Nor is it aimed at evaluating the impact of the globalization process of the last two decades on the developing countries or on whether globalization has been 'pro-poor'. Although these two subjects are examined in the book, the main focus is on the lessons from these experiences with respect to the policies that the LDCs should implement in the next wave of globalization.

The focus on the near and more distant future is critical because considerable changes have taken and are still taking place in a number of areas that shape the entire globalization process and these changes are likely to continue with the deepening impact of the trendsetting developments mentioned above. Many of the lessons from the experience of the 1980s and the 1990s may therefore be less relevant in the coming years and policies that proved to be effective in the past, including the policies that have driven the EA miracle, may not be equally useful for the LDCs in today's and tomorrow's global economy.

The book analyses possible scenarios and alternative policies which are likely to affect the LDCs and their poor population in light of the changes that are taking place in the globalization process, its organization and its institutions in order to draw lessons for future policies of the LDCs. This is also the background against which the lessons that can be drawn from the failure of the LDCs to integrate in to the global economy and accelerate their growth and the lessons from the highly successful policies of the EA countries are evaluated. The book examines these developments and policies against the backdrop of the changes in the global economy,

As a prelude to this analysis, Chapter 1 reviews the debate on globalization. A comprehensive survey of the different views on globalization is, of course, mission impossible, because of the wide range of topics covered and the even wider range of opinions expressed. The chapter therefore concentrates on the economic dimensions and polar views on the merits and demerits of globalization. To represent these polar views, the chapter concentrates on the writings of a small number of leading economists that have been the leading voices in this debate. The two books that are, perhaps, the most prominent representatives of the two polar views are the book of Joseph Stiglitz, *Globalization and its Discontents*, and the book of Jagdish Bhagwati, *In Defense of Globalization*. Since writing his book in 2002, Stiglitz wrote two more books, and one of them is also reviewed in the chapter, as well as many articles in the different arenas that the modern means of communication made available, and some of them are also icluded in this survey. Bhagwati has, undoubtedly, led the defence on globalization. Among economists outside the World Bank and the IMF, he was admittedly in a minority, since defence is always the more difficult and less rewarding task. Another prominent voice was that of Dani Rodrik from Harvard University who is one of the most prolific and knowledgeable writers on this subject. There is a clear common denominator that connect the writings of Stiglitz, Bhagwati and Rodrik: They are all concerned with the impact of globalization on the poor and are alarmed by the developments of the past decade that underscored the disadvantages of the poor countries and poor people in the global and intense competition that came to characterize the globalization process.

The debate on the impact of globalization on the poor and on the question whether globalization has been 'good for the poor' has become much more passionate after the publication of the famous article by Dollar and Kraay (2002), who presented a very thorough but controversial analysis on the impact of globalization on the poor. Dollar and Kraay came to the unequivocal conclusion that the number of the world's poor has declined and that globalization has indeed been good for the poor. Later studies challenged this conclusion and showed that the reduction in the world's poverty was primarily due to the steep decline in the number of the poor and in the share of the poor population in China, whereas in the vast majority of the other developing countries poverty remained unchanged or even increased. Chapter 2 analyses the effects of globalization in a multi-country framework and shows that some of the key conclusions that have been derived in the theory of international trade may not be valid when there are simultaneous changes in the income distribution between and within countries. Although a number of studies have demonstrated that there can be, and indeed has been, a decline in the global income inequality, there has also been an increase in income inequality between countries and within all countries. The second part of the chapter concentrates on the impact of globalization on the LDCs and on the reasons for the decline in their income in the last two decades that brought so many of them to the bottom of the global income ladder.

Chapter 3 provides a more general analysis that shows how the changes in the 'identities' of the poor as an effect of different growth rates of different countries and subgroups within countries affect both the measure of global poverty and the measure of global income inequality. The chapter also shows that a staightforward extension of the basic results of the classical analysis that international trade is a win-win strategy benefiting all trading countries and income groups may not be valid in the multi-country multi-income groups framework when different groups have different growth rates. Against this background, the chapter again examines whether globalization is 'pro-poor'. There have been different definitions of 'pro-poor' growth and the answer to the question on the impact of growth in the multi-country framework depends both on the rates of growth between and within different countries and on the specific definition that is used.

Chapter 4 reviews the development policies that have been promoted by the Washington Institutions and implemented within the framework of the structural adjustment programmes. The guiding principles of these programmes have been the very controversial set of policies of the 'Washington Consensus' and the World Bank itself went through an introspective re-examination of these policies in light of their failure to accelerate growth or reduce poverty. The chapter examines the lessons from the experience with these policies in light of the very ssuccessful experience of the EA countries that implemented policies that were diametrically opposed to the policies recommended in the Washington Consensus. This review emphasizes the very different social and political structures in the SSA and LA countries on the one hand and the EA countries on the other that explain a great deal of the differences in their economic performance. The lessons from the failure of the structural adjustment programmes must therefore examine not only the principles of the Washington Consensus, but also the lack of consideration in the design of the programmes for the conditions in the countries and the need to accompany the economic policies with additional measures, including measures to strengthen their system of governance; the lack of flexibility to adjust the policies from the outset so that they will be more suitable to the conditions in the countries, and the lack of flexibility to adjust the policies to changes in the conditions in the countries.

A key principle underlying the Washington Consensus that remains a fundamental guide for the policies promoted by the international development organization is the essence of free trade and the desirability of open trade policies for economic growth. The neoclassical economic theory emphasizes the importance and great benefits of open trade, and the very successful experience of the EA countries and later also India is brought as an undisputed proof of these benefits. Chapter 5 evaluates the benefits from open trade for an LDC against the backdrop of the successful experience of the EA countries and against the more general backdrop of the debate between inward-looking and outward-looking policies. In principle, this debate seems to have been decided in a knockout by the debt crisis and is no longer an issue in development economics. However, there are still a number of questions about the lessons from the trade policies of the EA countries: How outward-looking were, in fact, the policies of the EA countries? How suitable would these policies have been for the LDCs given the wide difference between the political and economic systems in these two groups of countries? How suitable are these policies today, given the changes in the global economy and the very strong influence of China and India on world trade, particularly in labour-intensive products? What are the lessons with respect to role of government in the economy given its very decisive influence in the EA countries? Could the government have an equally effective role and impact in the LDCs of SSA? What are the implications of the decision of these countries in the recent years to reduce government interventions and increase the role of the private sector?

Finally, Chapter 6 concentrates on the lessons for the countries of SSA. It discusses several characteristics that distinguish these countries from the other low-income countries and LDCs and that affected their growth in the last two decades. In light of their past experience and their social, geographic and political characteristics, the chapter examines alternative future scenarios and the effectiveness of alternative policies. This analysis takes into account the wider changes in the global economy that affect their future prospects. What are the implications of the institutional changes in world trade and the damage to the multilateral trade system and to the LDCs of the collapse of the trade negotiations in the Doha Round? What are the implications of the formation of regional trade agreements? Several projections of the recent years see dismal prospects for the future growth of the African countries and the potential for their integration into the global economy. On the other hand, the entire international community has recognized its obligation to help the poor in these countries and declared a joint commitment to come to their help in the Millennium Development Programme. The EU also initiated a far-reaching programme to help these countries in the EU Economic Partnership Agreement. The potential impact of these and other aid programmes are in grave doubt and there may be a need to reassess their design.

The growing divergence between the LDCs and the developed and emerging economies also symbolizes a divergence of the debate about the impact of globalization in the coming years. On the one hand, the majority of the LDCs are likely to remain embroiled in deep poverty and the entire international community will have to help them climb out of this abyss. On the other hand, in the developed and emerging economies, the rising income inequality may lead to internal conflicts that may cast a shadow on the major achievements of the globalization process. These potential effects of globalization must be taken into account as early as possible in order to ensure that most people will be able to benefit from this progress.

1 Diverging Views on Globalization

1 Introduction

In the last two decades, globalization has been the dominating force that shaped the world economy. Although there have been periods in the past when world trade grew at higher rates, most notably in the 19th century with the increase in trade between the 'new world' and the 'old world', the current process of globalization is unprecedented. World trade has reached nearly all countries in all continents; the most rapid increase has been the trade between Europe, the USA and the Far East. A century ago, trade with the Far East was only a trickle and provided mainly material for legends. The rise in world trade was made possible by a number of factors, such as the sharp drop in transport costs and transport time across sea and land, the vast improvements in the logistics of moving goods between and within countries, as well as the steep reduction in the costs and the huge improvements in the means and speed of communication. These changes are truly amazing, and today it is almost impossible to imagine how people and enterprises could function earlier. The rise in world trade was further accelerated by vast changes in the structure and organization of production and trade along the supply chain from producers to consumers. These changes were driven by unprecedented technological progress and many new scientific discoveries; in addition, the mode of operation in the management in large corporations changed and led to the growing dominance of transnational corporations that now control large segments of the production and supply chain. Finally, there have also been very significant changes in the rules and institutions that govern world trade, most notably the General Agreement on Tariffs and Trade (GATT) and the World Trade Organization (WTO) that reflect and represent the high level of multilateral collaboration that has been achieved.

Clearly, this is only a partial list of the changes that took place, and it is often not all that easy to distinguish if the changes were an effect or a consequence of globalization. In any case, globalization not only revolutionized the structure and methods of production and communication and transformed the lives of millions around the world, but also ushered in a new era of evermore rapid changes both in the methods of production and in the organization and management of production and supply that will continue in the coming years and will shape the world at large, and thus affect the life of each and every person. If it is difficult for us to recall how people and enterprises could function a generation ago; it is even more difficult to imagine how they will function in the future and how the world will look a generation from now.

All these developments had a profound impact on the world economy. Most notable was the rapid economic growth and the enormous improvement in the standard of living of a very large portion of the world population in both the developed and the developing countries. However, not all countries benefited equally from this growth, and even in countries that did grow, not all people were able to improve their standard of living. Particularly striking are the wide variations between different subgroups of developing countries with respect to their gains from global growth and their success in becoming integrated into the global economy. Although some developing countries managed to grow very rapidly and became the world's emerging economies, taking on the role of the driving force for the entire globalization process, other developing countries remained stagnant and drifted to the bottom of the global income ladder to become the world's least-developed countries (LDCs). As a result, there has been a sharp increase in income inequality between countries. Moreover, in many developing countries, including those that experienced rapid growth, a large segment of the population did not benefit from that growth and many remained chronically poor; consequently, there has also been a large increase in income inequality within these countries.

A recent World Bank report entitled *Global Economic Prospects: Managing the Next Wave of Globalization* highlights three features that are likely to be most prominent in the next wave of globalization: a growing significance of the developing countries, particularly the emerging economies, in the world economy; the rising productivity in all segments of production and along the global supply chain; and rapid technological changes in all industries and segments of production with an accelerated diffusion of advanced technologies to more countries. Moreover, the decline in costs and large improvements in the means of communication, changes in rules of property rights and the abolition of many restrictions will further strengthen the relations and the collaboration between enterprises in all countries. Considering this list, we would have to conclude that we are already in the midst of the next wave of globalization (Rodrik, 2002; Hausmann *et al.*, 2005a).

The trends that characterized the changing relations between countries and between population groups and regions within countries will continue to dominate the next wave of globalization: the standard of living of a large and growing share of the world population will continue to rise at a rapid pace, but income inequality will also continue to rise in and between countries. The most significant development will be the convergence between the emerging economies and the developed countries that started in the 1990s. That convergence will include larger segments of the population in the emerging economies that will benefit from improved education, more advanced methods of production and better technologies which will raise their productivity and consequently also their income. The other significant, and ominous, development will be the growing divergence between the LDCs and nearly all other countries, and, consequently, the deepening gap between their standards of living. This means that in a world of unprecedented affluence and unparalleled levels of production, hundreds of millions of people may remain undernourished, undereducated and in poor health (Hausmann *et al.*, 2005b).

The growth in the world economy will be driven by the exceptionally high-growth rates in the Asian economies, and it will accelerate as these countries create their own cadre of scientists and technicians that will help them to push forward their technological progress. Africa, with the exception of South Africa, is not likely to be integrated into the next wave of globalization, and therefore it is expected that most of Africa will continue to trail behind and remain the continent that will benefit least from globalization and the rise in the standard of living.

However, some African countries have made considerable progress in improving their system of governance and the functioning of their institutions and they may also be able to build an efficient and competitive industrial sector by taking advantage of their natural resources and abundance of cheap labour. But many other African countries may remain embroiled in internal conflicts and bogged down by ineffective systems of governance and rampant corruption. The core of global poverty in the coming years will be in these countries. They will have great difficulty in building local industries due to low levels of foreign and domestic investments, poor infrastructure and lack of managerial experienced and skilled workers; in addition, these difficulties are likely to be further aggravated by problems at the macroeconomic level due to inefficient and corrupt public institutions and the lack of the necessary private institutions, primarily an efficient financial system (Hausmann and Rodrik, 2003).

African countries will also have to face new challenges in their efforts to build an industrial base, since the initial stages of industrialization that traditionally started with labour-intensive textile and apparel industries are not likely to continue their familiar cycle from more developed to less developed countries and, in the foreseeable future, these industries are not likely to migrate from Asia to Africa. The reason is that the demand for low-wage workers will be met by a seemingly endless supply of low-skilled labour in the Asian countries themselves that will remain an inexhaustible source of cheap labour, and these industries will therefore remain in China, India, Pakistan, Bangladesh and the other countries in the region for many years to come. The African countries will also be negatively affected by a brain drain due to the migration of many of the trained and more educated people who will leave to the more developed countries in search of higher income and better employment opportunities.

2 The Background of the Debate on the Merits of Globalization

To gain a better perspective of the background for the conflicting views and the heated debate on the merits of globalization, Table 1.1 presents selected indicators of the standard of living in the main developing regions. The most striking observation that emerges is the very large difference between the welfare indicators in sub-Saharan Africa (SSA) and South Asia (SA), and between indicators in the other regions. The extremely low indicators of living standards in SSA, ranging from income per capita to life expectancy and child mortality, are the most outstanding; another discouraging indicator of the developments in the African subcontinent is the decline in the average income per capita during the 1980s and 1990s, while in all other regions incomes were rising – though at very different rates. Other indicators, not included in the table, also show that the living conditions in SSA deteriorated very significantly in the last two decades; in particular, life expectancy has dropped drastically, largely due to the spread of the AIDS epidemic, and child mortality has increased.

The table also highlights the large differences in the annual growth rates of income between regions. These differences reflect in part the negative impact of the debt crisis on the countries of SSA and LA. However, after the 'lost decade' during the 1980s most LA countries, particularly the larger ones, managed to overcome the crisis and returned to a more robust growth during most of the 1990s, whereas nearly all SSA countries had another decade of stagnation and decline that brought them to the bottom of the world income distribution and widened the income gap between them and the emerging economies in East Asia (EA) and the Pacific.[1]

In 1970, the average income per capita in China (measured in real prices and adjusted for the cost of living) was 30% *lower* than in SSA; in 2001, the average income per capita in China was more than 2.5 times *higher* than in SSA. The comparison between the stagnation in Africa and the rapid growth in most EA countries provides the most prominent example of the gains that countries can have by opening up their economy to trade and by adjusting the structure of their production and trade to the structure of world production and trade. Most African countries did not make these adjustments. As a result, Africa, which in 1970 was home to about 10% of the world's poor, and Asia, which was home to more than 75% of the world's poor, found themselves in a very different situation 30 years later: Africa had more than a third of the world's poor, while Asia reduced its share to just 15% despite its much larger share in the world population (World Bank Report, 2000, 2001).

Globalization can therefore be viewed from very different perspectives. On the one hand, globalization has changed and continues to change the way societies operate and interact and the way people work, think and communicate. For many countries and people, globalization offers dream opportunities for success and prosperity. With the unprecedented increase in trade and

[1] The group of the EA countries also includes the relatively more advanced countries of South Korea, Taiwan, Hong Kong and Singapore (the 'four tigers').

Table 1.1. Selected development indicators for the main developing regions.[a] (From World Bank, 2003.)

Region	GNI per capita, 2001 (US$)[b]	Average annual growth in GDP per capita, 1980–2000 (%)	Life expectancy at birth, 2001 (years)	Under-5 mortality rate, 2001 (deaths per 1000 live births)	Average annual growth in population, 2001 (%)
Tropical sub-Saharan Africa	271	−1.1	46.0	172.5	2.3
South Asia	449	3.3	62.5	95.3	1.7
Latin America	3669	0.5	70.5	32.7	1.4
East Asia and the Pacific	3710	6.4	70.2	38.3	0.8
Middle East and North Africa	2207	0.9	68.4	49.8	2.0

[a]Population-weighted averages of countries in the region.
[b]GNI, gross national income.

trade-driven growth, globalization provides huge benefits that include not only material gains but also greater scope for aspirations and inspirations. The wider world view and vision is due to much better education, greater openness and an exposure to far better and more varied information. On the other hand, however, globalization has also widened the gap between countries and between people, and it has been particularly damaging for the people in the LDCs, for whom it has exacerbated poverty, misery and hopeless stagnation.

In the debate about globalization, one of the fundamental notions is that globalization is identified with free trade and should be perceived as a multi-country realization of the familiar free-trade model of David Ricardo that shows the gains all trading countries can have if they open their economy to trade. Although import-competing industries may have losses, exporters who have a comparative advantage in production will gain and the country at large, as well as the global economy, will have net gains. Globalization is thus a multi-country realization of 'laissez-faire' economics in which all markets operate most efficiently and all individuals maximize their gains, whereas the protection of the local market by restricting trade when the local markets are non-competitive will lead to a reduction in efficiency which will also reduce the incentives for trade and globalization (Rodrik, 1997).

The principles of the pro-trade and pro-globalization strategy that has been espoused by the Bretton Woods institutions, i.e. the International Monetary Fund (IMF) and the World Bank, are based on the neoclassical vision of open and free trade that motivates countries to specialize in their most productive industries and achieve the most efficient allocation of their resources. Openness to trade also enhances the welfare of the trading countries' populations and reduces their poverty by raising incomes and by giving them a wider choice of quality products at lower prices (see, e.g. IMF, 2001). The prime achievement in the last 20 years that is attributed to globalization is the unprecedented improvement in the standard of living of many millions of people all over the world, made possible by the rise in world trade and the accelerated world growth that increased incomes and raised many of the poor from poverty, hunger and hopelessness, and enabled them to improve their living conditions and develop aspirations for a much better future.

However, these achievements have not been equally shared, and a large number of countries did not benefit from globalization, their populations were unable to improve their living conditions and many of them became even poorer. In many other countries, most of the gains of globalization were concentrated in the hands of a relatively small elite. These wide variations in the impact of globalization on different countries and different population groups provide the background for the heated debate among economists and policy makers on the trade policies and the development strategies that countries should pursue in order to benefit from globalization and enable their poorer population to share these benefits.

Of fundamental importance for this debate is the realization that, in practice, the model of laissez-faire economics has to be left to textbooks that often introduce a more realistic picture of the world only in later chapters. Indeed, in the 'real world' markets, production and trade do not work as smoothly and efficiently as the models envisage, due to unavoidable market imperfections and information limitations that give different agents different market power that they use to their advantage. A free market also cannot guarantee that the rules of law and order will be secured or that institutions will be effective. As a result, free markets do not really provide a level playing field, open trade may not avoid monopolistic controls over large segments of the market and trade protections are therefore alive and kicking. Open trade without any interventions cannot guarantee that the allocation of resources will be efficient and in accordance with a country's comparative advantage; nor can it secure the most effective use of the economy's productive capacity and resources.

For all these reasons, the government has an important role in the economy, and its involvement is essential to set the rules and monitor their implementation in order to minimize market distortions and put restrictions on individual producers and consumers that may try to take advantage of their market power and better information to the disadvantage of others; only when the government fulfils this role can it guarantee that markets would work more efficiently. Successful economic policy has never been laissez-faire, leaving the market forces to operate freely while the government remains on the sidelines. Quite the contrary, a successful economic policy has always been characterized by extensive efforts to ensure that profit-driven market agents will operate in ways that provide the highest possible benefits for all, even though it is clear that such efforts constitute an intervention in the market activities and thus reduce their efficiency. Therefore, these interventions may not secure the best possible outcome and a choice may have to be made between the distortions introduced by the government and those created by a non-competitive market (Rodrik, 2004).

In the trade between countries, there is no global authority that has the power across the entire world market that is equal to that of a government within a country's market; thus, there is no global authority that can set rules that will be binding for all countries and that can enforce their implementation. The rules and their implementation in the world market can only be trade agreements that the governments of sovereign states agree to accept and abide by voluntarily, and these agreements are therefore negotiated according to a quid pro quo approach. There are, of course, pressures that countries, especially the larger ones, can apply in order to set rules that are most advantageous to them, but even then it is not easy to enforce these rules and supervise their implementation.

Major progress on these issues has been achieved in the successive rounds of the GATT, and with the establishment of the WTO in 1995. The collaborative framework that has been established went a long way to secure that world trade will be beneficial to all participating countries. However, the stalemate that derailed the multilateral negotiations at the Doha Round demonstrates the limitations of this voluntary cooperation between countries when there is no authority that is accepted by all countries and that can enforce the rules of trade on all countries. In the multilateral agreements, the rules of world trade are therefore bound to be much more difficult to enforce, and since the efforts to reach an agreement are influenced by the interests of individual countries, they may not secure the degree of efficiency that trade models predict.

Many of the debates on the merits and drawbacks of globalization have centred on whether the basic tenets of free trade that view trade as a positive sum game, from which all trading countries will benefit, can be extended to the world economy where many countries are engaged in the global trading system, and the rules of world trade have to be negotiated and commonly agreed. Since different countries have vastly different shares in world trade, they also have different powers in determining the rules of world trade, and the potential to reach a collaborative agreement that will be equally beneficial for all, at least in the sense of securing a level playing field, is by no means guaranteed. For this reason, the progress that has been made in the GATT and the WTO was a huge step forward. However, even that progress had its limits, and the efforts to reach an agricultural trade agreement have so far failed.

The benefits that free trade is supposed to guarantee according to the classical and neoclassical economic theory have been the focus for the economists of the international development organizations, primarily the IMF and the World Bank, and, consequently, the set of policies they prescribed for the developing countries in order to achieve the optimal economic progress included free trade. The list of optimal policies became known as the Washington Consensus, and many of the debates on globalization did not concentrate on the merits of free trade, but rather addressed the entire set of policies in that list. Indeed, the Washington Consensus for many became anathema, and remains so until today. One of the points raised in these debates was that, although the policies in the Washington Consensus list were defined for an open economy, they were a 'one-size-fits-all' set of policies that were not adjusted to the specific conditions of a country or its position in the global economy. Thus, for example, a country's trade policy according to this prescription remains unchanged even if one of its trading partners is large enough to assume a monopolistic control over the market.

Another dimension of the debate on the merits of free trade in the global economy concentrated on the impact of globalization on a country's poor and sought to determine whether this process has been pro-poor. In principle, this analysis sought to determine whether the trade laws in the basic models of Ricardo and Heckscher–Ohlin, according to which the labour-abundant country specializes in the production of labour-intensive goods and thus raises the income of its working people, remains valid in the global economy. To answer this question, the relevant studies made a cross-country analysis to evaluate whether globalization has contributed to reduce poverty in the developing countries. This analysis did not examine the impact of different policies in a country or in its trading partners on the country's level of poverty. Therefore, the question remains whether trade will still be pro-poor if a country's trading partners implement very different policies, and whether trade that has been pro-poor in the past will remain so in the future if the structure and the rules of international trade change (Rodriguez and Rodrik, 2000; Irwin, 2002).

In the mid-1990s, the debate on free trade was renewed when it became evident that the policies of the EA countries were fundamentally different from the policies recommended by the Washington Consensus. These countries were the main beneficiaries from the globalization process and their economic performance was truly miraculous. In contrast, many countries that followed the guidelines of the Washington Consensus under the conditions established by the IMF and the World Bank for their structural adjustment programme went through a long period of stagnation and decline. In light of this very sobering experience, the debate centred on the lessons that should be drawn from the experience of the EA countries and their implications with respect to the policies that are best suited to LDCs given the current conditions of the global economy.

All these writings evaluated and analysed in great detail the impact of globalization and of different trade and macroeconomic policies on world poverty and income inequality. The most remarkable development was the growing gap between the two groups of developing countries: the emerging economies, mostly in EA, integrated very effectively into the global economy and experienced very rapid growth, while most other developing countries, particularly the LDCs in SSA and Central America, remained nearly stagnant. An analysis of the impact of globalization on the world's poor that combines these two groups of countries may fail to identify the specific factors that affected the poor and the relevant policies to ameliorate their poverty (World Bank, 2000, 2001).

The focus of this book is on the LDCs. One of the book's main concerns is an examination of the reasons for the failure of these countries to integrate into the global economy; in this context, it is important to understand why they had little success in their attempts to make effective use of their abundance of labour and their natural resources in order to develop competitive export industries of products in which they have a comparative advantage. Could these countries develop their exports if they had implemented an outward-looking trade policy despite their poor governance and continued internal conflicts? Could they develop their exports if they had been able to improve their system of governance and resolve their conflicts even with an inward-looking trade policy? What can these countries learn from the successful experience of the EA countries?

The answers to these 'what-if' questions are obviously impossible to validate conclusively and are therefore very much in the eyes of the beholder. The need to introduce reforms both in these countries' trade policies and their system of governance is undisputed, but the relative importance and the sequencing of these reforms are still debated. In the book, I consider these questions and discuss the empirical evidence that has been accumulated in the last decade, which may shed light on these issues. However, I consider the questions with two caveats. First, I focus more narrowly on the impact on the poor, and second, I examine these questions against the background of the far-reaching developments that took place in the global economy, in the globalization process and in the LDCs themselves.

The question about priorities between policies received a partial and unrepresentative, yet instructive, answer in recent years with the steep rise in the world price of several key export products of a number of African countries. Obviously, the policies of these countries themselves had nothing to do with the large windfall gains from their exports, but this was an opportunity to examine the impact of the increase in exports in countries that did not make changes in their system of governance. The results should not come as a surprise: in countries that did not improve their system of governance, most of the gains from exports were spent by their leaders to strengthen the army or were transferred to Swiss banks, and the share of the poor in these profits was very small.

An analysis of the lessons that can be drawn from the successful experience of the EA countries should also emphasize the fact that the policy options that the LDCs have today are very different from those that the EA countries had just two decades ago. In part, this is because today's conditions in the global economy are significantly different from those that prevailed two or three decades ago, and these changes themselves have a substantial impact on all countries and on the policies that can be implemented. A policy of open trade that may have been effective a decade or two ago is less effective and less beneficial today when world trade is much less free and competitive, but is increasingly dominated by transnational corporations and by regional trade agreements (RTAs) that divert the flows of trade. Another important change is the reduced impact of the multilateral trade agreements in the aftermath of the Doha Round and the shrinking power of the multinational development organizations.

Against this background, this chapter provides a survey of the different views of scholars and of common people on globalization. As noted earlier, a lot has been written on this subject in the last two decades. Although opinions have changed over the years and positions have been shifted, this is a much slower process than the changes in the globalization process itself, and heated debates are still conducted on issues that may well be less relevant today. Given the huge volume of books and articles that have been written on this subject, this rather brief survey cannot do justice to all the views that have been expressed and all the controversies that have been raised. In this review, I have therefore elected to focus on two polar views: the criticisms and the defence of globalization. I should hasten to emphasize that the critics did not oppose globalization itself and saw great potential in its contribution to world growth and to the developing countries. The main criticism was levelled against the policies that have been advocated and implemented to advance the globalization process and to help the developing countries take advantage of this process to accelerate their growth and ease the integration of their economies into the global economy. At the centre of this controversy were the policies promoted by the IMF and the World Bank, and, recently, by the WTO. It should also be noted that these policies, as well as the globalization process itself, are now in a transition period and, in the past 3–4 years, there have been fundamental changes that are likely to continue through the end of the decade.

3 The Critics of Globalization

Much of the criticism on globalization and its impact on the developing countries has focused, in fact, on the policies advocated by the IMF and the World Bank to help the developing countries to integrate into the global economy and thus also accelerate their growth. The main criticism was targeted at the group of policies prescribed in the 'Washington Consensus'. These policies were the guidelines for the structural adjustment programmes that were designed by the IMF for developing countries in need of the IMF concessional credit to meet their debt payments. These programmes are discussed in more detail in Chapter 4 with a specific focus on the LDCs; this review examines the backdrop of the 'pro-globalization' policies of the Washington Institutions and the grounds for the criticism of these policies.

Dani Rodrik (2002) highlighted the large difference between the policies advocated by the IMF and the World Bank and the highly successful policies implemented by the EA countries. As Rodrik emphasizes with some admiration, China's economic policies have violated virtually every rule of the structural adjustment programmes. Indeed, China continued to violate these rules also in later years:

> China did *not* liberalize its trade regime to any significant extent, it joined the WTO only in 2001; to this day, its economy remains among the most protected in the world, and the Chinese currency markets were *not* unified until 1994. China also did not open its financial markets to foreigners, again until around 2005, and it made the first steps towards adopting private property rights only in 2007. China thus achieved its miraculous growth and integration with the world economy even though (and perhaps because) it ignored all the rules of the Washington Consensus.

Rodrik noted that India also regulated capital flows well into the 1990s and maintained one of the world's most protectionist trade regimes. India's comparatively mild import liberalization in the 1990s came a decade *after* the onset of higher growth in the early 1980s. None the less, he emphasized, India managed to increase its growth rate through the adoption of more pro-business policies. Rodrik suggests that the fact that India had not yet opened its financial market was the reason it emerged unscathed from the Asian financial crisis of 1997. It should be pointed out, though, that India's growth during the 1980s was the result of the Green Revolution, whereas its industrialization started only in the early 1990s, following the reforms of 1991 in which it took first steps to liberalize imports.

In contrast to the successful experience of China and India, Rodrik highlighted the bad experience of countries that *did* open up to trade and capital flows but had financial crises and disappointing performance. The large Latin American (LA) countries adopted the globalization agenda in the 1990s, but suffered rising inequality, enormous volatility and economic growth rates significantly below those of the post-World War II decades. Argentina is perhaps the most outstanding example of a country that implemented a strategy that was grounded in the theories expounded by the economists of the World Bank and the IMF. After several years of very successful performance, it suffered a deep crisis in the late 1990s and the early 21st century. The developing countries, Rodrik argues, caught between WTO agreements, World Bank strictures and IMF conditions, forcing them to implement far-reaching institutional reform, and their need to maintain the confidence of financial markets, have effectively lost their sovereignty and are no longer free to devise their own policies.

The World Bank itself also grapples with the lessons of the 1990s, and its report *Economic Growth in the 1990s: Learning from a Decade of Reform* (World Bank, 2005) provides one of the more critical and introspective evaluations of the Washington Consensus. The failure of countries that implemented the structural adjustment programmes, particularly the SSA countries, to accelerate their growth despite their comprehensive policy reforms, motivated this critical evaluation. The report compares the policies of the Washington Consensus with the highly unconventional policies pursued by China and India, which continued to maintain high levels of trade protection, avoided privatization and implemented extensive industrial policies but rather lax fiscal and financial policies. As the report notes: 'Indeed, had they been dismal failures instead of the successes they turned out to be, they would have arguably presented stronger evidence in support of Washington Consensus policies' (World Bank, 2005, p. 6).

The report concludes that the

> principles of . . . 'macroeconomic stability, domestic liberalization, and openness' have been interpreted narrowly to mean 'minimize fiscal deficits, minimize inflation, minimize tariffs, maximize privatization, maximize liberalization of finance,' with the assumption that the more of these changes the better, at all times and in all places – overlooking the fact that these

> expedients are just *some* of the ways in which these principles can be implemented.
>
> (World Bank, 2005, p. 11, emphasis original)

Perhaps the main criticism in the World Bank report is on the tendency to exaggerate the advantages of rules over discretion in government behaviour. Rules were meant to discipline the governments, but experience shows that they cannot come at the expense of government discretion. Thus, for example, enhancing private investment incentives may require improving the security of property rights in one country, but enhancing the financial sector in another.

In a review of this report, Rodrik (2006) discusses the different views highlighted in the report with respect to the conflict between rules and discretion and provides another example to emphasize the need for discretion: 'Argentina's currency board, which removed monetary policy from the hands of the government, worked well when the binding constraint was lack of credibility, but led to disastrous outcomes when the binding constraint became an overvalued currency.'

The debate between rules and discretion is, of course, an old one and is far from being resolved. Its relevance for globalization may not be immediately obvious, but it is relevant in the context of the structural adjustment programmes which included a wide variety of countries. However, while the World Bank report concludes rather cautiously that 'government discretion cannot be bypassed', it is by no means certain that discretion always leads to better outcomes. Some governments are more capable and have more qualified staff than others to apply discretion, and these differences must also be taken into account. There is also a delicate balance with respect to the question when government discretion should be the deciding authority and when a government should abide by the rules. In other words, the international development organizations themselves must also apply their discretion in this entire process.

Easterly and Levine (2003) showed that policies (i.e. trade openness, inflation and exchange rate overvaluation) do not exert any independent effect on long-term economic performance once the quality of domestic institutions is included in the regression. The potential limitation of these findings is that the quality of institutions may be defined in retrospect on the basis of certain criteria, but the article does not make clear how the government can ensure that the quality of the institutions is high enough or what criteria the government should apply in order to improve the quality of the institutions before it implements other policy reforms. Easterly and Levine (2003, p. 11) note that institutions are by their very nature deeply embedded in society, but the political system and the country's leaders are no less important in this context. Very frequently, leaders establish and structure institutions so that they will serve their needs or goals, including the goal of securing the leaders' rule over society. In many developing countries that conduct democratic elections (with or without quotation marks), business people openly admit that they avoid making investments a year or so before the elections because they are not sure which party will come to power and whether and how the rules will be changed.

In his criticism of the globalization process itself, Rodrik (2002) emphasizes that the regulations which developing nations confront in the world markets are highly asymmetric. Import barriers tend to be highest for manufactured products of greatest interest to poor countries, such as garments. The global intellectual property rights regime tends to raise prices of essential medicines in poor countries. His main criticism in this article is directed against the rules of international labour mobility. The rules on cross-border labour flows are obviously determined unilaterally by the developed countries and are highly restrictive. In Rodrik's view, even a small relaxation of these rules would produce huge gains for the world economy and for poor nations in particular (see also D. Rodrik, 2005). He recognizes that in the developed countries both trade liberalization and labour mobility are highly unpopular. He quotes a public opinion poll which shows that when asked about their views on trade policy, fewer than one in five Americans reject import restrictions. In most advanced countries, including the USA, the proportion of respondents who want to expand imports tends to be about the same or lower than the proportion who believes that immigration is good for the economy.

So why are trade flows nearly unrestricted, whereas restrictions on labour mobility remain so tight? In Rodrik's view, the main difference is that the beneficiaries of trade and investment liberalization are the multinational firms and financial enterprises, which have been successful in setting the agenda of multilateral trade negotiations. Cross-border labour flows, by contrast, usually have not had a well-defined constituency in the advanced countries. This, however, is a somewhat rosy picture. In fact, in practically all the developed countries, there is a strong constituency against labour migration or any other form of cross-border labour flows, and the rules are getting tighter and implemented ever more harshly.

In the debate about the merits and drawbacks of globalization, another leading voice critical of globalization has been that of Joseph Stiglitz. It is probably not a coincidence that Stiglitz titled his widely quoted book *Globalization and its Discontents* (2002). Just as Freud in his 1930 treatise, *Civilization and Its Discontents,* discusses the clash between the instinctive demands of selfish individuals and the need for civilization and the rule of law to curb those aggressive urges, globalization can also be seen as a process of curbing the conflicting interests of countries that seek to structure their trade with other countries so that they can maximize their own self-interests. Stiglitz does not see the failures of world trade to cater to needs of the less-developed countries as endemic to globalization. On the contrary, he states clearly (Stiglitz, 2006, ch. 9) that the origin of all problems is the way globalization has been managed, rather than globalization per se. The problems and failures were, in his view, primarily the results of actions and policies taken by the major governing institutions and the powers behind them which set the rules, the IMF, the World Bank or the WTO. This, however, is a very narrow view of the inherent limitations of globalization and of the world governing institutions. In fact, the failures of world trade *are* endemic to globalization, and the limitations of the power of the governing institutions, more than the policy decisions that they have made, are the primary reasons for the failures.

The most obvious example of the limitations of the multilateral governing institutions is the failure of the successive rounds of the GATT to extend the trade agreements to agricultural products. This has been demonstrated most manifestly in the very last round of negotiations of the Doha Round in mid-2007, which was derailed due to the failure of the representatives of the EU and the USA to reach an agreement on the subsidies to agriculture. In the end, these subsidies were left and no agreement could be reached for reasons that had nothing to do with rational economics.

Another important point noted by Stiglitz is that free markets, left to their own devices, do not necessarily deliver the positive outcomes promised by textbook economic reasoning where people are assumed to have full information, and suppose they can trade in complete and efficient markets, and depend on satisfactory legal and other institutions. Hence, he concludes, whenever information is imperfect and markets are incomplete, which is to say always, *and especially in developing countries*, then the invisible hand is bound to work imperfectly, and under these circumstances, the government can greatly improve the performance of the economy by well-chosen interventions.

The conclusions that should be drawn from this analysis are therefore clear: the first step that should be taken in the implementation of development policies is to renounce the free-market ideology; instead, there should be recognition of the need to see individual circumstances in their complexity and distinctiveness. Instead of a fixation on ideology, Stiglitz argues that 'the most fundamental change that is required to make globalization work in the way that it should is a change in governance' (Stiglitz, 2006, p. 226). In a world of imperfect markets and incomplete information, effective development policy, as Stiglitz demands from the international development organizations, requires both good information on local conditions and a broad consensus on aims and a sense of the legitimacy of the strategy recommended to achieve them.

However, the conclusions that Stiglitz draws from his analysis contain ingredients of a 'catch-22': On the one hand, effective governance of the IMF and the World Bank requires these institutions to set the rules and monitor their implementation, since globalization without governance is likely to lead to chaotic results that will be especially devastating for the less-developed countries. However, when the governing institution sets the rules, there are no exceptions and everyone has to obey them. With the global governance by enlightened and compassionate international bodies recommended by Stiglitz, the government of each individual country should in fact be free to act as it judges appropriate. How can this system of governance work if each member is free to choose whether or not it accepts that the rules are appropriate and whether it is therefore willing to obey them? So, are privatization and liberalization attractive policies, or is that only the case if they guarantee extra jobs at fair conditions? How can a system of governance work under these conditionalities if the decision whether or not to obey the rules is in the eyes of the beholder? Hence, countries should embrace globalization, but only on their own terms, rather than following a 'cookie-cutter' approach and an arbitrary obedience that the system of law and order sometimes demands. But can any system, let alone a global one, operate under these terms? Can any multilateral agreement be reached if it is not binding and cannot be enforced?

Stiglitz's book only partly discusses the remedies to the system and is primarily focused on whom to blame for the failure to help the developing countries in the process of globalization. The culprit, according to him, is the IMF that followed 'market fundamentalists' and enforced on the developing countries the set of policies specified in the Washington Consensus. Stiglitz considers, and rejects, the view that the choice of the IMF to promote the Washington Consensus was some sort of a conspiracy between the IMF and the other members of the trinity of global financial governance and development together with powerful interests in the richer countries, primarily the US Treasury, to govern the process of globalization. He does, however, hold the view that the IMF 'was not participating in a conspiracy, but it was reflecting the interests and ideology of the Western financial community'. Arguably, blaming the IMF for being, unwittingly and unwillingly, the 'Sanctus Simplicitas' in this 'conspiracy' is no less insulting. He also does seem to agree, though, that the international development organizations do set the rules of the game of international development and cooperation in ways that largely serve the interests of the more-advanced countries.

As noted earlier, Stiglitz also claims that the three leading global financial governance and development organizations have set the rules of the game of international development and cooperation according to policies based on the views of economics that are referred to as 'market fundamentalism', which posit that markets by themselves work efficiently and are self-adjusting. What drove the IMF particularly to push these policies was, in his view, a perspective, an ideology and interests. The IMF was the target of his criticism because, he argued, it applied a nearly unvarying agenda of deregulation, trade liberalization, privatization, deflationary monetary policy and the other ingredients of the 'Washington Consensus', even when they made little economic sense.

Although in hindsight many economists in the IMF and the World Bank were in agreement with many of the criticisms that Stiglitz made, it should also be noted that the actual policy making obviously had to proceed without the benefit of hindsight and had to face many uncertainties; leaving the policy decisions to the countries themselves would not have guaranteed that the optimal decisions would have been made even by the country itself, and countries' policies also involve a great deal of arbitrariness, bad judgement, ideology and interests. In this context it should also be noted that the great success of the EA countries was because they chose to ignore the policy advice of the IMF, and, as Stiglitz claimed, governed globalization, took globalization on their own terms and chose what part of globalization they found appropriate at which pace. At least an equally important reason was that these countries were not drowned in debts like the SSA and LA countries, which the IMF had to bail out because, during the 1960s and 1970s, these countries were controlled by dictatorial regimes that suffocated economic development cut relations with the outside world and made any foreign debt impossible.

I share many of Stiglitz's criticisms of specific policies that were recommended and sometimes required by the IMF, primarily on fiscal austerity, trade liberalization and the liberalization of the capital markets, but leaving these rather complex policies in the rather inexperienced hands and heads of the policy makers in many of the developing countries is by no means a panacea for efficient policies. I also agree that, at the level of the national economy, governments should be allowed to fight recessions and depressions by using expansionary monetary and fiscal policies to spur the demand for goods and services. At the microeconomic level, governments can also use a variety of devices, ranging from job creation to manpower training to welfare assistance, to put unemployed labour back to work and cushion hardships. In fact, Jagdish Bhagwati made the same recommendation. There is a risk, though, that if governments have no restrictions and no supervision when they implement these policies, they may use them excessively and, in the end, may cause more damage than good. Nevertheless, some macroeconomic policies that were part of the Washington Consensus proved to be valuable later on: fiscal discipline, spending caps, certain privatizations effected with proper oversight, the institution of regulatory bodies for public services, the establishment of central banks with a degree of autonomy and the simplification of tax schemes. It took tough measures to make these policies an integral part of proper government policies, but now most governments accept them.

Stiglitz's criticism highlights the two non-congruent goals that the conditions specified by the Washington Consensus seek to achieve at the same time. On the one hand, they try to achieve the goals and priorities of the aid donors that are mostly concerned with policy reforms that would accelerate a country's growth and reduce its poverty, thus boosting the effectiveness of their aid. On the other hand, they attempt to meet the demands of private creditors that are mostly concerned with the safety of their loans and their rate of return, and require the conditions of the Washington Consensus to provide reliable signals of the soundness of a country's macroeconomic policies and its capacity to meet these demands. Private creditors are not concerned with a country's social goals, but mostly with the rate of return on their loans. The seal of approval of the IMF that a country's economic policies are sound and meet the conditions of the Washington Consensus would enhance its credibility in the global financial markets and increase its capacity to mobilize more credit.

Another important aspect of the debate about globalization is pointed out by Stiglitz and Charlton (2006) in their book *Fair Trade for All: How Trade Can Promote Development*, which identifies the great weaknesses of the existing world trade system that provides a very biased playing field that works to the disadvantage of the developing countries, and offers some solutions that would ensure that the world trade system contributes more to promote development.

Highlighting the great asymmetry of power between North and South, Stiglitz and Charlton argue that, as a result, the current system of international trade disadvantages, even victimizes, the developing countries in virtually all respects. When the book was published, it was not yet clear that the Doha 'Development Round' would fail, thus increasing the disadvantages of the South. The book argues that 'fairness' requires to empower the developing countries rather than to insist on reciprocity and agreements through negotiations. Instead, the authors suggest, the WTO should substitute the negotiations by a centrally designed economic analysis of costs and benefits. No doubt, however, that a trade agreement designed by a committee is bound to be far more controversial. The failure to reach an agreement in the Doha Round after nearly 6 years of intensive negotiations shows the prospect of an agreement of the kind suggested by Stiglitz and Charlton.

The book also emphasizes that economic theory and empirical studies have so far been unable to establish the benefits of open trade, especially those associated with liberalization, due to market failure, government failure and high adjustment costs. The benefits of competition are often marred by the dominating role and influence of multinational corporations. On these grounds, the book argues for 'policy space' to allow developing countries to experiment with industrial policy. This will be done by giving countries that liberalize their trade substantial amounts of assistance under a 'principle of compensation'. I was not clear, though, how this principle would work, how the compensations would be determined and how long they would be given.

Again the book suggests an arrangement that might work well on paper, but is likely to encounter considerable stumbling blocks in the 'real world'. Market failures due to incomplete information may also work to the disadvantage of arrangements designed by committees that are also handicapped by incomplete information. I also do not think that by lumping all the developing countries together this arrangement would secure 'fairness', and I certainly do not think that China, India and many other emerging economies should be rewarded in any way for liberalizing their trade. In fact, the countries that would be rewarded under this arrangement are those that kept their markets closed until now, and the countries that would be punished are those that under the structural adjustment programme opened their economies for trade.

Another committee-designed and implemented arrangement that the book proposes is that all rich countries should commit to unrestricted imports from the LDCs (which the book identifies as countries with both a smaller per capita and smaller absolute GDP). Again, incomplete information of the committee that designs, implements and supervises this arrangement may create a nightmare of paperwork and piles of complaints from both the developed and the developing countries.

4 In Defense of Globalization

The rapid growth of the Asian countries, most notably China and India, was driven by the rise in their exports; by contrast, the countries of SSA and, to a lesser degree, the LA countries, adopted inward-looking policies in order to protect and promote their local industries. In practice, the main impact of the inward-looking trade policy was to dampen trade and preserve inefficient industries that slowed down growth. In his book *In Defense of Globalization*, Bhagwati (2004) argues that the differences between the trade policies of the countries in these regions and their growth rates underscore and strengthen the case for pro-trade policies that contributed to raise the rate of growth in the Asian countries, while the countries that opted for policies that restricted their trade and closed their markets for competing imports experienced a considerably slower or reduced rate of growth.

The reason for the favourable impact of globalization on the poor is, according to Bhagwati, straightforward: by increasing a country's trade, globalization contributes to accelerate its economic growth, and the higher rate of growth is increasing people's incomes and reducing poverty. Indeed, Dollar and Kraay (2002), who conducted an empirical test in a large sample of developing countries on the impact of greater openness, or 'globalization', on growth, poverty and income inequality, found that greater openness accelerated the growth of income per capita and did not increase income inequality within countries or disproportionately hurt the poor. True, Bhagwati would admit, in many developing countries poverty did in fact increase, but many social, political and economic factors contributed to this and, he would perhaps continue along the line of Tina Turner's song, 'What's globalization got to do with it?' After all, one cannot blame all the ills of the world on globalization.

On the other hand, to show that globalization eases poverty, Bhagwati offers a tale of two continents. In 1970, average African incomes were 30% higher than average Asian incomes. About 30 years later, African incomes had remained stagnant and were then half of Asian incomes. Although there were undoubtedly many causes for this reversal, Bhagwati documents that a primary one was that Asia opened itself and adapted to external markets while Africa did not. As a result, Africa, which – using standard measures of the absolute level of poverty – in 1970 was home to about 10% of the world's poor, found itself 30 years later with more than a third of the world's poor, while Asia's share in the world's poor, which in 1970 was more than 75%, had declined 30 years later to just 15%. However, the question 'What's globalization got to do with it?' may also be appropriate here. After all, one cannot blame all the ills of the world on the lack of globalization. Africa went through decades of incessant conflicts and wars, which, in addition to dictatorial leaders and corruption, were arguably the main reasons for the continent's impoverishment. Moreover, under the structural adjustment programmes of the IMF, the African countries actually opened their economy to trade more than the Asian countries, as we will see below.

In this book, I have therefore taken a much more cautious and qualified approach, both in praising globalization and in condemning its ills. First, I will argue that there is a strong connection between some of the ills that quite a few countries, particularly the LDCs, suffer from, and globalization. Second, and perhaps more important, there is also a strong connection between the prospects and the potential of these countries to remedy these ills and globalization. Third, not all the blessings that benefited many other countries, primarily the highly successful emerging economies in EA, can be attributed to the policies of free market and open trade that, according to Bhagwati, are the policies that ushered in globalization. This book emphasizes that the attributes of globalization that contribute to accelerate growth and the policies that the developing countries should implement in order to take advantage of these attributes must also take into account the potential pitfalls of globalization, and must take measures at an early stage in order to protect their economies and their people from these ills.

The experience of China and India are undoubtedly the most noteworthy examples: both China (since the late 1970s) and India (since the early 1990s) took gradual and very measured steps to liberalize their trade and cut down tariffs; it cannot therefore be concluded that they adopted outright an open-trade policy. In fact, both countries remained heavily protected during all these years, the share of foreign trade in their GDP is still relatively small[2] and a large part of their imports that were liberalized more extensively consist of raw materials rather than final consumer goods. Their imports were therefore driven and expanded in part by the rapid growth of their exports that increased, in turn, their demand for raw materials, including oil. Both countries continue to protect many of their local industries even today, in large measure in order to protect their employment and shield their regimes from internal political pressures, and they did not reduce by much the tariffs on imports of products that compete with local production. In addition to relatively high tariffs, they also maintain other trade barriers and a selective approach to foreign direct investment (FDI).

With high levels of trade protection, limited privatization, extensive industrial policies, as well as lax fiscal and financial policies through the 1990s, these two countries and most other EA countries hardly look like exemplary practitioners of the Washington Consensus. In his *Defense of Globalization*, Bhagwati does not take a clear position on the role of the government in a globalizing economy. At one point, he discusses with approval the Indian government's decision to install agricultural price supports to prevent a collapse in food prices during the green revolution (Bhagwati, 2006, p. 56), whereas at another point he urges policy makers to ensure that 'bureaucrats are replaced by markets wherever possible' (Bhagwati, 2006, p. 58). The Washington Consensus, in contrast, was very clear on this issue: other than measures essential for coherent and consistent macroeconomic policies aimed at securing economic stability, the government should maintain free and open markets and take measures to reduce its influence over the price system that it maintained through subsidies, taxes or direct controls.

The market-friendly approach of the Washington Consensus was an integral part of the view of globalization that praised the benefits of trade liberalization and promoted export-led growth by cutting tariffs and removing trade barriers. However, Bhagwati recognizes the need to accompany

[2] The share of trade in their GDP is small, particularly in India, even after taking into account the large size of their economies.

trade liberalization with assistance programmes for those who lose their jobs as a result of the adjustments that the economy undergoes, and he thus acknowledges the significance of social policies and institutions to protect the poor. Indeed, rich countries implemented comprehensive assistance programmes when they opened their markets to trade and allowed the imports of products in which their industries did not have a comparative advantage; as expected, the cheaper imports bankrupted many producers, left many workers – mostly unskilled – unemployed and reduced the wages of many others when they were forced to compete with low-wage workers in the developing countries.

The poor countries that are just beginning to liberalize their trade are not able to implement similar adjustment assistance programmes due to their limited resources. These countries would need institutional support from the international development organization and from the donor countries to help them during the adjustment phase in establishing adequate safety nets in order to help those who may lose their jobs and be impoverished by the policy of open trade. This is also why a developing country should be careful in opening up its economy and should be watchful in determining the optimal speed of adjustment in order to manage the transition very carefully, taking into account the difficulties that certain sectors and population groups may encounter.

Rodriguez and Rodrik (2000) also emphasized that trade reforms, by themselves, are not the panacea for growth and poverty reduction in a developing country and that these reforms cannot substitute for other policy and institutional reforms that these countries must implement. The priority that is given in the Washington Consensus to greater openness may therefore be misguided, since equal emphasis must be given to establishing the necessary conditions that would guarantee the success of the reforms, including effective governance, well-functioning institutions and the rule of law. If the required institutional reforms are not implemented and effective governance is not secured, the trade reforms are more likely to go astray and their goals will not be achieved. Free-trade advocates tend to believe that the rising tide of the growing world economy with international specialization and investment will lift all boats. But even in a fully competitive and free world trade, the rising tide would not lift all boats at the same pace, and people who lack the capacity to adjust, retool and relocate with the changing market conditions would be at a disadvantage for quite some time. The benefits of specialization can materialize only in the longer run, but during the adjustment period, until people can obtain the necessary training and resources, they are bound to suffer great pain. During that period, there will also be a rise in income inequality.

According to Bhagwati, the measures to open a developing country to trade must therefore be conditioned on the measures that the country needs to take to put adequate safety nets in place in order to provide protection to the sectors, regions and people who are likely to be hurt when the economy is opened up to trade; these conditions are the constraints that determine the pace of implementing the adjustment programme. In practice, however, the guiding principles of the World Bank and the IMF for the trade reforms that have been summarized in the Washington Consensus did not include any such conditions or constraints on trade liberalization, and the list of common 'conditionalities' in the IMF/World Bank structural adjustment programmes also did not include these restrictions. Instead, these conditionalities include not only requirements to reduce import tariffs and eliminate all other impediments on imports that protect inefficient domestic producers – except, perhaps, for infant industries – but they also include requirements to reduce government expenditures, eliminate the more wasteful social programmes and reduce the size of other social programmes. Obviously, these conditionalities are in contradiction to the measures that Bhagwati describes as necessary for the trade reforms. Only towards the end of 1999, the IMF and the World Bank introduced far-reaching changes in the structural adjustment programmes and endorsed a programme to strengthen the focus on poverty reduction in their Poverty Reduction and Growth Facility (PRGF), but even then the actual changes were quite limited.

The conclusions drawn in the World Bank study on *Globalization, Growth and Poverty* (2001) raise a number of questions that are also relevant to some of the conclusions of Bhagwati. The study attributed the good economic performance of the last two decades in the countries that were defined as 'globalizers' to their open trade policy and their integration into the global economy. In that study, China and India were included among the globalizers. However, according to the common criteria of the World Bank and the IMF both countries were highly protected during all these years and should be regarded as protected economies even today. The inclusion of China and India among the globalizers and their rapid growth and the steep decline in their poverty were, in fact, the reason why the countries that were defined in that study as the group of globalizers had much better economic performance and a steeper decline in poverty than the countries that were included in the group of 'non-globalizers' – even though quite a few of the countries that were included in the latter group actually had more open economies.[3] The World Bank report *Attacking Poverty* (2000, 2001, p. 36) recognized, in contrast, that there is not necessarily a causal connection from opening up the economy and allowing more free trade to more rapid economic growth; however, significant trade liberalization may help to increase production efficiency and reduce the waste of resources.

China is undoubtedly the most successful EA country, but it is still very much centrally planned and tightly controlled, despite the far-reaching measures taken in recent years to gradually liberalize the markets, trade and foreign investments, and to privatize many of the state's assets. Although China is loosening up its central controls over the economy and allowing more foreign investments, the interprovincial commerce is still regulated, the trend towards market liberalization is very gradual, foreign investments are closely monitored and the markets have opened to trade only very carefully

[3] The list of 'globalizers' in the World Bank study also includes countries like Nepal, Cote d' Ivoire, Rwanda, Haiti and Argentina.

and gradually. All these measures were taken in a deliberate effort to find a proper balance in order to avoid the political and economic chaos of the kind that came in the wake of the 'big bang' approach to economic reforms in Russia, and to avoid the political and social unrest that these adjustments may trigger.

The approach of China and India to the trade reforms and the restrictions that they left on trade and capital flows may also qualify the conclusions drawn by Bhagwati by comparing their successful trade policies and rapid growth with the poor performance of post-colonial India and of other countries that implemented protectionist, inward-looking policies and failed to deliver the prosperity that was expected when the market system replaced the bureaucratically determined rations of goods and services. To some extent, these conclusions seem to be based on a number of assumptions that are quite similar to the assumptions made in the World Bank study *Globalization, Growth and Poverty* (2001) in its division of countries into 'globalizers' and 'non-globalizers'. More important is the fact that this comparison does not provide adequate answers to a number of key questions that may raise doubts whether this comparison can be taken far enough and remain valid for many developing countries:

- Can the trade policy in China be characterized as 'pro-trade' and 'market-oriented' and as free of protectionism?
- To what degree can the economies of China and the other EA countries be categorized as free-market economies? What was the role of the Chinese government and its bureaucratic administration in controlling and navigating the economy?
- To what degree have the SSA and the LA countries been closed to trade throughout this entire period? Have their trade policies remained inward-looking since the early 1980s, when their economies were guided by the IMF and the World Bank and their policies were determined by the conditionalities set by the structural adjustment programmes?
- How rapidly were the LDCs able to liberalize their trade and open their economies given their resource constraints and budget limitations to set up adequate social safety nets to protect their poor?

These questions are discussed in the book and the discussion is focused, in particular, on the experience of the LDCs. But to some extent the questions are tautological and the implicit critiques that they express also reflect some of the thoughts I had when I read Bhagwati's book, as well as his and others' comments in subsequent discussions on the book. My overriding impression was that Bhagwati wrote about globalization not the way it actually shaped up to be, but the way it perhaps ought to be: had countries taken all the necessary measures to prepare adequate safety nets and/or had the donor countries and the international development organizations given sufficient assistance to the developing countries so that they would have had the necessary resources and would have been able to help those population groups that were impoverished by the adjustment, *then* globalization would have been much more beneficial and would likely have raised the living standards of all people, while the negative effects of the economic adjustments could have been mitigated or prevented.

Similar questions can be raised with respect to the assumptions that have implicitly been made in the analysis of globalization with respect to the production side. The theory is straightforward: had all industries been competitive and all markets been free, *then* countries would have been able to allocate their resources more efficiently with open trade and consumers would have been better off with globalization. In practice, markets are not entirely free, both because *all* governments remain actively involved in the market and often impose restrictions on trade, and because many industries are dominated by monopolistic transnational corporations. The gains from trade when all these interventions and restrictions are taken into account are therefore considerably different from the gains that the textbook predicts if markets are free. Given these constraints, the policies that governments in different countries should actually implement in order to maximize their gains from trade may therefore be different from policies that advocate the minimization of government interventions and an unfettered free market.

However, that is not to say that the way trade between countries is actually practised makes globalization or open trade undesirable. But it does change the costs and benefits from trade, and it also changes the trade policies that different countries should implement in order to maximize their gains. The most obvious examples are the gains from trade with and without bilateral or RTAs, the gains from trade for a small and a large country, and gains from trade for a country when its corporations have a monopolistic power over the world market of certain products. As Bhagwati notes very critically in his book: 'Pharmaceutical and software companies muscled their way into the WTO and turned it into a royalty-collection agency simply because the WTO can apply trade sanctions.' Again, we could ask 'What's globalization got to do with it?' A lot, and Adam Smith's early warning that when 'three people of the same trade meet together, the conversation ends in a conspiracy against the public' tells us most clearly why.

The forces that shape the structure of globalization, wittingly or unwittingly, have also got a lot to do with preferential trade agreements (PTA), and the massive proliferation of these agreements in the last decade is testimony to that connection. In his book *Free Trade Today*, Bhagwati (2001) already points out that the procedural problems created by the overlapping trading blocks are enormous, especially with respect to the 'rules of origin' that are primarily aimed at preventing non-members from gaining access to the markets of the member countries through the member with the lowest applicable tariffs. With the reduction or removal of the tariffs for member countries, they increase the obstacles for non-members with a maze of producers that in practice increase their protection. For the LDCs that are not members, the free-trade agreements (FTAs) increase the obstacles for their exports. The PTA s actually reduce efficiency and welfare not only worldwide, but also among member states as an effect of the trade diversion by giving incentives to member states to divert their imports from the lower-cost non-member countries to the more expensive, but tariff-free, imports from a member country within the trading block.

Trading blocks that raise tariffs against non-members can reduce the efficiency and welfare for member states; in addition, Bhagwati recognizes that the RTAs inevitably politicize trade. Moreover, with the formation of a trading block, import-competing industries under pressure from lower-cost producers in member nations may demand to increase the tariffs on the imports of these products from the lowest-cost producers of non-member states, thus diverting trade to higher-cost producers among the member countries. Such tariff increases are prohibited by GATT, but there are many ways to impose trade restrictions through administrative means, and with the weakening of the WTO after the stalemate in the Doha Round, these agreements are gaining force. The free-trade regions may thus spread discrimination against producers in non-member countries, and Bhagwati rightly fears that these complex discriminatory trading block agreements may weaken and gradually displace the multilateral agreements and their central tenet of non-discrimination.

The large countries take advantage of their market power even without a trade agreement. In China, import tariffs continue to protect entire industries. Thus, for example, the domestic market of shoes and other leather products is protected by a 27% surcharge. Furthermore, Chinese currency policy acts as a massive subsidy for exports by keeping prices of goods sold abroad artificially low, while the intellectual property of foreign firms is taken without payment and without compunction.

India is another prominent example of a country that had large benefits from trade and globalization, even though its trade policies were, and in part remain, quite different from those advocated by Bhagwati and promoted by the IMF and the World Bank. Despite the parallels that can be drawn between the trade policies of China and India in 2007, there were clear differences between them in the past and some of these differences still remain today. The Indian government took more active measures to liberalize the economy only in the reform programme that started in 1991. Even then, the changes in its trade policies were very gradual, and today, a decade and a half later, there are still strict restrictions on trade in many sectors. Nevertheless, since the introduction of the liberalization programme and more active measures to promote exports, India managed to accelerate its economic growth very significantly. In the 12 years from 1995 to 2007, the annual growth rate rose to over 6.5%, and since the early 2000s, it even exceeded 8%. The economic dynamism that has been brought by this growth has imparted the confidence that India will be successful in building one of the world's leading economies. The country that was sometimes referred to as 'the sick man of Asia' has become a credible contender for a major role in the balance of power in Asia. However, Fig. 1.1 shows that India is still far from being an open economy, and its imports are highly regulated and controlled by means of high tariffs and many administrative

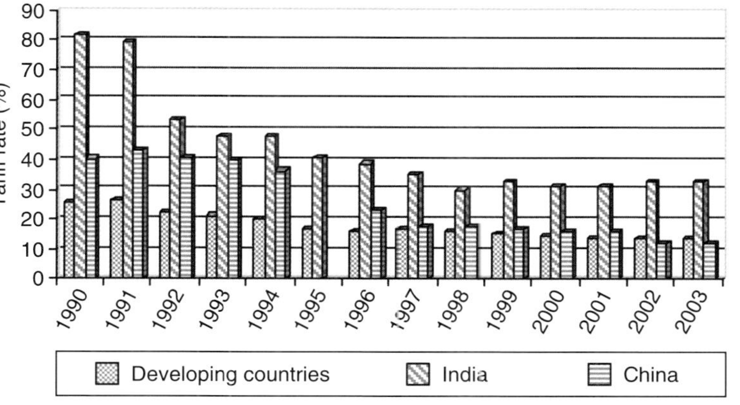

Fig. 1.1. Average tariff rates in India, China and the developing countries.

restrictions.[4] Nevertheless, India managed to achieve an unprecedented rate of growth and the country became a major attraction for foreign investors.

Despite initial concerns about major losses of jobs when India opened up its economy, the reduction in import tariffs did not slow down the rapid growth of its local industries that continued to grow at a rate of over 7% since the mid-1990s, and it did not create unemployment. Early concerns about the impact of foreign investments on its capacity and sovereignty to develop its industry according to the government's plans and priorities also proved to be unjustified and, in practice, these investments stimulated the development of its industrial sector. The multinational corporations that came to India became an integral part of the local economy and proved to make a significant contribution to its long-term and increasingly more ambitious industrialization plan. This plan was gradually extended to include an expansion beyond the domestic market to the wider world economy and to expand the country's exports by establishing subsidiaries and by making direct investments abroad.

Poverty is still a dominating fact of life in India and most of the benefits from globalization have been reaped by the top 10% of its population, primarily the skilled workers. Yet, despite the benefits globalization has brought for India, each step forward in opening up the Indian economy to global trade was associated with great concerns in different quarters about the impact on their lives and their businesses: will the Indian economy be dominated by the transnational corporations? Will people be free to choose their

[4] There are still limits on imports of hundreds of products – from bicycle parts to electronic equipment – aimed at protecting the small-scale industries. India's prohibitive import tariffs on cars are another example: These high tariffs are forcing Volkswagen to make large investments and build a new factory to serve the Indian market, even though its factories in Europe are operating below capacity and the company is even considering closing some plants – a move that would result in mass layoffs.

place of work? Will they be able to choose their own and their children's education? Will India remain merely a service provider for the rich countries? These concerns notwithstanding, poverty declined from 55% of the population in the 1970s to 26% in 2000 and the economy continues to grow at unparalleled rates. The number of the poor still remains high and in some states, particularly the Hindi heartland of Uttar Pradesh and Bihar, it has barely declined. Moreover, in all states and all sectors the pursuit of better standards of living has clearly gained momentum with migration to the urban centres, people's savings and investments and the hunger for better and higher education. A large percentage of the population did manage to improve their standard of living, poverty declined, new jobs were created and new industries were built by local and foreign companies.

The Government of India has taken a very different approach to protect its local industries and local workers than the one advocated by free-trade economists: instead of establishing safety nets for those who might lose their jobs, the government maintained high protection on many local industries and on the local agriculture. What's globalization got to do with it? In my view, a great deal: for one, the problem of creating social safety nets in the developing countries is due not only to the lack of resources, but also to a widespread abuse of the system. Even in Sweden, they have come to realize that false claims for social security benefits are a considerable percentage of the total claims. In a developing country, where there is no way to implement effective supervision, it is not at all clear that social safety nets are more cost-effective than the protection of inefficient industries with high tariffs. Another point to be considered is that internal political pressures often play a major role in the decision regarding whether or not to keep the protection of certain industries. Indeed, political pressures played a central role in preventing the members of the WTO to achieve a trade agreement in the Doha Round. Given all these considerations, a complete cost–benefit analysis may lead to the conclusion that in some industries and some countries trade restrictions may well be a sensible solution. The conditions for a successful trade liberalization that guided the principles of the Washington Consensus and the conditions of the structural adjustment programmes of the IMF and the World Bank were, in most cases, determined in total disregard to the political and social needs and constraints. This, in my view, was the main weakness of these conditions that, in the end, also led to the rather turbulent demise of these programmes.

5 The Opposing Views on Globalization in Retrospect

Sections 2 and 3 presented two polar views on globalization and its impact on the developing countries, focusing on the positions put forward by Bhagwati, Rodrik and Stiglitz. They are undoubtedly the leading and most influential voices in this debate, but they represent only a small sample of the very intense controversy that developed and came to involve a large number of scholars who addressed a wide variety of other subjects and other aspects of globalization and its impact on the developing countries. This section highlights the two main 'camps' of the debate because their arguments provide a good overview of the subjects that will be discussed in this book. With the hindsight of 20:20, I would like to add some comments about the opposing views on globalization that define the debate.

Stiglitz's book *Globalization and its Discontents* is undoubtedly the more polemic of the three, and it clearly represents an unabashed criticism, though not on globalization per se, which Stiglitz does not see as inherently harmful or undesirable, but on the IMF and the two other leading multinational organizations, the World Bank and the WTO, that governed and guided the economic policies and processes that took place in the developing countries as an effect of globalization. Rodrik also attacked the structural adjustment programmes of the IMF and the World Bank. The policies that these institutions promoted and essentially enforced on the developing countries through a set of conditionalities determined the structure of the structural adjustment programmes and their principles that were summarized in the Washington Consensus. In Stiglitz's view, these policies are rooted in a misguided ideology and they were harmful and damaging for the developing countries. In Rodrik's view, at least part of the success of China and India to achieve such miraculous growth should be attributed to the fact that some of the policies they adopted were diametrically opposed to the policies advocated at the time by the Washington institutions.

Although this review is written 5 years after Stiglitz published his book, and many changes have taken place since then in the global economy and in the developing countries themselves, Stiglitz has in the meantime published many articles and another book on globalization that express essentially the same criticism. But how relevant are these subjects and this criticism today? Reading Stiglitz's book and his many articles, as well as publications of other authors who take up these themes, one dominant impression is that these writings often focus on a close-up image of the IMF. But with the passage and the perspective of time, the close-up focus on the wars that the IMF waged during the 1980s and part of the 1990s in the context of the structural adjustment programmes has become much less relevant, particularly in the last decade, largely due to the shrinking role and influence of the IMF in today's global economy. Against this background, Rodrik entitled a recent review article 'Goodbye Washington Consensus, Hello Washington Confusion?' (2006). The much criticized programmes lost their function and importance for several reasons: as the world economy became more stable and growth was spreading across more countries, including many of the LDCs, as interest rates were reduced, inflation rates fell and the exchange rates became more stable, and as the global capital market grew at exponential rates, reaching more and more countries, a growing number of developing countries were able to borrow from the international private financial institutions and cover all their credit requirements at rather low interest rates. Their needs for loans at concessional interest rates that the IMF could offer were therefore much

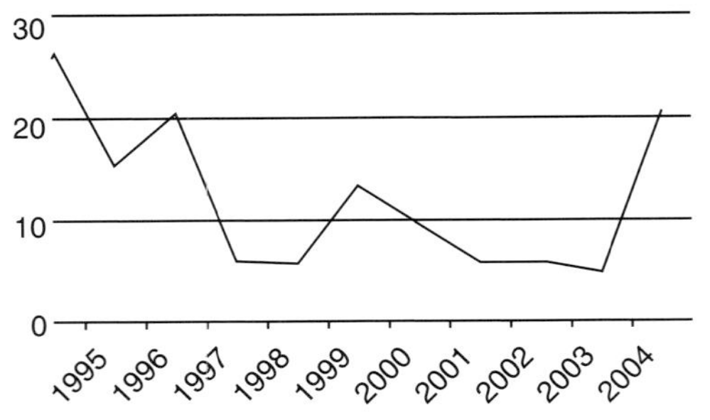

Fig. 1.2. The share of IMF lending to low-income countries (%).

smaller, and the IMF conditionalities further reduced their appetite for that credit. Figure 1.2, which shows the rather meagre share of the low-income countries in the total lending of the IMF since 1995, offers a more succinct summary of this point (IMF, 2002).

Another point that needs to be noted is that Stiglitz sometimes overstretched his criticism, because over the years, the IMF and the World Bank actually made very significant changes in their basic approach to the macroeconomic policies of the developing countries, and this was also reflected in their conditionalities. In the PRGF, which is the Fund's concessional lending arm in support of low-income members, far-reaching changes were made. The PRGF programme was designed to reflect more closely the needs of the low-income economies for pro-poor growth and the link between macroeconomic policies, growth and poverty reduction policies. In 2002, an internal review of the PRGF programme design emphasized that PRGF-supported programmes should focus on budget execution and on the efficiency and targeting of public spending to ensure that resources devoted to poverty reduction effectively reach their intended uses. An important goal in the design of the PRGF programme was to reflect national ownership and priorities of the underlying economic policies. The conditions were based on programmes that were owned by the country and were drawn up in close cooperation with the national authorities. The conditionality was also aimed to reflect genuinely country-owned programmes and the country's main concerns in the structure of the reforms (Independent Evaluation Office, 2005).

Most LA countries were already able to pay back their debts when the debts of Argentina (US$10 billion) and Brazil (US$15 billion) were repaid to the IMF either from their own resources or with generous loans that Hugo Chávez gave them, largely in order to release the LA countries from their dependence on the IMF. With these payments, the IMF no longer has such a significant role to play in the region. In SSA, the heavily indebted poor countries (HIPCs), which continued to have difficulties in paying their external debt-service obligations, received aid to reduce their external debts to sustainable levels. The HIPC Initiative was first launched in 1996 by the IMF and the World Bank, with the aim of ensuring that the poor countries can meet all their debt payments. In 1999, faster, deeper and broader debt relief was provided, and in 2005, the HIPC Initiative was supplemented by the Multilateral

Debt Relief Initiative that allowed for 100% relief on eligible debts for countries that completed the HIPC Initiative process.[5]

With these developments, the role of the IMF as a source of credit to the developing countries has been diminished very significantly, and with it also the impact and the significance of the Washington Consensus. Today, the Washington Consensus is effectively passé and the IMF itself is a large administration in search of a new role. The International Financial Institution Advisory Commission (2000), the Meltzer Commission, recommended in its final report that the IMF should restrict its lending to the provision of short-term liquidity. The current practice of extending long-term loans for poverty reduction and other purposes should end. The Overseas Development Council (ODC, 2000) report also underscored the need for the IMF to discontinue its role in the PRGF in favour of the World Bank; Birdsall and Williamson (2002) and Bergsten (2005) also supported an outright move of the PRGF, the Fund's concessional lending arm in support of low-income members, to the World Bank, on the ground that the latter is better equipped to deal with this group of countries.

Bhagwati's book provides an extensive overview of globalization that leaves the reader with a strong sense of its great potential. However, in some parts of the book, I could not avoid the impression that Bhagwati is holding a mirror up to what might be called a 'virtual reality' image of globalization and is writing on the reflection of that image. In other words, sometimes neither his praise nor his criticism are on globalization as it really shaped up to be, but on some kind of textbook globalization: if world markets had been free and competitive, if multinational corporations did not have the monopolistic power they have, if multilateral trade agreements had not been influenced as much by the large trading countries and if the RTAs had not distorted world trade with trade diversion, *then* the policy of open trade would have been most advantageous for the developing countries and would have guaranteed the most efficient allocation of their resources.

Even then, however, since opening the economy to trade involves a difficult adjustment process, neither Bhagwati nor Stiglitz favours a 'shock therapy' of the kind recommended by the IMF in the past, which failed so miserably in the case of Russia and earlier also in Latvia and Lithuania.[6] It took time for the IMF to adjust its 'ideology', but the great disadvantages of a 'shock therapy' that was supposed to rapidly rearrange all markets to a new general equilibrium have now been realized. The adjustment of the markets takes time, and the process of opening the markets for trade must take this into account.

Bhagwati as well as Stiglitz realize that countries would have to give assistance to those segments of the population and to those sectors that are

[5] However, even if all the external debts of these countries were forgiven, most would still depend on significant levels of concessional external assistance (Collier and Dollar, 2001).

[6] Today, the EU's pursuit of rapid liberalization with Africa, Caribbean and Pacific nations through Economic Partnership Agreements (EPAs) may have similar effects and is contrary to their development needs.

harmed by the adjustments: workers who lose their jobs would need welfare payments and training programmes so that they can find new jobs; enterprises that must change their line of production should get support to carry out these changes. In many countries, the damage caused by the adjustments affected entire regions, leaving few workers or entrepreneurs unharmed. In all these cases, the adjustments may be more painful and more expensive. The developed countries usually have well-organized welfare programmes and the central or local authorities are able to provide the necessary assistance. In the developing countries that rarely have adequate resources, Bhagwati calls on the international development organizations and the donor countries to provide the resources for this assistance to prevent people from becoming hopelessly impoverished. This indeed is what happened in many LA and SSA countries when entire industries were transferred to China or India. The actual implementation of such programmes, however, is not that simple due to several reasons:

- At present, there are no well-organized assistance programmes, either at the international level or at the country level in any developing country, to provide assistance to people who lose their jobs as an effect of structural adjustments in the wake of the globalization process. Thus, in several African countries, large numbers of people lost their jobs when the local textile and apparel manufacturing moved to China.
- Since aid donations are highly fungible, there is no guarantee that the aid provided by donor countries to help a developing country to set up assistance programmes during these adjustments will indeed reach the needy people. A necessary condition for this to work is that the countries themselves will recognize the needs and organize an effective and non-corrupt support programme. There are obvious doubts about that. Thus, for example, in recent years quite a few developing countries had large profits as a result of the sharp rise in the world price of some of their export products, primarily oil and minerals; yet, only a small share of these profits actually reached the poor and the needy.
- The prospect of most poor developing countries, especially in SSA, to provide alternative sources of employment in new industries in which they would be able to compete in the world market are very small. In the absence of PTA s, for textile and apparel in particular, most of these enterprises were transferred to China and India when the Multi-Fiber Agreement entered into force in early 2005, and for reasons that will be discussed in subsequent chapters, these industries are not likely to move back to Africa in the coming years despite the low wages of local workers.

Bhagwati also argues that the multinationals' capital should not necessarily be allowed to flow freely wherever and whenever it is deemed most profitable, and he condemns the lobbying of multinationals for intellectual property protection through the WTO; in his view, trade and intellectual property should be separate issues, and he regards their forced union in the Trade-Related Aspects of Intellectual Property Protection Agreement as a triumph of lobbying over logic: 'Pharmaceutical and software companies muscled their way into the WTO and turned it into a royalty-collection agency simply because the WTO can apply trade sanctions.' Intellectual property is today one of the three legs of a tripod – the other two being trade in goods and trade in services – on which the WTO is supposed to rest. Now other lobbies in rich countries want to inject their own agendas into the WTO. The trade unions want a social clause that would trigger trade sanctions against countries if Western labour standards are not met, and the environmentalists want the same for their causes for protecting nature.

Given that globalization has not followed a set of theoretical ideals and thus leaves much to be desired, how can a developing country still manage to take advantage of the process? What can be done if the world market is not competitive and transnational corporations do control the production or the supply of the products in which the country has a competitive advantage? What can be done if the multinational corporations manage to muscle their lobbying power and shape the Intellectual Property Protection Agreement to further increase their monopolistic power? What can be done if the developing country does not have the resources to provide assistance to those segments of its economy that are hurt when the market is opened to free trade? How should the WTO act when a country is essentially free to determine the pace at which its trade liberalization would be implemented? As discussed above, India is one obvious example: although the country has been one of the main beneficiaries from globalization, its integration into the world economy has been, and remains, very limited and guarded.

The impact of RTAs on the LDCs is mentioned by Bhagwati without being fully discussed. What is important to note in this context is that, in recent years, it has become evident that the large trading countries are using regional and bilateral trade deals to attain concessions from their trading partners that they cannot get at the WTO, and this has serious implications for the development of poor countries.[7] As many as 25 developing countries have already signed free-trade deals with developed countries, with more under negotiation. In total, there are more than 250 regional or bilateral trade agreements in force, governing 30% of world trade. These agreements require enormous irreversible concessions from developing countries and demand almost nothing in return from rich countries. These agreements envisage much faster liberalization and stricter intellectual property rules than the WTO and prohibit measures that could be imposed in order to protect the poorest people. The implications for development are significant. In the first 10 years after the NAFTA, Mexico lost 1.3 million agricultural jobs. Manufacturing jobs were initially created, but the Agreement on Textiles and Clothing (ATC) that abolished all trade quotas on textile products and apparel led to the loss of 200,000 jobs.

The impact of globalization on wages and on farmers' incomes has been perhaps the most controversial subject among scholars, policy makers and the

[7] These effects are discussed in the Oxfam report: *Spread of Free Trade*, 20 March 2007.

general public. Despite the rapid growth of the emerging economies, no one should expect a rapid rise in wages in the Far East or even in Eastern Europe. Although their growth brought a very large increase in employment, millions of farmers and slum dwellers in these countries are still waiting for industrial employment, which continues to create a downward pressure on wages. As a result, wages of low-skilled workers in the Far East are increasing at a very low pace and that rise does not slow down the inflow of foreign investments in new industries. It has been estimated that if wages of unskilled workers in the Far East and India continued to rise at the same pace in the next 20 years, they would rise by only around 80%. Given the training that these workers will have during these years and the improvement in the industrial infrastructure in these countries, the rise in productivity of local industries would more than compensate for the rise in their wages.

In all countries, the decision of entrepreneurs to keep their industries and the jobs that they provide in a given country is ultimately determined by their profits and the rate of return on their investments. Simply stated, this decision depends on whether the invested capital can be turned into more capital. Capitalists would invest only in countries where they have profits and where their capital can grow, and as the restrictions on the flow of capital would be gradually removed, capital movements would rise steeply. In 1980, total FDIs amounted to only US$500 billion. Today, worldwide direct investment has jumped to US$10 trillion, an increase of almost 2000% in only 25 years.

Workers would, in principle, make similar decisions: if they were able to migrate freely, they would go to the place or to the country where they can expect to have decent wages and a good standard of living. But whereas capital is free to move and is welcome almost everywhere, workers are not. For high-skilled workers, the more developed countries now offer special incentives to attract them, but practically all the developed countries are now closing their borders to low-skilled workers. Commodities are traded with relatively few restrictions, and outsourcing or offshoring enable people in low-income countries to take the jobs of people in the developed countries; capital can also move freely across national borders, but the low-skilled workers themselves are not allowed to cross.

The globalization process effectively expands the global labour market through trade and through FDIs, but the restrictions on the movements of labour are becoming increasingly tighter. The production process and international trade are generating trade in all factors of production, including capital goods and in human labour either through trade in commodities or through transfer of capital and, to a more limited extent, migration of labour. In countries that take active part in global trade, many workers have new opportunities and wages are gradually rising. For these workers, the globalization process gives great hope. But with the movement of working places as an effect of more open trade and foreign investments, millions of other workers in both developing and developed countries are losing their jobs, the wages of many others are declining, their work places are no longer secure and working conditions are often worsened.

Since the early 1980s, as the EA countries joined the global labour market, and later, when India, the East European countries and the countries of the former Soviet Union also joined the world trade system and the global labour market, the global labour force increased by some 1.2 billion workers who flooded and overwhelmed the West's 350 million well-trained, but expensive workers, who until then dominated global production. This process has squeezed down not only the wages in the developed countries, but also their trade unions as millions of workers were willing to underbid the workers organized in the trade unions.

In the last decade, wages started to rise in Eastern Europe and also in some of the countries of South and EA, but they are falling in the developed countries. Wages of low-skilled workers remained low in China and India under the pressures of the tens of millions of rural workers who were eager to join the labour force. In Western Europe, nearly 20 million people were unemployed and millions of others left the labour force, bringing the total number of unemployed workers in Europe to about 30 million. These major changes in the global labour market and in the wages and well-being of millions of people had a major impact on their position for or against globalization, and in the last decade, these developments led to significant changes in the public opinion vis-à-vis globalization.

The most significant and potentially more commanding changes in public opinion took place in the developed countries and reduced the willingness of people there to accept free-trade policies. The gradual de-industrialization in these countries with the transfer of millions of jobs to the developing countries has created a new underclass of unemployed or underpaid. They have become a serious opposition to globalization and a rising power in all developed countries. The people whom we can categorize as 'modern-day proletariat' are not hungry, they have a roof over their heads and they have some health insurance, thanks to the benefits of the welfare state. These benefits are no longer as generous, however, as they once were and they are no longer secured. Since they are not organized and do not have a class identity, they have a much smaller impact on the political struggle. But the rise in their numbers increases the pressure on the political parties and on the political process itself, and even parties that were in the past pro-trade – like the Republican party in the USA – are cooling to the multilateral trade agreement that has been negotiated at the Doha Round, and there are stronger pressures in the US congress to raise certain tariffs or to strengthen RTAs.

6 The Human Face of the Next Phase of Globalization

Two effects of globalization are of particular concern to all people in all countries, researchers, policy makers and the general public alike: one is the impact of globalization on poverty and the other is the impact on income inequality. The impact on poverty has been, and remains, the subject of research, extensive discussions and heated debates in the context of the developing countries, but it is becoming also a source of growing concern for workers, farmers and other people in the developed countries. The very different effects of globalization on the different groups of developing countries were not evident until the mid-1990s, because for a long time, the gap between the group of developing countries that became the world's emerging economies and the group of developing countries that became the world's LDCs was not as wide and noticeable.

The more visible impact of globalization in the EA countries at the initial stage of their industrialization was the widening gap between the urban and the rural population. This development corroborated essentially the first part of the hypothesis put forward by Kuznetz according to which in the first stages of industrialization income inequality is rising. According to this hypothesis, however, at later stages, income inequality would decline, whereas in China, the rise in income inequality still continues and even accelerates after nearly three decades of rapid growth. The EA countries also benefited from a rapid decline in poverty that was particularly exceptional in China.

The developed countries were, during these years, the main beneficiaries from globalization due to the decline in the prices of many consumer goods, and the impact on their employment was not yet noticeable. In India, the decline in poverty during the 1980s was mainly in rural areas owing to the Green Revolution; only in the early 1990s, a more rapid process of industrialization started as an effect of the economic reforms, and, as a result, the income of the urban population also started to rise.

The findings in the studies of Dollar and Kraay (2002), Chen and Ravallion (2001) and others,[8] which showed that the size of the world's poor population was falling, were therefore very exciting and seemed to provide clear evidence that the process of globalization is working and reaching also the world's poor. These trends also provided support and encouragement for the efforts of the international development organizations to promote policies that would widen the scope and accelerate the process of globalization by showing that an increase in world trade and policies of opening the markets of the developing countries to free trade can help them to achieve a more rapid reduction in their poverty. These results were taken as a clear proof that the policies recommended in the Washington Consensus were not merely a textbook recipe for the correct ingredients of the right economic policies, but also a proven prescription to remedy the developing countries' malaise and reduce their poverty.

With the passage of time, when more results from household surveys became available and more experience was gained, it became increasingly more evident that the reduction in poverty concentrated in China and India, whereas the majority of the developing countries were hardly affected by the global economic growth and by the rapid rise in world trade. These countries remained essentially stagnant, their share in world trade declined and the number of their poor actually rose. Many of these countries became the world's LDCs.

[8] These studies are discussed in detail in Chapter 2, which also includes a complete list of references.

Several studies were conducted following the studies of Dollar and Kraay on the trends in global poverty and income inequality. These studies also aggregated the entire world population in order to evaluate the changes in global poverty after dropping the artificial national borders and thus removing all differences in the identities of different individuals and treating them as equals. The results continued to show that world poverty as well as global income inequality were declining.

Studies at the regional and the national levels, however, painted a different picture: they highlighted the marked and growing inequalities between individuals within many of the countries and the growing gap between regions and between countries. The main differences were between the two subgroups of developing countries: the emerging economies and the LDCs. Whereas China and India as well as most other EA countries were leading the group of emerging economies with growth rates in the 1990s registering world records, in the LDCs, particularly in SSA, average per capita income declined.

The majority of the world's poor population is still concentrated in India and China despite their rapid growth. The sharp drop in the number of their poor and the unparalleled increase in their national income during the last decade are clear indications, however, that these two countries are on their way to resolve their poverty problem. Not all segments of their population and not all regions in these two countries benefited equally from their countries' growth. In China, the income gap between urban and rural households and between provinces has grown very significantly. According to most estimates, mean per capita income in urban China is more than triple than in rural areas, giving China one of the highest urban–rural income ratios in the world. (The gap is likely to be smaller, however, after adjustments are made for the differences in purchasing power.[9]) The rapid rise in income inequality slowed down the reduction in poverty and it became a source of great concern for the Chinese authorities that prompted them in recent years to make larger investments in the country's rural and more remote regions in order to accelerate their growth.

In India, the large income gap between population groups was partly due to the gap between the urban and the rural populations and partly due to the gap between skilled and unskilled labour; correspondingly, this accelerated the growth in the southern states to 7–10% a year, whereas the northern states were stagnating with an annual growth rate of 3–4%.[10] Despite the large and growing income gap, the political ramifications are still rather muted, to some degree, because these differences are quite closely correlated with caste differences that are still accepted as a preordained fate, but mostly because India's democratic regime gives an outlet to the poor to express their discontent. The

[9] Ravallion and Chen estimated that the Gini measure of income inequality in rural areas increased from 25.7 in 1985 to 36.5 in 2001; in urban areas from 17 to 32.3, and at the national level from 26.5 to 39.5.

[10] Devarajan, S. and Nabi, I. (2006) Economic Growth in South Asia: Promising, Un-equalizing,... Sustainable? World Bank.

Indian government has also been making determined efforts in recent years to integrate the poor and to raise their incomes by making large investments in education.

With these developments in China, India and the other emerging economies, the gap between these countries and the LDCs deepened since the mid-1990s despite the financial crisis in EA in 1997–1998 and in Latin America in the early 21st century, and the differences between their growth and standard of living became increasingly more evident. At one end of the global income distribution, the rapid growth of the emerging economies accelerated the convergence between their economies and the economies of the developed countries and gave these countries a leading role in the world's economy. At the other end, the stagnation of the LDCs led to a widening divergence between the LDCs and the rest of the world's economy. While global economic growth was soaring, Africa continued to lose ground and the continent's share of world exports fell from 4.6% in 1980 to 1.8% in 2000, and its share of world imports declined from 3.6% to 1.6%. Africa's share in the world's FDI also fell from 1.8% in 1986–1990 to 0.8% in 1999–2000, largely because Africa did not share the exploding rise in the flow of investments from the developed to the developing countries.

These opposing trends of convergence between the emerging economies and the developed countries and divergence between the LDCs and emerging economies were not clearly visible in the 1980s and early 1990s, because at the start of the globalization process, per capita income in SSA was 30% higher than in the EA countries. In the early 1990s, the differences between these two groups of countries were still quite small and they were all lumped together as 'developing countries'. The differences between their per capita incomes increased at a rapid pace during the 1990s, but the practice of lumping them all together as one group of developing countries still persists. This was most evident in the negotiations during the Doha Round, when the group of 90 developing countries (the G-90) joined together to present their demands despite the very large differences between the interests of the emerging economies (the G-20) and the interests of the other developing countries that belong mostly to the group of LDCs.

In the last decade, the rising poverty in the LDCs, primarily in the countries of SSA, has become the core of the world's poverty problem that should be the focus of the development programmes. Until the past few years, nearly all these countries remained stagnant and the size of their poor population continued to rise. In the early years of the 21st century, quite a few of these countries started to grow more rapidly as an effect of the sharp rise in the world price of their main export products, particularly oil and minerals. None the less, in the majority of these countries, only a small percentage of the profits from their exports trickled down to the poor population. Inadequate governance and rampant corruption has been, and remains, the central problem in these countries. As a result, the large amounts of aid and the comprehensive structural adjustment programmes that the IMF coordinated had only a rather meagre effect on their economies and practically no effect on poverty.

The PRGF of the IMF and the World Bank correctly recognized the need that the countries themselves should own and take responsibility for the implementation of these reforms. This could be a long process, however, and the international development organizations would have to plan carefully the ways and means of securing that this process would indeed take hold and produce the expected results. The gradual spread of democracy in a growing number of SSA countries and the successful example of the process in Ghana and later also in Kenya offers a great promise for a more orderly political process in which the government will be at the helm and take full responsibility not only for receiving the money, but also for spending it. One should expect this to be a rather long process, however, and the more urgent and immediate problems of rising poverty are likely to remain in most African countries. Internal conflicts still persist in quite a few countries, despite a noticeable decline compared to the 1990s, and the tragic events in Darfur that led to genocide show that fundamentalism remains a great threat to the continent.

In the global economy, the collapse of the trade negotiations at the Doha Round on the one hand and the expansion of the EU, the efforts to expand and strengthen RTAs, particularly in Latin America and South-east Asia, and the decline in both the economic and the political power of the USA on the other symbolize and usher in the next phase of globalization. While in the second half of the 20th century the global economy was dominated by the USA and the EU, and the main efforts at the global level were to expand the multi-country organization of world trade in a series of trade agreements within the framework of GATT and the WTO, the global economy is now facing the possibility, and the risk, of fragmentation into a multipolar organization that is more competitive and less cooperative. The danger is particularly grave for the LDCs: whereas the multilateral agreements and organizations always recognized the special needs of the LDCs and all the developed countries jointly agreed on the need to help these countries to become integrated into the global economy, a multipolar and competitive structure will weaken that commitment and the willingness to share the burden. Instead, the LDCs are less likely to become equal members of the RTAs and the large trading powers are less likely to reach an agreement on sharing the burden and the costs that will be involved in helping the LDCs to develop their economies and strengthen their integration into the global trade system.

6.1 People's views on globalization

The popular discourse on globalization has tended to blur the lines between the different dimensions of globalization and to discuss its merits and drawbacks as if it were a homogeneous, undifferentiated phenomenon. Indeed, in recent years, many polls on attitudes towards globalization have been conducted – some of which are discussed below – and many of them are marred by a failure to specify which aspect of globalization the poll is addressing.

Findings from polls conducted by the Center on Policy Attitudes of the University of Maryland concluded that, overall, Americans tend to see globalization as more positive than negative and appear to be growing familiar with the concept and more positive about it. A large majority favours moving with the process of globalization and only a small minority favours resisting it. Another poll by the PEW Global Attitudes Project of 38,000 people in 44 countries found that majorities in every nation surveyed are saying that growing business and trade ties are at least somewhat good for their country and for themselves. But it may be too optimistic to go by these polls, as they may also reflect changed circumstances in national economic performance. Good times dampen anti-globalization attitudes, while bad times deepen them.

The Voice of the People 2006 poll of Gallup International, which covers public opinion in 64 countries, found that on average 38% of the population in these countries considers globalization to be good for their countries, 15% believes it is bad, and 31% says it is neither good nor bad. The interviews were conducted between July and September 2006.[11] This public opinion poll found that seven out of ten people in Africa view globalization favourably, while less than a third of Americans and Europeans believe that globalization helps their country. Attitudes vary widely between continents and countries. Poorer countries, particularly in Africa, tend to see the benefits of international integration, while richer countries put greater emphasis on the costs. The common view in all regions is that the process tends to favour the wealthy, and nearly two-thirds of those surveyed think that the rich benefit more. This feeling is especially strong in Western Europe, where three-quarters of the people surveyed have this opinion.

Among the African respondents, nearly three-quarters said that globalization is a 'good thing' for their country. Africans also tend to look favourably on foreign investment. As many as 75% say that attracting such funds is 'necessary and positive'. Only 20% say that it could be 'dangerous'. None the less, nearly two-thirds of the Africans say globalization helps the rich more than the poor, and most of the rest think that rich and poor benefit equally.

An early international poll conducted between December 2003 and January 2004 by GlobeScan and the World Bank also found that Africa was the region most favourably disposed towards globalization. This poll was conducted in eight (relatively more affluent) African countries. About two-thirds of the Africans surveyed believed that globalization had a positive effect on their lives. Although over three-quarters of the African respondents believed that democracy was the best system of government, less than half felt that their own government represented 'the will of the people'. Africans tended to pin high hopes on integration with the global economy, but they did not generally think that poor countries benefit as much as rich countries from free trade and many of them had resentments because they felt treated unfairly by the wealthy countries.

[11] Global Envision. Available at: World Public Opinion.org

In Western Europe, the most common reply is that globalization is neither good nor bad. In the 16 Western European countries surveyed, 40% of the respondents took this neutral view, while 30% said that globalization is good and 20% said that it is bad. Majorities look favourably on globalization in Denmark, Iceland and Sweden, while majorities reject it in Greece, France and Austria. These differences are not very surprising since among the European countries, Greece, France and, to a lesser degree, Austria are threatened the most by competitive imports of agricultural products and textile and apparel, while Denmark, Iceland and Sweden are threatened the least. Foreign investment is generally regarded positively and only one-third of Europeans see it as dangerous. Nevertheless, three out of four Western Europeans say that the rich profit more from globalization than the poor, and most of the rest say that both gain equally.

Opinions in the USA and Canada on globalization are also divided. The most common view is that it is neither good nor bad. Canadians look a bit more favourably on international integration than the other NAFTA partners: about one-third considers globalization as good, and one in five considers it as bad. In the USA, only one in four considers globalization to be good and the same proportion thinks it is bad. A third believes that it is neither good nor bad and the rest say that they do not know. Americans, however, tend to view foreign investment negatively. Nearly half of them say that it is 'dangerous' compared to one-third who view it as 'necessary and positive'.

In Latin America, globalization is a passionately debated topic. With the rise of leftist and socialist movements in many countries and the election of Hugo Chávez as Venezuela's president in 1998, many leaders of leftist parties have won elections, including Luiz Inácio Lula da Silva in Brazil in 2002, Néstor Kirchner in Argentina in 2003, Tabaré Vásquez in Uruguay in 2004; 2006 saw victories at the ballot box for Evo Morales in Bolivia, Michelle Bachelet in Chile, Daniel Ortega in Nicaragua, Rafael Correa in Ecuador, and the re-elections of Chávez in Venezuela and Lula in Brazil. LAs are divided on globalization: 35% say it is good, 15% call it bad and 35% think it is neither good nor bad. Chile, Colombia and the Dominican Republic are the most positive. The ten countries polled are more favourable towards foreign investment. A slim majority (56%) calls it positive, while a third considers it dangerous. Nearly six in ten LAs say the rich are the primary beneficiaries of globalization, while most of the rest say both rich and poor benefit equally.

Nevertheless, the process of globalization seems inevitable to the LAs and the policy debate focuses instead on how it will be shaped in the coming years. It is recognized that globalization provides not only greater economic opportunities but also a remarkable resilience to events that in the past would have proven highly disruptive. Moreover, most of the investment opportunities are related to the global markets more than to the local market, so closing the economies to the outside world is impossible and all countries have to preserve an open economy, increase foreign trade and, in particular, create a good environment for investment – domestic and foreign – into the country. Globalization is also clearly irreversible, even though certain policies of the WTO, the World Bank or the IMF could be changed. It is also necessary to include the underclasses, even though in practice the political and economic system of capitalism favours the few in the privileged elite.

The poll recorded the greatest support for globalization in Chile and Brazil, whereas in Mexico, Bolivia, Ecuador and Argentina less than half of the people surveyed had a positive view on globalization. The leftist governments generally have an anti-globalization and strongly anti-American platform. In the 1990s, in contrast, the dominant political agenda was to open markets and increase trade liberalization. The sharp rise in income inequality in the entire region and the deepening poverty in many countries as an effect of the series of economic crises brought to power the leftist movements. However, Lula in Brazil and Bachelet in Chile have remained committed to open market economics despite being swept into power by populist sentiment.

Eastern and Central Europe are the most sceptical regions on globalization: less than one-quarter of the respondents believed it is good, while one in five said it is bad and the rest did not express an opinion. The East and Central Europeans are divided on the issue of foreign investment, but most of them hold the view that the rich tend to benefit more from globalization.

In Asia and the Pacific, about half of those polled look favourably on globalization. The most positive among the 13 countries surveyed are Taiwan (78%) and Vietnam (75%). In addition to these two countries, Gallup International polled China, India, Indonesia, Japan, Malaysia, Pakistan, Philippines and Thailand.

Another survey in 44 developing countries conducted by the Pew Research Center in 2003 found a common desire for democracy and free markets, and an acceptance of globalization.[12] People worldwide have become aware and are highly supportive of the impact of increasing interconnectedness on their countries and their own lives. In almost all countries, three-quarters or more of those interviewed think children must learn English in order to succeed in today's world. People generally view the developments associated with globalization as good for themselves, their families and their countries, although their approval is often guarded.

Despite wide support for the changes associated with globalization, people are also concerned that some aspects of their lives which are affected by globalization are actually getting worse as they lose many of their unique traditions; many are also worried about losing their jobs. Majorities in 34 of the 44 countries surveyed say that the availability of well-paying jobs has become worse, the gap between rich and poor is growing, the affordability of health care has diminished and the ability to save for old age is much lower. But people mostly blame domestic factors, including their government policies. This is especially true in economically faltering countries in Africa and in Latin America. At the same time, they are inclined to credit globalization for conditions they see as improving, such as increased availability of food in stores and more modern medicines and treatments.

[12] Available at: Views of a Changing World 2003.

The anti-globalization movements have not convinced the public that globalization is the root cause of their economic malaise, but the public does share the critics' concerns about eroding national sovereignty and the loss of national cultural identities. Large majorities in 42 of the 44 countries believe that their traditional way of life has to be protected against foreign influence, but there is less agreement whether consumerism and commercialism represent a threat to their culture and traditions. The rather abstract concept of the 'global economy' is perceived as a great threat, whereas the more specific and strictly economic aspects associated with interconnectedness, trade and travel are viewed in a much more positive light.

People in most countries have a positive view on the symbols of globalization. Large corporations from other countries get a favourable review in much of the world, as do international organizations. In Africa, people express highly favourable opinions about foreign corporations, while in the Middle East opinions are more divided. The most negative opinions of foreign firms are expressed by people in the developed countries of Western Europe, the USA and Canada. However, even in these countries the positive evaluations of multinationals outweigh the negative.

Most people have a negative view of the anti-globalization protesters. In the European countries, where large demonstrations were held, the French and the Italians give higher ratings to the multinational corporations than to the protesters. The main reason, in my view, is that the anti-globalization protests are generally concerned with the environment and with poverty, and most people are less concerned about these issues.

Most people feel, however, that their way of life should be protected from foreign influence, and a large majority of the people in nearly all countries favour tougher restrictions on foreign migrants entering their countries. Overwhelming majorities in the Western European countries surveyed support tighter border controls. The Western Europeans take a much less favourable view of foreign workers from Eastern Europe, the Middle East or North Africa and more than half of them say that the foreign workers are bad for the country.

A BBC World Service poll of 32 nations that was conducted prior to the meeting of the World Economic Forum in Davos in 2006 found highly divergent economic perceptions in countries around the world. In 15 countries, majorities or pluralities saw conditions in their country getting better, while in 17 countries, respondents saw conditions getting worse. Perceptions of the world economy were also divided, with 14 countries seeing it getting better and 15 seeing it getting worse (with 3 divided). Among the 19 countries that were polled in 2005 as well as in the poll of 2006, there were also diverging trends with some growing more optimistic and some less so.[13]

The BBC also found that despite the widespread criticism it has received, the World Bank is widely seen as having a positive influence in the world

[13] The poll of 37,572 people was conducted for the BBC World Service by the international polling firm GlobeScan together with the Program on International Policy Attitudes (PIPA) at the University of Maryland. The 32-nation fieldwork was coordinated by GlobeScan and completed between October 2005 and January 2006.

and was viewed most positively among a list of global economic actors, including the IMF and global companies. On average, 55% rated the World Bank as having a positive influence on the world, while just 18% rated it as having a negative influence. Among the 32 countries polled, in 30 a majority (17 countries) or a plurality (13) rated the World Bank as positive.

The opinions expressed in the polls are certainly different from those one finds in economic journals. Although opinions in these polls may be shifting like quicksands and reflect a tendency to give recent events the greatest influence on the responses, an important opinion that was expressed in all countries and all polls concerns the perception that rich people and rich countries gain more from globalization than the poor. In some countries, this perception is very strong, but it also expresses a mixture of anti-Americanism and more abstract anti-capitalism than a clear view about the rich people in their own countries. In fact, the rather positive opinion in most developing countries (particularly in Africa) of the transnational corporations (which are associated more with work opportunities than with 'exploitation' or corruption) counterbalanced the negative views about the rather amorphic concept of 'capitalism' that Chávez is expressing. Perhaps the most explosive issue that people associate with globalization is international migration. On this issue, the opinions in the developed countries are becoming increasingly more extreme, and the importance of this issue in the national elections in all European countries is an indication for the rise of hostility towards foreign migrants.

6.2 Globalization and the least developed countries

The concern of the book with global poverty and food insecurity makes it necessary to focus the analysis on the LDCs. Although the size of the poor population in these countries is less than half of the world's total poor population, and the majority of the world's poor population is still concentrated in the emerging economies, primarily China and India, the size of the poor population in these countries is declining very rapidly, which is promising a solution to their poverty problem in the coming years. It is also certain that these countries will meet the Millennium Development Goals (UN Millennium Project, 2005), and China, in fact, has already met these goals. However, in the LDCs, long stagnation and decline has led to an increase in the number of poor people, and these countries have now become the core of the world poverty problem. As the world is now proceeding to the next phase of globalization, both the LDCs and the international development organizations are going through a very trying transition period and are facing a number of central decisions about the shape of development strategies for the coming years. The key for these strategies in the next phase of the globalization process is to help the LDCs to integrate their economies into the global economy. Although the process of globalization is now going through a transition that will change its structure, organization and global institutions, it is evident that all countries must become part of the global economy in order to be able to develop their economies and increase the standard of living of their population.

One dimension of poverty that is becoming more specific to, and more prevalent in, the LDCs, as it becomes less common in the other developing countries, is food insecurity. Nearly all low-income developing countries went through a phase in which a large segment of their population suffered from food insecurity and malnutrition, often leading to outright hunger. Until the 1970s, chronic food insecurity was primarily an extreme form of poverty due to insufficient food supplies and relatively high food prices resulting from low productivity in agriculture, and low incomes and purchasing power made it impossible for the poor to provide all their food needs.

Temporary food insecurity afflicted from time to time large areas due to droughts or floods that led to shortfalls in food supply. However, these shortfalls were not the only reason for temporary, but often extreme, shortages. As Amartya Sen (1983) already noted in his classic *Poverty and Famine*, some of the extreme famines occurred without a significant fall in the supply of food. In 2005, some countries in the Sahel were in the grip of a food crisis that was reminiscent of the 1968 and 1973 famines in the entire region, although this time it affected mostly Niger. A sharp rise in food prices due to high foreign demand prompted large sales of the local food for exports, thus raising local prices and aggravating the food scarcities in the home country by steeply reducing the purchasing power of the local pastoralists. As Sen observed in his book, 'In the fight for market command over food, one group can suffer precisely from another group's prosperity' – an observation which has become much more pertinent and prevalent in today's global economy, with the increase in the size of the urban population and with the rise in income inequality.

The Green Revolution in SA during the 1980s and the industrialization in the EA countries also increased their food supply and the income of large segments of their poor population, and while poverty remained prevalent, the poor managed to provide their most essential food needs. The rising levels of food production in the world, most notably in the developed countries, and the growing world trade led to a reduction in food prices and enabled more developing countries to cover their temporary food scarcities when they were afflicted by droughts or floods. The LDCs, particularly in SSA, were not able to gain from these developments; moreover, their food insecurity was increasingly an outcome of their macroeconomic, trade and exchange rate policies, and of the impact of globalization on their economies. The rise in world production, the fall in world prices of many agricultural products and the collapse of many of these prices in the 1990s, led, paradoxically, to growing chronic food insecurity as a macabre outcome of the fall of farmers' incomes. Temporary food scarcities were increasingly caused by price instability of agricultural products on the world markets as an effect of the greater and rapid price transmission with trade liberalization that made it more difficult for the LDCs to import these products due to their balance of payment difficulties, and for the local poor it became more difficult to purchase these products due to the rise in their price in the local markets.

The central question that these developing countries are now facing is how to structure their trade policy and development strategy so that they can benefit from trade and be able to integrate into the global trading system. They have several major difficulties in answering these questions. First, there have been fundamental changes in the structure and organization of production and trade of agricultural products in the recent years that are likely to become more pronounced in the coming years. These changes are due to the rapid diversion of agricultural production into the production of maize and sugar, which are used for making ethanol. The second reason is the huge increase in the demand of China for grains and other food products with the steep increase in local demand. The third, and undoubtedly more critical, factor in the coming years is the growing scarcity of water and the climatic changes that may turn out to be most damaging for the developing countries.

Another factor that is likely to change the structure and institutions of world trade is the failure to reach a trade agreement in the Doha Round. This failure may be a prelude for large changes in the organization of production and trade with the collapse of the multilateral trading system and the trade agreements under the WTO and the fragmentation of that system into many separate regional or bi-national trade agreements. This change will significantly reduce the power and weight of the low-income developing countries, which may be effectively excluded from many of these trade agreements. The impact of all these changes is not clear, but when the LDCs and the international development organizations are planning the development and poverty reduction strategies for the coming decade, they must carefully consider all these changes and the potential impact on the economies.

References

Bergsten, F. (2005) Reform of the International Monetary Fund. Testimony before the Subcommittee on International Trade and Finance, Committee on Banking, Housing, and Urban Affairs, US Senate, Washington, DC.

Bhagwati, J. (2001) *Free Trade Today*. Princeton University Press, Princeton, New Jersey.

Bhagwati, J. (2004) *In Defense of Globalization*. Oxford University Press, Oxford.

Birdsall, N. and Williamson, J. (2002) *Delivering on Debt Relief: From IMF Gold to a New Aid Architecture*. Centre for Global Development and the Institute for International Economics, Washington, DC.

Chen, S. and Ravallion, M. (2001) *How Did the World's Poorest Fare in the 1990s? Methodology*. Global Poverty Monitoring Database, World Bank, Washington, DC. Available at: http://www.worldbank.org/research/povmonitor/method.htm

Collier, P. and Dollar, D. (2001) Can the world cut poverty by half? How can policy reform and effective aid meet international development goals. *World Development* 29(11), 1787–2002.

Dollar, D. and Kraay, A. (2002) Growth is good for the poor. *Journal of Economic Growth* 7(3), 195–225.

Easterly, W. and Levine, R. (2003) Tropics, germs, and crops: the role of endowments in economic development. *Journal of Monetary Economics* 50 (January), 1.

Hausmann, R. and Rodrik, D. (2003) Economic development as self-discovery. *Journal of Development Economics* 72, 603.

Hausmann, R., Pritchett, L. and Rodrik, D. (2005a) Growth accelerations. *Journal of Economic Growth* 10(4), 303–329.

Hausmann, R., Rodrik, D. and Velasco, A. (2005b) *Growth Diagnostics*. John F. Kennedy School of Government, Harvard University, Cambridge, Massachusetts.

Hausmann, R. and Rodrik, D. (2007) Discovering El Salvador's production potential, *Economia*.

Independent Evaluation Office (2005) *Evaluation Report on PRSPs and the PRGF*. International Monetary Fund, Washington, DC.

International Financial Institution Advisory Commission (2000) *Final Report*. US Congress, Washington, DC.

International Monetary Fund (IMF) (2001) *World Economic Outlook*. IMF, Washington, DC.

IMF (2002) *Evaluation of Prolonged Use of IMF Resources*. IMF, Washington, DC.

Irwin, D.A. (2002) *Free Trade Under Fire: An Economist's Look at the History and Complexities of Free Trade*. Princeton University Press, Princeton, New Jersey.

Overseas Development Council (ODC) (2000) *The Future Role of the IMF in Development*. ODC, Washington, DC.

Rodriguez, F. and Rodrik, D. (2000) Trade policy and economic growth: a skeptic's guide. In: Bernanke, B. and Rogoff, K. (eds) *NBER Macroeconomics Annual 2000*. MIT Press, Cambridge, Massachusetts.

Rodrik, D. (1997) *Has Globalization Gone Too Far?* Institute for International Economics, Washington, DC.

Rodrik, D. (2002) Globalization for whom? Time to change the rules – and focus on poor workers. *Harvard Magazine* 104(6), 29–31.

Rodrik, D. (2004) *Getting Institutions Right*. Harvard University, Cambridge, Massachusetts.

Rodrik, D. (2005) Why We Learn Nothing from Regressing Economic Growth on Policies? Harvard University, Cambridge, Massachusetts.

Rodrik, D. (2006) Goodbye Washington consensus, hello Washington confusion? A review of the World Bank's economic growth in the 1990s: Learning from a decade of reform. *Journal of Economic Literature* 44(4), 973–987.

Sen, A. (1983) *Poverty and Famines: An Essay on Entitlement and Deprivation*. Oxford University Press, Oxford.

Stiglitz, J.E. and Charlton, A. (2006) *Fair Trade for All: How Trade Can Promote Development*. Oxford University Press, Oxford.

UN Millennium Project (2005) *Investing in Development: A Practical Plan to Achieve the Millennium Development Goals*. United Nations Millennium Project, New York.

World Development Report (2000/2001), *Attacking Poverty*. Oxford University Press, Oxford.

World Bank (2001) *Globalization, Growth and Poverty. Building an Inclusive World Economy*. World Bank Group, Washington, DC.

World Bank (2005) *Economic Growth in the 1990s: Learning from a Decade of Reform*. World Bank, Washington, DC.

2 Globalization and the Marginalization of the Least Developed Countries

1 Introduction

Since the 1980s, the development of the world economy and international trade is heavily influenced by immense changes in the structure of world production and trade and by the changes in the private and public organizations and institutions that govern the global economy. The proliferation of world trade did not lead to a 'convergence' between countries and working people as the early textbooks predicted, but to increasing inequality between and within countries and a divergence both in the rate of growth and in the standard of living between different regions and countries. Even the reduction in poverty was, in most countries, far lower than could have been expected and, in light of the rapid growth, more could have been achieved. Moreover, despite plentiful production of food, the rise in the standard of living in the LDCs was disappointingly slow and the number of the world's malnourished people remained practically unchanged.

The objective of this chapter is to provide an overview of the developments during the last two and a half decades as a background for the discussion of specific issues and policies in the subsequent chapters. This overview also examines how the developments in the global economy and their implications for the design of economic policies were interpreted and debated in the economic literature in the early discussions on economic growth and international trade, and how these developments are interpreted and debated today. In light of the diverging trends in different regions and countries, this overview must include a brief description of the developments in selected developing countries that may explain some of the reasons for this divergence. The subsequent chapters seek to draw more general lessons from these developments and to explain their main driving forces.

The overview in this chapter concentrates on four regions in which the majority of the world's poor and low-income people concentrate: Latin

America and the Caribbean (LAC), SSA, South Asia (SA) and EA. These regions experienced very different courses of development in the last two and a half decades, but the reasons for this differed from region to region:

- In LAC, the economic developments in most countries were overshadowed by the political developments that augmented the differences between countries and led to considerable instability within many countries at different time periods.
- SSA has become the core of the world's poverty problem, but in the 1950s and 1960s, when they shook off colonial rule, most countries there were far more affluent than the EA countries. In those days, the African countries offered considerable promise for future growth due to their wealth of natural resources, their abundance of low-wage workers and their proximity and close ties to Europe.
- SA and EA were in those days the world's poorest regions, deeply scarred by World War II and the Cold War conflicts and ravaged by internal wars and social revolutions. Practically all the EA countries were controlled by brutal dictatorial regimes that imposed suffocating economic, social and political reforms. The vast majority of their populations lived in rural areas and depended on agricultural production. Only in Japan in the early 1950s, and few years later also in the four Asian countries that later became known as the 'Asian tigers' – South Korea, Taiwan, Hong Kong and Singapore – there were early buds of an industrial sector.

Few economists would have predicted at that time the course of development of these countries two decades later, and the term 'Asian miracle' indeed reflects how unimaginable the economic achievements of the EA countries were. In order to draw the lessons from the region's miraculous economic development, it is necessary to adopt an approach that avoids broad generalizations and instead pays attention to factors that set countries or regions apart.

In SA, economic developments until the early 1990s concentrated in the rural sector and only in the past decade, due to the far-reaching economic reforms in India, the industrial sector started to grow. However, the majority of the population still concentrates in the rural areas, and the rapid growth in India affected only a narrow segment of the urban population. As a result, the equally miraculous growth in India since the early 1990s widened the divide between the two-thirds of the population that live in rural areas and the one-third that lives in the urban centres, and between the top decile of the country's income distribution that reaped most of the gains from their country's rapid growth and the rest of the population. In addition to the growing gap between the urban and the rural sector, there are also rising inequalities within the rural and the urban populations that further increased the country's overall income inequality. Over 300 million people in the rural areas still live in dire poverty, and the rich–poor divide reflects less the urban–rural divide and more the growing gap between states, between districts within states and between skilled and unskilled workers.

In EA, the developments in the entire region since the early 1980s were dominated by the developments in China, but many other countries also experienced high growth. Since the pertinent details are widely available in many sources, this chapter will focus on examining the common denominators of the region's development and the lessons that can be drawn for today's LDCs.

Among economists, there are heated debates concerning not only the course of development of the global economy, but also the likely impact of these developments on the future growth of different countries, as well as different economic sectors and population groups. The controversy over the impact of globalization on world poverty has galvanized the debate over the gains of the developing countries from globalization and, more importantly, over the policies that enabled the developing countries to benefit from globalization.

Another debate concerns the impact of globalization on income inequality. The data of the 1980s through 1997/98 did not provide a conclusive answer, but the accumulation of more extensive, more detailed and more recent data from the developing countries themselves shows how income inequality has been affected. On the one hand, the majority of the developing countries benefited very little from global economic growth, their income increased very slowly and their income gap vis-à-vis the developed countries increased very significantly. On the other hand, a subgroup of developing countries – including both China and India, and therefore the majority of the population in the developing countries – grew very rapidly in the last two decades and became the world's 'emerging economies'. The gap between this subgroup of developing countries and the developed countries has narrowed notably, but within most countries, both developing and developed, income inequality has considerably increased. A combined measure of global inequality that lumps all these different trends together is therefore providing a misleading summary of these different trends.

The impact of globalization on the world's poor has also long been controversial, but recent data are shifting that debate in a different direction. On the one hand, it is now clear that the number of poor people in the world and their share in the world's population has indeed declined in the last two decades. On the other hand, recent data also highlight the fact that this is primarily due to the steep reduction in the number of the poor in China; given its large share in the world's population and its high incidence of poverty in the 1970s, the developments in China were the main reason for the decline in the total number of the world's poor. In most other developing countries, including India, the decline in poverty has been much slower. As a result, the goal of the international community – declared in 2000 in the UN Millennium Development Goals (MDGs) – to accelerate the eradication of poverty and hunger in a world of unparalleled affluence is not likely to be realized.

The current debate about globalization centres on two questions: the first concerns the impact of globalization on the LDCs, and, more specifically, the contribution of globalization to the reduction of poverty; the second concerns likely future scenarios – will globalization and free trade contribute to reduce poverty and food insecurity in the LDCs in the future?

The objective of this chapter is to review the evidence of the last two and a half decades and examine in more detail the policies implemented in selected developing countries in order to reduce the number of the poor. This review serves as a background for the subsequent chapters that discuss the large differences of the impact of globalization on different groups of developing countries and on different economic sectors within these countries.

2 Has Globalization Been 'Good for the Poor'?

The question whether globalization has been 'good for the poor' by contributing to a reduction in world poverty is highly controversial, even though the positive impact of globalization on poverty in different subgroups of developing countries is not in dispute. The impact of globalization on the poor is a much studied subject, and many multi-country and individual country studies have been conducted to shed light on this important question. Most multi-country studies have concentrated on the changes in world poverty since the early 1980s, but a number of studies present a long-term perspective and evaluate the changes and the trends in poverty in the last century. The central questions in these studies remain relevant: Have the benefits from global economic growth trickled down to the world's poor? Has global growth reached the developing countries, where the world's poor are concentrated?

The studies generally confirmed that the accelerated global growth of the last two decades reduced the number of the world's poor and had a positive impact on the countries where the majority of the world's poor population concentrated. As noted earlier, these results were affected by the very impressive impact of globalization in China and the consequent reduction of the large share of the poor in the country's total population during the 1970s. The impact of globalization on today's LDCs and on the poor population in the majority of the developing countries is much less impressive. This section provides a summary of the main findings of these studies as a background for an analysis of the different views in this debate, the reasons for the different conclusions and the methodology used in these evaluations.

In this context, it is important to note that the reduction in poverty cannot be attributed only to the globalization process; other factors, including the various policies pursued by governments and developments that were outside the economic domain must also be taken into account. Ravallion (2004, p. 3) argued: 'One reason why such different views [on the impact of globalization on the poor] persist is that it is difficult to separate out the effects of globalization from the many other factors impinging on how the distribution of income is evolving in the world.' An evaluation of the factors that determined the changes in global poverty with the benefit of the hindsight that we have today should allow a more accurate analysis of the different factors that affected these changes and make it possible to identify the impact of the factors that were driven by globalization.

In a series of multi-country studies, Dollar and Kraay (2001, 2002) found strong evidence that global poverty indeed declined during the period from 1980 through 1997. Moreover, the average income of the world's poorest quintile moved almost one-for-one with average income overall, so that the percentage change in incomes of the poor was on average equal to the percentage change in the average income of the general population. Hence, the population in the bottom quintile of the income distribution gained as much from globalization as the top quintile, so that there was no change in the global income inequality. From these findings, *The Economist* concluded: 'In short, globalization raises incomes, and the poor participate fully' (27 May 2000). There are, however, potential fallacies in the methodology that Dollar and Kraay used to arrive at this conclusion, and the next section, focusing on the methodological aspects of the measures of poverty and inequality, shows why they can sometimes be biased and even outright misleading. The study of Collier and Dollar (2002) '*Globalization, Growth and Poverty:* Building an Inclusive World Economy' draws heavily on the methodology developed by Dollar and Kraay, but, as is shown in Section 2.1, the bias embedded in these measures may lead to dubious and possibly even incorrect conclusions.

Their conclusion that there was a decline in the incidence of poverty in the developing countries has been supported by many other studies and is by now widely accepted, although the specific impact of globalization on this decline is still debated. However, the second conclusion of Dollar and Kraay (2002) that, by and large, and despite wide variations between countries, the poor participated fully in the world's growth and hence global income inequality did not change, is still controversial. Many other studies showed that there was a steep rise in income inequality both within most countries, including many developed countries, and between countries.

Ravallion (2004) suggested another possible reason for the continuing debate about the conclusion of Dollar and Kraay (2002) that global income inequality remained, on average, unchanged. Ravallion noted that there are no basic disagreements about the data on incomes, prices and so on; the disagreements are therefore on the value judgements about the criteria for a just distribution of the income gains that are inherent in the different measures of inequality. However, in Section 2.1 it is argued that the conclusion of Dollar and Kraay (2002) is more likely due to a fallacy of identification, and their findings can have a very different explanation and lead to different conclusions.

Chen and Ravallion (2001) concurred in their own multi-country study with the findings of Dollar and Kraay (2002) that during the 1990s, the incidence of extreme poverty in the developing countries fell from 28% to 21%, and growth during these years was *pro-poor* and contributed to shift the global income distribution, however modestly, in favour of the extreme poor. Their conclusion that there was a decline in global income inequality was also supported by a thorough study of Sala-i-Martin (2006), but it still remains a highly controversial issue. In another empirical study, Ravallion concluded:

> The poor typically do share in the benefits of rising affluence, and they typically do suffer from economic contraction. However, there is a sizable variance around the 'typical' outcomes for the poor . . . [due to] differences in initial inequalities between countries and between regions within countries that create sizable differences in how much the poor participate in aggregate growth or contraction.
>
> (Ravallion, 2000, p. 21)

2.1 The experience in the main regions

The empirical analysis of the trends in poverty has some limitations because in many countries results are still based only on a small number of household income and expenditure surveys. In most countries in SSA and Central Asia, these surveys were conducted on a regular basis only since the mid-1990s and there are often inconsistencies between surveys in the same country at different time periods.

Nevertheless, the available surveys provide a clear picture of the developments in the last two decades. In the early 1980s, over 40% of the population in the developing countries lived in extreme poverty, on a daily income of less than US$1 per person (in 1993 prices, with incomes adjusted to the cost of living in the individual countries – i.e. purchasing power parity [PPP] adjusted), and some two-thirds of the population lived in moderate poverty, on a daily income of less than US$2 per day (in 1993 prices and PPP adjusted). In 2001, the incidence of extreme poverty in the developing countries declined quite sharply to 21%, and the incidence of moderate poverty declined to 53%.

The most dramatic decline took place in EA, primarily in China, where three-quarters of the population were starving and millions died in the 1960s. In 1981, the incidence of extreme poverty in China was still 64%, but in 1991, extreme poverty declined to less than one-third of the population, and in 2001, it fell to 16%. These numbers epitomize the 'East Asian miracle'.

In SSA, in contrast, the incidence of extreme poverty rose from 42% to 46% from 1981 to 2001, and the number of people who lived in extreme poverty increased from 164 million in 1981 to 316 million in 2001.

In SA, primarily in India, the reduction of poverty during the 1980s from 382 million people in 1981 to 357 million in 1990 occurred mostly in rural areas and as a result of the 'Green Revolution'. During the 1990s, there was another cycle of decline in the incidence of poverty in India, though there are some disagreements about the numbers. However, it is widely recognized that the decline was rather small in spite of the country's rapid and accelerating growth driven by India's rapid industrialization following the economic reforms in the early 1990s, but that growth concentrated in the urban centres and in sectors that mostly benefited the high-skilled workers. In the late 1990s and early 21st century, an increasing number of poor people also benefited from industrialization, some by migrating to the urban areas, others due to the spread of low-skilled labour-intensive industries to more remote regions and nearly all by the flow of remittances to the rural areas from family members who managed to settle in the urban centres.

In other SA countries and in many other developing countries as well, the decline in poverty was marginal and their poor did not share the benefits of globalization and the rising affluence in the world or even in their own countries. It is the sharp fall in the number of the poor in China that provides the statistical explanation for the decline in the number of the poor and in the incidence of poverty in the last two decades despite the very small decline and often even the rise in poverty in many other developing countries, particularly in the LDCs. The concentration of the decline in

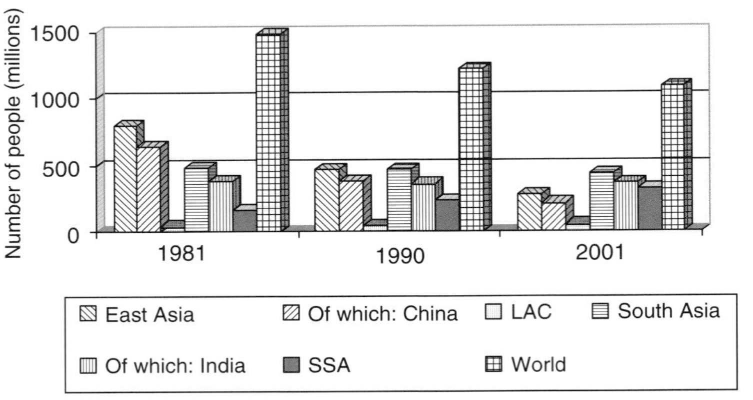

Fig. 2.1. The number of people living in extreme poverty. (From Chen and Ravallion, 2004.)

poverty in a relatively small subgroup of developing countries and the concentration of the poor in other developing countries, primarily in SSA, contradicts the rather sweeping conclusion that 'globalization raises incomes, and the poor participate fully', or the conclusion that the poor shared the benefits of the rising global affluence. Figure 2.1 shows the diverging trends between the sharp decline in poverty in China and the rise in poverty in SSA.

In a paper published in 2004, Ravallion made the following observation:

> The 1990s saw reasonable growth in the developing world as a whole; the overall rate of growth in real per capita private consumption for the low- and middle-income countries over 1990–97 was 2.6% per year. Yet 24% of the population of the developing world lived below $1 per day in 1998, which was only 3–4 percentage points lower than 10 years earlier. The total number of poor by this standard was about the same in 1998 as in 1987, when roughly 1.2 billion people lived below $1 per day.[1]

However, the findings of Chen and Ravallion themselves do not seem to agree with this conclusion. The large differences in the growth rates of different countries in different regions, the decline in the per capita income in most SSA countries and the rise in the number of the poor there do not square with the conclusion that 'the 1990s saw reasonable growth'. This conclusion is based on an average growth rate for the entire group of the developing world that was calculated for a large number of countries. Given the very large differences in the growth rates of different countries within this group during this period, the decline in the income in most SSA countries and the extremely rapid rise in the incomes of the poor in China and many other EA countries, the average growth rate is, in fact, meaningless.

[1] These figures are based primarily on the household surveys and are therefore somewhat different from the figures noted earlier that use also the income surveys.

3 The Changing Faces of the World's Poor

There is another reason to be sceptical about sweeping interpretations of the results of the multi-country studies: the reduction in poverty that these studies find is due not only to the concentration of the decline in poverty in a small number of countries, but also to a consistent bias in the methodology used to calculate the changes in poverty, and while this bias does not affect the headcount measure of poverty – though, as we have seen, it may affect its interpretation – this methodology may bias the other poverty measures that are also affected by the income gap of the poor below the poverty line.

In the common measures of the change in poverty, the change in the incomes of the poor is calculated by comparing their mean income in the base year, say 1981, with their income in, say, 2001. The annual growth rate in the incomes of the world's poor is calculated on the basis of this comparison, and the poor can be defined by an absolute poverty line, as done by Chen and Ravallion (2001, 2004), and as done commonly in studies on the developing countries, or as the lowest percentile – quartile or quintile – in the world's income distribution, as done by Dollar and Kraay (2002). In their study, Dollar and Kraay found that the average income of the world's lowest quintile grew during the 1980s and 1990s at an average annual rate of 3.2%, whereas the incomes of the world's highest quintile grew at a rate of 1.9%. On these grounds they concluded that global growth had been *pro-poor*, by tilting the global income distribution in favour of the poor, and therefore their conclusion was that global growth has been 'good for the poor'.

This measure of the change in the incomes of the poor does not take into account the changes in the *composition* of the world's poor during these years as an effect of the highly skewed distribution of the growth in income across regions and countries. By ignoring the changes in the composition of the world's poor, the measures of the changes in their income and income inequality may bias the estimates of the changes in the poverty gap and income inequality.

To illustrate these effects, consider the estimates in Table 2.1 of the gross national product (GNP) per capita in the main regions of the developing

Table 2.1. GNP per capita in main regions.[a] From Arrighi *et al.*, 2003; calculations based on World Bank, World Tables, 1984; World Development Indicators, 2001.

Region	1960	1970	1980	1990	1999
Sub-Saharan Africa	5.2	4.4	3.6	2.5	2.2
Latin America	19.7	16.4	17.6	12.3	12.3
West Asia and North Africa	8.7	7.8	8.7	7.4	7.0
South Asia	1.6	1.4	1.2	1.3	1.5
East Asia (without China and Japan)	5.7	5.7	7.5	10.4	12.5
China	0.9	0.7	0.8	1.3	2.6
Developing countries	4.5	3.9	4.3	4.0	4.6

[a]As % of developed countries' GNP per capita.

countries. These estimates were calculated as percentage of the income per capita in the developed countries.

Consider the change in income inequality in the subgroups of developing countries: the income gap between the poorest and most affluent subgroups of developing countries increased from 21.9 in 1960 to 22.0 in 1980 and then declined to 9.5 in 1990 and 5.7 in 1999. However, during these years, there were fundamental changes in the identity of the poor: until 1990, the lower quintile mostly included the poor in China. Only in the early 1990s, that quintile included a larger share of the population in SSA that dropped to the bottom, while a growing share of the EA population rose to higher quintiles. Moreover, by 1999, the population in the group of EA countries, which did not include China and Japan but included the four 'Asian tigers', became the most affluent group among the developing countries.

3.1 The impact on the measure of income inequality

Dollar and Kraay (2002) also concluded that the income of the bottom quintile in the income distribution of the population in the countries in their study gained as much from globalization as the top quintile, so that there was no change in global income inequality, and the poor participated fully in global growth. In fact, the measure of income inequality that is based on the income gap between the top and the bottom quintiles regardless of their identity is quite meaningless in this case and is the reason for the results.

The main problem in using this measure for analysing the income changes during the 1980s and 1990s is that it ignores the large differences between the countries that had been at the bottom of the world's income distribution and the changes in the national identity of the population at the lowest quintile. From Table 2.1 we can see that this measure neither conveys, for example, the rise in the income gap between LA and SSA from 3.8 in 1960 to 5.6 in 1999, nor shows the rise in the income gap between EA and SSA from 1.1 in 1960 to 5.7 in 1999. In 1980, the mean income in China was less than a quarter of the income in SSA; in 2000, the mean income in China was nearly 20% higher than the mean income in SSA.

Figure 2.2 offers one way of observing why the conclusion that income inequality remained unchanged is, in fact, misleading. In Fig. 2.2, countries are defined by their mean income in 1980 and 2000 relative to the world's mean income in 2000, and show their distribution according to their growth rates in 1980–2000. Since countries are defined only by their income, anonymity is ensured, except that the 'anonymous dots' on the diagram are population-weighted. The figure also shows the regression lines that were calculated on the basis of these income ratios. In the upper diagram, the regression line is lower than the 'zero' line at low ratio and rising above 'zero' line at higher ratios; in the lower diagram, the opposite is the case: the regression line is higher than the 'zero' line at low ratios and is converging to the 'zero' line at higher rations. How can we explain the diametrically opposed behaviour of the regression line?

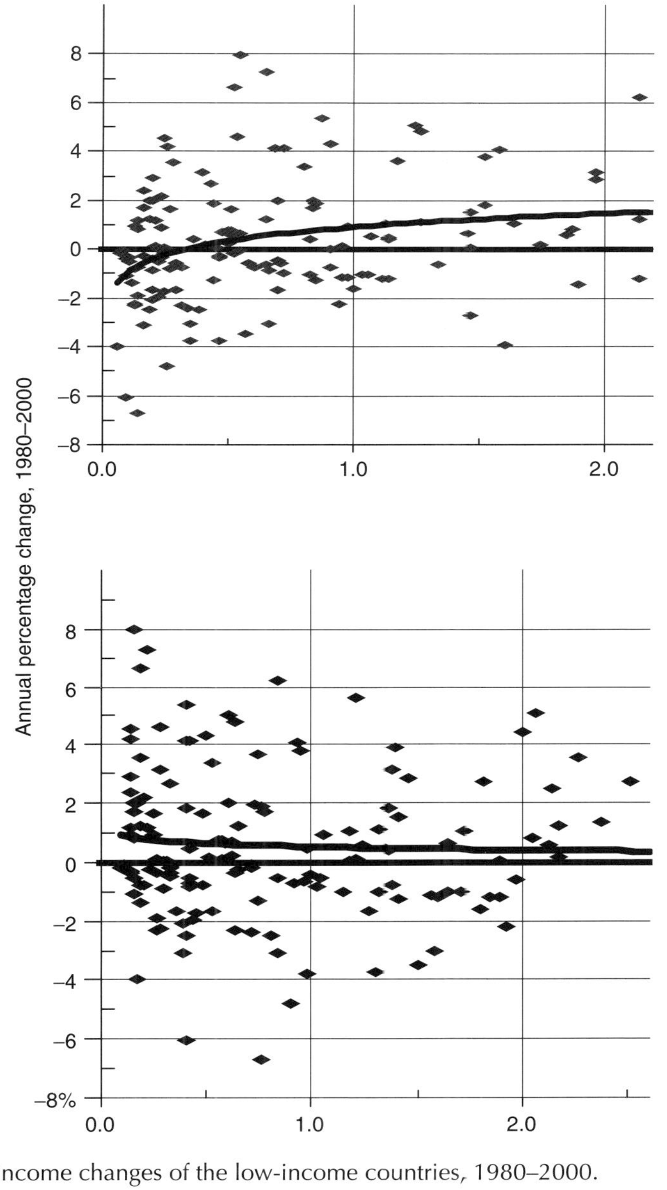

Fig. 2.2. Income changes of the low-income countries, 1980–2000.

If the 'dots' were randomly distributed, then the regression line would have been parallel to the 'zero' line. But, in fact, we know that this was not the case: the reason is that the countries that had been the world's poorest in 1980 (and were thus distributed to the left of the vertical 1.0 line) – including China and most other EA countries – grew during these two decades at a higher rate than most of the affluent countries that included most of the SSA countries. The incomes of the population at the bottom of the world's income distribution in 1980 grew during these years at an average annual rate of 2.3%. The larger population in these countries 'pulled' the regression line above the 'zero' line at low incomes in the lower diagram, but the regression line was negatively sloped and was converging towards the 'zero' line.

In the upper diagram, most of the world's poorest countries in 2000, with mean incomes below the world's mean income during that year (and thus distributed to the left of the vertical 1.0 line) were the SA and SSA countries. The incomes of the population in these countries declined at an average annual rate of −1.9%, whereas the incomes in most other countries were rising. The smaller weight of the population in these countries in the world's total population 'pulled' the regression line in the upper diagram below the horizontal 'zero' line at low incomes, but the regression line was positively sloped and was rising above the 'zero' line.

We can now use these findings to evaluate the conclusions that Dollar and Kraay (2002) came to: the growth rate of the individuals that constituted the poorest quintile was higher during the 1980s, when China was still at the bottom of world's income distribution, than the growth rate of the mean income of the world's population. The world's mean income was slowed down by the decline in the income in most SSA countries and the slow growth in many SA and LAC countries. In the 1990s, the trend was reversed: the lowest quintile in the world income distribution in that decade included an increasing share of the poorest individuals in SSA, India, other SA countries and later also in Central Asia. The decline in the incomes in the SSA countries and the slow growth in the other poor countries slowed down the rise in the incomes of the lowest quintile. The higher quintiles and the world's mean income, in contrast, then included a larger share of the population in China and the other emerging economies, and their rapid growth also accelerated the growth of the world's mean income. A trend line of growth rate of the lowest quintile over the two decades that ignores these changes in the identity of the poor 'averages out' the differences in the growth rates of the different population groups that were included among the poor at different time periods. This is the reason why the trend line of the growth of the lowest quintile was more or less parallel to the trend line of the growth rate of the mean income, and the corresponding income elasticities therefore remained nearly the same.

The reason for going into these details is to determine whether the results of Dollar and Kraay (2002) were coincidental and represent the very unique circumstances during the last two decades, or whether one can draw a more general conclusion from the experience of the last two decades about future developments. Can we expect that future growth, which may well include many of the SSA countries, will indeed be shared fully and the poor will benefit equally, or is future growth more likely to increase income inequalities between and within countries? Will future growth reach the rural poor in SSA, SA, LA and even China, or will these poor be excluded, and should we expect diverging trends between the rapid growth of the developed and the emerging countries and near stagnation and deepening poverty in the other countries? Obviously, the results of Dollar and Kraay (2002) and of Chen and Ravallion represent observations that are relevant for the specific periods and the specific countries covered in their studies. The next step is to define a more general model that offers an explanation for a wider set of countries.

Finally, Table 2.1 does not give information on the poverty line income, but it can be assumed that in the illustration absolute poverty remains unchanged over time except for adjustments to the price index. The ratio between the poverty line income and the GNP per capita in the developed countries is therefore changing over time with the growth of these countries. Consider, as an illustration, a poverty line income equal to 2.3% of the GNP per capita in the developed countries. In this case, the poor population included until 1990 the populations in China and SA; in 1999, however, there was a steep decline in the world's headcount measure of poverty when the mean income of the Chinese population rose above the poverty line, despite the decline in the mean income of the population in SSA below that line.

These examples are only illustrations that highlight the changes in the identity of the world's poor due to the very different effects of globalization on the poor in different countries. Can we conclude from these results that globalization has been good for the anonymous poor or should we rather emphasize that globalization has been good for the poor in China and few other emerging economies, but has been bad for the majority in the other developing countries?

Do the measures of the income gap between the poorest and the most affluent segments of the population, regardless of their identity, or the measure of the poverty gap of the (anonymous) poor below the poverty line provide any useful information? Can we understand better the convergence in the incomes of the EA population with the incomes of the developed countries, the divergence in the incomes of the other developing countries and the incomes of the emerging economies?

3.2 Evaluating the changes in poverty and income inequality

To provide a more general evaluation of this effect, define the 'poor' as the segments of the population whose per capita income falls below the poverty line, given by: $\Psi_t(p) = \{\varphi{:}y_t(\varphi) \leq z\}$, where z is the poverty line. The incomes of these poor are given by: $Y_t(p) = \{y_t(\varphi){:}y_t(\varphi) \leq z\}$, where $y_t(\varphi)$ is the income of the φ-percentile and their share in the general population determines the headcount measure of poverty, given by:

$$H_t \equiv w_t(p) = \Sigma_{\varphi \varepsilon \psi_{t(p)}} w(\varphi); \ (0 < H_t < 1) \tag{2.1}$$

where $w(\varphi)$ is the share of that percentile/subgroup in the population. When the poor are defined as the lowest 'quantile' (quintile, quartile, etc.), they include all the percentiles on the (ranked) income ladder with lower incomes and their share in the general population is equal to that quantile.

The *poor*'s mean income is a population-weighted mean of the incomes of all the poor percentiles:

$$y_t(p) = \Sigma_{\varphi \epsilon \psi_{t(p)}} [w(\varphi) / w_t(p)] \bullet y_t(\varphi) \tag{2.2}$$

The *change* in the *poor*'s mean income is thus a weighted average of the income changes of all the poor individuals or poor percentiles, and the weights are their shares in the *poor*'s total income:

$$\hat{y}(p) = \Sigma_{\varphi \epsilon \psi_{t(p)}} [S_t(\varphi) / S_t(p)] \bullet \hat{y}(\varphi) \tag{2.3}$$

where a hat over a variable indicates its percentage change over time; $\hat{y}(\varphi)$ is therefore the percentage change in the income of the φth poor percentile, given by $\hat{y}(\varphi) = \{[y_{t+1}(\varphi)/y_t(\varphi)] - 1\}$, and $S_t(\varphi)$ and $S_t(p)$ are the shares of the corresponding percentile and of the entire poor population in the national income at time t.

The ratio $[y_{t+1}(\varphi)/y_t(\varphi)]$ is calculated regardless of the identity of the individuals classified in that percentile at the two time periods, although over time their identity may change if the incomes of the individuals who were ranked at the φ percentile at time t increased at a more rapid pace and at time $(t + 1)$ other individuals dropped to that rank on the income ladder. Hence, over time, the ratio $\{[y_{t+1}(p)/y_t(p)] - 1\}$ estimates the percentage change in the poor's mean income, even though the incomes at the two time periods $y_t(p)$ and $y_{t+1}(p)$ may estimate the mean per capita income of two possibly different groups of individuals that were classified as poor at each of the two time periods, instead of comparing the change in the incomes of the *same* individuals.

The changes in the composition of the poor are disregarded in these comparisons under the ethical prerequisite established by the *anonymity* (or *symmetry*) axiom which is stipulated in order to guarantee that all individuals are treated impartially. Poverty and inequality should therefore be a function of the levels of incomes irrespective of the identity of the individuals who earn these incomes or any of their personal characteristics.

In the measures of poverty or income inequality at a given point of time, anonymity is an obvious requirement aimed at securing equal treatment for all and avoiding any possibility that the aggregate measure and the policies guided by this measure will be affected or biased by the personal identity of the poor. Thus, the goal of this axiom is to guarantee that the representation of individuals in the aggregate measure would not depend on their place of residence, their country of origin or any of their individual characteristics such as gender, skin colour or ethnic group. Instead, the only criterion that should be relevant for allocating benefits and targeting poverty alleviation policies should, as a matter of principle, be the level of the individuals' incomes.

In an a posteriori evaluation of changes in poverty, the ethical considerations underlying the anonymity axiom may, however, be more complex. There may still be reasons to make the evaluation under the veil of anonymity and compare the mean income or the mean poverty gap regardless of the identity of the poor in order to evaluate the success or failure of past policies to raise incomes regardless of the personal identity of the poor. However, a more complete evaluation of past policies could determine their effects on different subgroups among the poor, and this can be done only by explicitly identifying the subgroups that gained and rose out of poverty and the subgroups that lost and fell into poverty. By identifying these subgroups, especially those that lost, future policies can be targeted on these poor and thus be more effective.

This can be seen most clearly by introducing different notations to identify the reference population and determine the time period at which incomes were measured. In the present context, the reference population is determined according to the time period at which these individuals were classified as *poor*, or ranked in a given percentile. Thus, for example, the set $I_t \equiv \Psi_t(p) = \{\varphi: y_t(\varphi) \leq z\}$ consists of all the individuals/percentiles that were poor at time t. Under the veil of anonymity, these individuals are identified only by their rank on the income ladder. In the following time period, however, some of the individuals in I_t remained poor, although their position on the income ladder may have changed, and some others managed to escape poverty. However, some individuals who were not poor earlier drifted into poverty at time $(t + 1)$. The group of poor individuals at time $(t + 1)$, denoted by I_{t-1}, can therefore be different from the group at time t, I_t. In addition, the general population continues to grow, but different population subgroups grew at different rates and their shares in the population also changed.

The growth rate of the mean income of the individuals in I_t is calculated by comparing their mean incomes at the two time periods and given by $\{[y_{t+1}(I_t)/y_t(I_t)] - 1\}$. This growth rate is likely to be different from the growth rate calculated by $\{[y_{t+1}(p)/y_t(p)] - 1\}$, which compares the incomes of the individuals who were poor at time t with the incomes of the (possibly other) individuals who were poor at time $(t + 1)$. Using the above notations, the latter ratio can be written as $[y_{t+1}(I_{t+1})/y_t(I_t)]$, and this notation clarifies that we compare, in fact, the incomes of two different groups of individuals.

Consider how these calculations can bias the estimation of the growth rate: the method of identifying the poor and calculating their mean income implies that at time t: $y_t(I_{t+1}) > y_t(I_t)$. In other words, the incomes of those who later drifted into poverty were higher at time t than the incomes of those who were poor at that time. The rate of growth calculated by the ratio $[y_{t+1}(I_{t+1})/y_t(I_t)]$ is therefore larger than the rate of growth of the individuals in I_{t+1} but smaller than the rate of growth of the individuals in I_t. In fact, this is exactly the result of the ratio $[y_{t+1}(I_{t+1})/y_t(I_t)]$ when we compare the mean income of the poor in 1981 with the mean income of the poor in 2001. This ratio compares the mean income of the lowest quantile in 1981, and most of the people in that quantile were the poor Chinese, with the mean income of the lowest quantile in 2001, when most of them were from SSA.

In 1981, the mean income of the poor in SSA was 7% higher than the mean income of the poor in China; in 2001, the mean income of the Chinese poor was 36% higher than that of the African poor. During these years, the mean

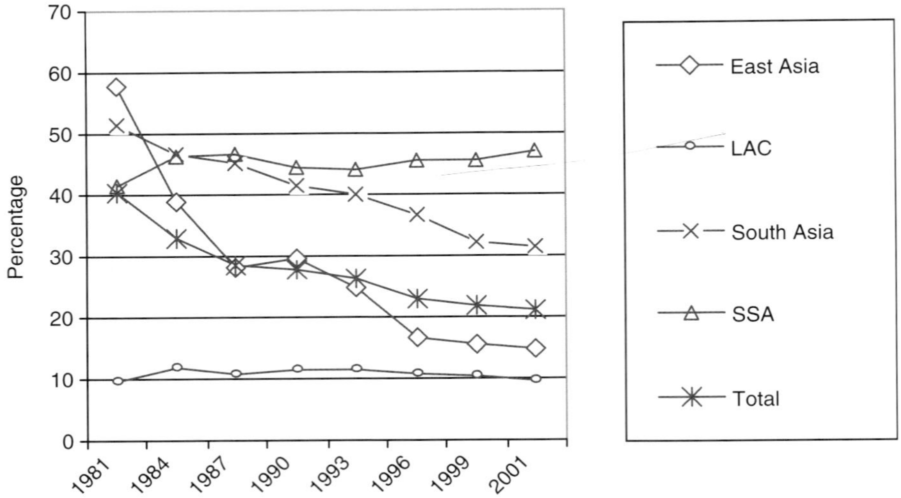

Fig. 2.3. Headcount measures of poverty: percentage of the population living below US$1 per day.

income of the Chinese poor increased by 42%, while the mean income of the world's poor increased by 32% and that of the African poor declined by 3%. Obviously, the change in the mean income of the world's poor does not provide any useful information for the analysis of the changes in poverty.

Figure 2.3 shows the changes in the composition of the poor population between 1981 and 2001. In 1981, the headcount measure of poverty in China was nearly 60% and the Chinese poor constituted the majority of the world's poor population. In fact, in the early 1980s, less than 15% of the world's poorest quintile was from the SSA countries, and the average per capita income in the SSA countries was more than 20% higher than in the EA countries. During the 1980s and 1990s, the average per capita income in the EA countries grew very rapidly, while it declined in the SSA countries by an average of 8.3%.

The headcount measure of extreme poverty in China declined during the 1980s at a dramatic pace from 64% to 33%, while it increased in SSA from 42% to 45% of the population. During these years as well as during the 1990s, a large share of the population in SSA and many segments of the population in SA, primarily in the rural areas, dropped to the lowest quintile of the global income distribution, and by 2001, they were the majority of the world's poor, while significant portions of the EA population rose to higher quintiles and out of poverty.

If the growth rate of the average income of the lowest quintile is calculated regardless of the changes in the identity of the poor, it will show neither the rapid growth in the incomes of the Chinese who had been the world's poorest population in the early 1980s, nor the fall in the incomes of the Africans, who became the world's poorest in the late 1990s. Under the veil of anonymity, that measure of growth also does not show the rise in poverty in rural India and the fall in poverty in urban China (Fig. 2.4). In fact, this does not measure the actual

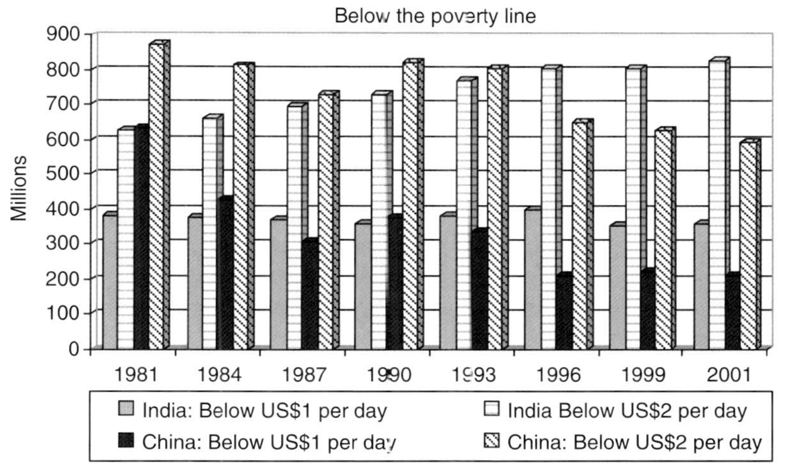

Fig. 2.4. Number of people in India and China who live below the poverty line.

growth in incomes, but rather the difference between the mean income of the Chinese poor, many in urban areas, and the mean income of the African and the poor in rural areas that existed already in the 1980s.

In his book *Imagine There's No Country*, Surjit Bhalla (2002) deliberately disregarded the national identity of the world's population in his analysis of global poverty, income inequality and growth. To some extent, the division of the poor population according to their national identity is obviously arbitrary, and similar problems exist in the analysis of the changes in poverty *within* countries. However, the purpose of identifying the national identity of the subgroups among the poor is not to contest the noble desire to 'imagine that there is no country', but rather to identify the target groups and the geographical regions on which poverty alleviation policies should concentrate. At the country level, effective targeting requires additional criteria for identification: the geographical region within the country, the sector of employment, rural or urban, the gender of the head of the household, etc. These criteria are not intended to create division or discrimination between population groups, but rather to determine consistent measurements that will provide a solid basis for policy making.

The World Bank (2001) study divided the developing countries into 'globalizing' countries (those that integrated well into the global economy) and 'non-globalizing' countries (those that were and remained only marginally integrated into the global economy). In 1980, the majority of the world's poorest people were concentrated in countries that later became the globalizing countries, among them China and India. In 2000, a large and growing share of the world's poorest people were concentrated in the LDCs (even though the majority of the world's poor were still in India and China). An evaluation of the change in income inequality that is based on the change in the income gap between the world's 'anonymous' *poor* and the world's *rich* thus compares the income gap between different groups of people at different time periods: in 1980, this gap is the ratio between the incomes of the poorest people – who were then in the future 'globalizing' countries – and the world's rich people; in 2000, this gap was the ratio between the income of poorest people – who were then in the 'non-globalizing' countries – and the world's rich people. The change in that ratio does not show the large increase in the gap between, say, the mean income of the African poor and the income of the world's rich that year, or the considerable decline in the gap between the mean income of the Chinese poor and the income of the world's rich that year.

In the studies of Dollar and Kraay (2001, 2002) and Sala-i-Martin mentioned earlier, the changes in the composition of the income quantiles were not taken into account when the income gaps between the quintiles and the deciles were calculated. In Fig. 2.5, the income gap between the top and the bottom deciles measured in the 1970s is the income gap between the Chinese poor and the American rich, whereas in the 1990s, this gap is the income gap between the African and the Indian poor and the American rich. This measure of change in inequality therefore shows neither the *rise* in the income gap between the African poor and the American rich between 1970 and 1998, nor the *decline* in the income gap between the Asian poor and the American rich during these years. The decline in income inequality that is shown in the figure during the 1980s and possibly also during the 1990s is due to the growing share of the African poor in the bottom decile and to the fact that the income of the African poor in the 1970s was higher than the income of the Asian poor in that decade. With the rise in the share of the African poor in the bottom decile, the average income of that decile therefore increased, even though the income of the African poor remained essentially stagnant.

But for the majority of the developing countries, particularly the LDCs, the process did not have the expected effects. Although these countries are labour abundant and wages are low, their industrialization was very minimal and incomes remained stagnant. There were other reasons why these countries effectively drifted to the bottom of the global income distribution, and the subsequent chapters discuss these reasons in great detail, but the empirical conclusions are unequivocal: the income gap between these countries and the developed countries increased very significantly.

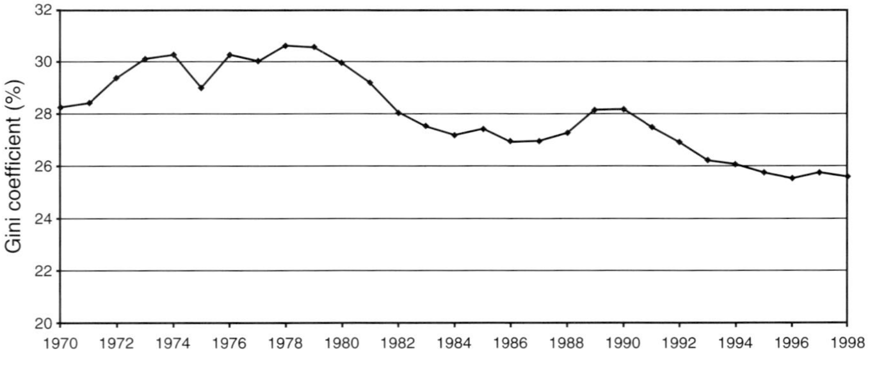

Fig. 2.5. World income inequality. (From Sala-i-Martin, 2004.)

The statistical results of Dollar and Kraay (2002) and Sala-i-Martin reflect the weighted mixture of these two contradicting processes in these two groups of countries: the rising incomes in the emerging economies that were at the bottom of the income distribution in 1970s and the stagnation of the LDCs that, in the 1970s, were ranked in higher deciles. They also reflect the much larger weight of China in the world's population that tilted the distribution in the direction described in the figure. In the early 21st century, the weight of the Chinese poor in the bottom decile has been much smaller than three decades earlier and, moreover, their income has increased during these years. The share of the African poor in the bottom decile has increased and, as a result, the income gap between the African countries and the developed countries has increased.

The main outcome of this process has been a decline in the income gap between the bottom and the top income deciles despite the rise in the income gap between the African and the developed countries. To clarify this outcome, consider the following numerical example: assume that in 1970 the mean per capita income of the lowest decile was US$700 and of the top decile was US$10,000. In 1970, the majority of the lowest decile included the Chinese poor, while the mean per capita income of the African poor was US$1,000 and most of the population in these countries was at higher income deciles. In 2000, the mean per capita income in China rose significantly while the African countries remained stagnant and their poor became even poorer. As a result, the mean per capita income of the lowest decile, which mostly included the impoverished African poor, was US$900, while that of the top decile increased to US$11,500. The result of these changes was a decline in the income gap between the bottom and the top deciles, and a decline in this measure of global inequality, even though the African poor became poorer and the rich developed countries became richer. In fact, this illustration has a much stronger implication: it is possible that income inequality within all countries and between all countries will rise and yet the measure of global inequality will show a decline – due to the changes in the composition of the income deciles (see also Chapter 3, this volume).

4 Has Globalization Been Good for the LDCs?

The conclusion that globalization has been good for the poor, drawn by Dollar and Kraay (2002), Chen and Ravallion (2004), Sala-i-Martin (2004) and others, is based on statistical estimates that compare the number of the poor, the incidence of poverty and the incomes of the poor at the initial stage of the globalization process in 1980 and two decades later; however, these estimates are dominated by the dramatic decline in poverty in China and the large share of the Chinese population. In most other developing countries, the gains from globalization have been very meager. Even in India, where the economy was booming after the reforms in the early 1990s, the majority of the poor population, most notably the nearly 700 million people living in rural areas, was barely affected by or benefited from their country's growth and the unparalleled development of high-tech industries and services. In most of the other developing countries in SA, Central America and SSA, the majority of the population was actually impoverished.

A significant development during these years was the growing divergence between two subgroups of developing countries: one subgroup includes many EA countries, India and a number of LAC countries, particularly the larger ones; these countries grew and continue to grow very rapidly, and they have become the world's *emerging economies*. The other subgroup includes the developing countries that had a long period of stagnation, drifted to the bottom of the global income distribution and became the world's LDCs. Thus, for example, in 1973, the income per capita in China was 11% *lower* than the average income per capita in Africa (including North Africa). In 2000, income per capita in China was 380% *higher* than the income per capita in Africa. In 1973, the income per capita in China was less than one-tenth of the average income per capita in Western Europe; in 2000, it was nearly one-third.[2]

The diverging trends between these two extreme subgroups of developing countries suggest that an analysis of the developments in the entire group of developing countries can no longer be conducted along the traditional lines that combine all these countries together and compare, for example, the growth rate of the developing countries with the growth rate of the industrial countries. Instead, this analysis must clearly distinguish between these subgroups, particularly between the LDCs and the emerging economies. An analysis of the trend in the entire group of developing countries on the basis of the average growth rate of their per capita income would be dominated by the rise in income and the decline in the number of the poor in China, even though the trend in the LDCs was in the opposite direction.

Another significant development was the rise in income inequality between the developing countries, which was due not only to the deepening gap between the emerging economies and the LDCs, but also to large disparities in the growth rates of countries within the same group. Table 2.2 pro-

Table 2.2. GDP per capita in the four large developing countries relative to the world average. (From World Bank Report, 2000/2001.)

	1980	1990	2000	% Change 1980–2000
India	0.164	0.206	0.246	+8.2
China	0.169	0.307	0.602	+43.3
Indonesia	0.364	0.495	0.527	+16.3
Brazil	1.029	0.869	0.822	–20.7
		The ratio of the GDP per capita[a]		
China/Brazil	0.16	0.35	0.73	
India/Brazil	0.16	0.23	0.30	
China/India	1.03	1.49	2.45	

[a]GDP per capita measured in 1995 PPP dollars.

[2] There is a third subgroup between these two that comprises many low-income developing countries, but most of them also grew very slowly in the last two decades.

vides selected indicators on the changes in income inequality between the leading countries within the group of emerging economies. The most notable observation is the rapid rise in the income per capita in China. As a result, China managed to considerably narrow the seemingly unbridgeable gap that it had with Brazil in 1980.

The most dramatic decline in poverty occurred in EA, and the incidence of poverty in the region's countries declined from over 60% in the early 1980s to 15% in 2001. Despite the rise in income inequality in most EA countries, most notably in China, the majority of the poor benefited from their countries' rapid growth. However, that growth did not reach the large provinces in the western part of the country due to the poor infrastructure. In other EA countries, the pace of growth slowed down very sharply or even entered negative territory at the end of the 1990s as a result of the deep financial crisis, but most of the affected countries started to recover in the early 21st century, and since then they have experienced several years of rapid growth. In Indonesia, growth was lower due to the political crisis that followed the financial crisis and slowed down the country's recovery. By 2015, the region's extreme poverty is expected to be drastically reduced, but the political instability afflicting some of countries raises doubt about the feasibility of achieving this goal.

The growing income in the emerging economies, however, was distributed very unevenly between regions, sectors and income groups within these countries, and income inequality rose quite significantly in most of them. This was also one of the reasons for the large difference in the rates at which poverty declined in these countries. Neither the decline in income in Brazil nor the rise in income in the EA countries were equally shared between the population groups, particularly in urban and rural areas.

In the SA countries, economic growth during the 1980s was driven primarily by the success of the Green Revolution and the introduction of more productive technologies in agriculture that increased the incomes of rural households and led to a reduction in their poverty. Since the mid-1990s, the region's growth has been driven by the policy reforms and trade liberalization primarily in India that precipitated industrialization, whereas in rural areas the levels of poverty have remained high. India was the most rapidly growing country in the region, but its growth concentrated in services and was driven by the country's high-skilled labour. In Bangladesh and Pakistan, growth was driven by industrialization in labour-intensive industries, primarily textile and apparel. That growth trickled down to the poor, either by migration to the urban centres in search of work and higher income, with the rise in income from agriculture in rural communities that send their products to the urban centres, and, mostly in India, with the introduction of large-scale milk production. By 2015, the incidence of extreme poverty is expected to decline to around one-third of its level in 2001, but the stalemate at the Doha Round may slow down that decline.

Since the countries where poverty declined included China and India, and since these countries have the majority of the population of the developing countries as well as the majority of the world's poor population, the sharp decline in poverty in China and later in India resulted in a reduction in the number of the world's poor. However, in the majority of the developing countries in SSA, Central America, Central Asia and SA, poverty increased, their growth was very meager and in many of them the average per capita income even declined. These countries benefited very little, if at all, from globalization. Given the growing concentration of the world's poor in the LDCs, the relevant questions are: Has globalization been good for the LDCs? Has it been good for the three-quarters of the population in India who live in rural areas? Has it been good for the poor in Central America or in the Caribbean? Has it been good for the poor in Central Asia? How will globalization affect the future trends in world poverty?

4.1 The changes in poverty in the LDCs

The answer to these questions cannot be based on a simple comparison of the incidence of poverty or the number of the poor in these countries in 1980 and in 2000, since these comparisons reflect not only the effects of globalization; in fact, most of these countries were not even affected much by globalization. The rapid growth in world trade, the unparalleled technological progress and the improvements in the standard of living in the developed countries and emerging economies did not reach the majority of the population in the LDCs. For the rural population in most of these countries, neither their living conditions nor their production or cultivation methods changed by much, and they still trade mainly in traditional agricultural crops, like coffee, tea, bananas and cotton.

The total exports of the African countries still consist primarily of their exports of natural resources and traditional agricultural products, and their export earnings were highly volatile during these years, partly due to unstable world prices, but mostly because of their own governments' unstable and often quite damaging trade and exchange rate policies. Erratic government policies had a negative impact on their economies, but their prolonged stagnation and their inability to integrate into the global economy were primarily due to unstable political regimes that were established in many countries after a military coup, and the affected countries suffered from continued internal conflicts between rival groups of their population. Many countries are still embroiled in civil strife that often deteriorates into regional conflicts that drain their resources and ruin their economies. As a result, the incidence of poverty in most SSA countries remained high or even worsened during the last two decades, and the number of their poor continued to climb. The structure of most of the region's economies changed very little, and their share in global trade declined by the year 2000 to half its level, two decades earlier.

It is by no means certain that the growth process of the past decade in the developing countries that focused quite narrowly on the emerging economies and was driven by their rapid industrialization represents a model of growth that will also be suitable for the LDCs; it is also unclear whether this would be their best strategy to combat poverty. Even in

Table 2.3. The share of the poor in developing countries. (From World Bank, 2005.)

Number of people living on less than US$1 per day (millions of people and the head-count index in %)

	1980	2001	2015[a]
East Asia and Pacific	472 (65.6%)	271 (14.9%)	19(0.9%)
South Asia	462 (41.3%)	431 (31.3%)	216 (12.8%)
SSA	227 (44.6%)	313 (46.4%)	340 (38.4%)
Total	1218 (27.9%)	1039 (21.1%)	662 (10.2%)
Excluding China	(844) (26.1%)	(877) (22.5%)	(606) (12.9%)

[a]Projected

China, the rural poor did not share much in the country's rapid growth during the 1980s and through the mid-1990s. Since then, the government made considerable investments in rural areas to improve the infrastructure, education and health, and to introduce advanced agricultural technologies. In addition, the growing demand of the population in the more affluent urban centres for consumer goods enabled farmers to increase their income by selling more of their products to the urban population, by diversifying their agricultural production, and also by gradually increasing the value-added of their products with local processing. The driving force for growth was rapid industrialization which was targeted mostly for the export markets and was accelerated by the flow of FDIs. The diverging trends between the different subgroups of developing countries are summarized in Table 2.3.[3]

From the mid-1960s to the mid-1970s, the relatively rapid growth of the developing countries was dominated by the rapid growth of the African and the LA countries despite the stagnation in the EA countries. Since the mid-1970s, income per capita in the African and LA countries started to decline and they entered into a prolonged stagnation and a deep debt crisis, while the EA countries started to grow more rapidly. The growth rate of the entire group of developing countries declined very sharply despite the moderate growth in India, and only in the early 1980s, the Chinese economy started to pick up at a more rapid pace, though its industrialization was still at its initial stage.

The gradually more rapid growth of the EA countries during the 1980s and the recovery of the large LA countries in the 1990s deepened the division of the developing countries into two very distinct groups: one group became the world's 'emerging economies' and their rapid growth in the last two decades narrowed down the income gap between them and the developed countries; the other group remained stagnant, and there has been a growing

[3] These estimates do not yet take into account the impact of the escalating world prices of oil, natural gas and minerals.

divergence between the per capita income there and in the emerging economies and developed countries.

4.2 The dilemma in policy decisions

Despite their rapid growth, the EA countries were not ardent 'globalizers' at the initial phase of their growth process, in the sense that they did not open their economies to trade and did not pursue many of the other policies that have been prescribed by the International Monetary Fund (IMF) and the World Bank as sine qua non for rapid growth and successful globalization. As the LDCs must now choose their growth and development strategies, they face a dilemma whether to select a package of policies that will be designed along the lines of the heterodox policies that the EA countries pursued with so much success, or the more orthodox policies that are promoted by the Washington-based institutions. These latter policies have three main goals: accelerate growth, reduce poverty and integrate into the global economy. In the short run and, possibly also in the medium run, these policies may not be complementary, and that will require the setting of clear and perhaps difficult priorities that may make the implementation of these policies politically very complex.

Moreover, in order to carry out the necessary reforms, the LDCs would have to make comprehensive structural adjustments in their economies, and these adjustments would have to be designed to take into account the changes in the structure of world production and trade, as well as the changes that transformed the institutions governing world trade in the last decade. The changes in the global economy are still evolving and therefore, the reforms should not have a rigid and inflexible structure, but should also include rules for making additional adjustments in the future.

To make the required choices, it is necessary to carefully examine the experience of the developing countries that went through a very challenging process of structural adjustments during the 1980s and the 1990s; the experience of other countries, primarily the EA countries, is obviously also very important. However, the lessons that can be drawn have only limited relevance for the LDCs for two main reasons: First, these lessons can only draw on the experience during the 1980s and 1990s, and they cannot therefore properly reflect the effects of the decisive changes in the structure of world production and trade since then. Second, the policy of each country is necessarily built to suit its specific conditions and constraints, and the lessons from the experience of the EA countries may not be fully relevant for the African countries.

The analysis of each policy would have to be based on the experience that has been accumulated and summarized in the relevant studies. However, the findings of many country studies with respect to the impact of alternative policies on growth and poverty often produced inconclusive and sometimes contradicting results, and there are large differences between the findings in different regions. This also raises obvious questions about the use of the conclusions drawn in multi-country studies. Take, for example, the study of Dollar and Kraay (2002) which shows that, on average, there was no significant correlation between economic growth and income inequality. The most obvious question is whether a conclusion that applies, on average, to all countries, can be of any use in the analysis of similar developments in a specific subgroup of countries. In other words, are the differences between the results for all countries and the results for a subgroup of countries purely random, or are there persistent and consistent differences between the 'average' or the trend in a certain subgroup of countries and the 'average' in all countries? Under what conditions will the growth in a country's mean income have largely the same effect as the growth in the mean income of the country's poor population?

4.3 The choice between pro-growth versus pro-poor policy

Neither multi-country studies nor specific country studies provide conclusive answers to key questions about the desirable structure of the reforms:

- Should countries opt for rapid economic reforms and swiftly cancel inefficient government programmes or should the government implement the reforms only gradually in order to minimize the harm that will be caused to workers and enterprises and give them time to make the adjustments?
- Should the government open the country's markets to trade and remove all trade restrictions as rapidly as possible in order to increase the efficiency in the allocation of the country's resources and increase productivity in order to accelerate growth and become more competitive in the world markets, or should the country continue to protect its industries and its workers by restricting trade and open up the economy more gradually?
- How should a country deal with the rising income inequality that may also have economic justification by reflecting the higher returns to skilled workers, entrepreneurs or capital?
- Could progressive income taxes reduce the incentives of the more productive workers to work or the incentives of the rich to invest?
- What should be the guidelines for the social safety nets that, on the one hand, mitigate the negative impact of the structural reforms on the more vulnerable segments of the population and, on the other, reduce efficiency and thus slow down economic growth?
- Will growth lift all boats? Will the free market provide a level playing field or will it introduce rules that would tilt the playing field against those who have greater difficulty to find their niche in the new economy and may therefore fall behind?

In theory, a free-market economy accelerates growth by concentrating production in sectors and regions where it is most effective, by making investments in sectors where the capital is most productive, and by using the more efficient and high-skilled workers where their productivity is the largest. In a free-market economy, these goals are achieved through the price system, by remunerating capital and labour at the highest rates in the sectors where they are most productive. This system of allocating resources inevitably increases income inequality between the high-skilled and the low-skilled workers, between owners of capital and those who do not own capital, between entrepreneurs and employees.

In the EA countries, that rise in inequality was an outcome of their process of industrialization and it deepened the divide between the urban and rural sectors. Although the expenditures of all urban consumers increased, the rise in expenditures of the top decile, that mostly includes skilled workers and entrepreneurs, was more than two and a half times higher than that of the median decile and nearly six times higher than that of the poorest quintile. In rural areas, the rise in inequality was slightly less rapid, but average consumers' expenditures in urban areas were nearly double the average expenditures in rural areas.

The neoliberal free-market approach assumes that, even though these differential remunerations may not be fully justified in the short run and during the transition period, in the long run this organization of the economy will be beneficial to everyone, including the low-income groups and the low-skilled workers that will have incentives to acquire higher skills. These principles are well known from the basic courses in microeconomics, but the questions whether this is indeed what happens, particularly in the developing countries, must be asked again and again, because the market system is also open to widespread abuse that raises the remunerations of certain groups, entrepreneurs or individuals well beyond the level that can be justified by their contribution to production.

The following examples illustrate the dilemma that governments in developing countries must face when they weigh alternative policy options:

- In India, the rapid growth of the last decade was higher than at any time in the last 50 years, but that growth was not pro-poor and it concentrated in the urban centres and in high-tech industries and services. This growth increased income inequality since the rise in the rural sector was very slow and the majority of the rural population remained poor. Government investments also concentrated on improvements of the urban infrastructure and the share of the rural sector was very small.
- In China, growth concentrated in the industrial centres in the country's eastern provinces and thus contributed to a rise in income inequality. Government investments in infrastructure also concentrated in the urban centres in an effort to attract new industries, and only in recent years more public investments have been targeted on rural infrastructure and on the education and health of the rural population.
- Should the government in a developing country maintain price controls over local food crops in order to help the urban poor despite the distortions created by these controls, and despite their negative impact on agricultural growth and rural incomes?
- How should a government that plans to carry out land reforms proceed, given that the division of large estates to many small land owners thwarts the returns to scale in agricultural production, reduces productivity and prevents the use of more advanced methods and machinery in production? The alternative is to consolidate the lands of many small land owners,

establish cooperative rural supply chains and provide subsidized credit for the improvement of the production methods. The dilemma is clear: land reform is the main policy instrument to reduce income inequality and provide a source of income to the landless rural population; that reform must be combined, however, with the creation of local institutions in order to establish efficient cooperation between the small landowners in production, in the supply chain and in the use of concessional credit; otherwise, productivity is bound to fall and reduce rural incomes. Many ex-centrally planned economies still grapple with this dilemma.

- In some countries, developed and developing, but most notably in the EU and the USA, a significant component of agricultural policy includes targeted subsidies for specific crops that are largely politically motivated. These subsidies are economically wasteful, highly distorting and concentrate the agricultural lands in the hands of fewer and larger farmers or agricultural corporations. The difficulty to find a satisfactory solution to remove or replace these subsidies was the main reason for the failure to reach an agreement at the Doha Round.
- A reorganization of the entire agricultural sector is often part and parcel of the reforms that are necessary for a country's integration into the global market and for meeting the WTO conditions. That reorganization requires thorough structural changes, introduction of new and advanced technologies and the diversification of agricultural production to increase the share of the products in which the country has a comparative advantage and for which there is demand in the world market. The dilemma is that these reforms require large government investments in the rural infrastructure, in the country's ports, in the organization of efficient supply chains, etc., that are necessary to make agricultural production profitable.
- When a country seeks entry to the WTO, it is required to make significant policy changes that include the removal of subsidies and price supports to selected sectors as well as the removal of tariffs and administrative restrictions on imports. These reforms are highly damaging for sectors that in the past were able to produce under the protection of these restrictions that shielded them from the high tariffs. How should a country proceed with these reforms given their high social toll?

The rising inequalities that countries risk when implementing economic reforms and the policy choices that governments face in the short run are due to an inevitable trade-off between policies that are pro-growth and policies that are pro-poor. The significance of these policy choices is due not only to their impact on a country's economic growth and the people's welfare, but, perhaps more importantly, to the political context in which these decisions are made. The rise in inequality and poverty are bound to foster strong political opposition to the entire reform programme, which may lead to open resistance and confrontations that could either topple the government, as happened in many LA countries, or force the government to retract from large parts of the reform programme, as happened in quite a few developed countries. Where, then, should the limits be and what should be the guidelines for policies that are pro-growth but, in the short run, raise the country's income inequality? The fierce debate on the reforms implemented in many developing countries under the guidance of the IMF and the World Bank in the process of their structural adjustment programmes is testimony to the conflict between policies that, in the short run, are pro-growth and policies that are pro-poor.

5 Growth, Poverty and Income Inequality in the Main Regions

A multi-country analysis of the global changes in poverty and income inequality during the last two decades is bound to 'average out' not only the differences between regions and countries, but, more importantly, the design of the policy reforms. The structural adjustment programmes designed by the IMF and the World Bank have often been blamed for being 'one-size-fits-all' that 'average out' and therefore ignore the differences between countries. The multi-country studies discussed in the previous section that came to the conclusion that globalization is good for the poor 'averaged out' and therefore ignored the differences between countries and the fact that in most developing countries, poverty remained nearly unchanged or actually increased.

The main conclusion from the experience with the structural adjustment programmes is that policies and structural reforms must be tailored to the specific conditions and constraints in the countries in which they are implemented. To highlight the differences as a background for the discussions in the subsequent chapters, this section provides a brief overview of the trends in growth and poverty in the main regions and the conditions that determined the structure of these trends. This overview is necessarily brief; the main developments are summarized in figures and tables and the focus of the overview is on the social and political conditions that were the backdrop for these developments.

5.1 Sub-Saharan Africa

Today, the majority of the SSA countries are the world's LDCs, but in the 1950s and the 1960s, many African countries seemed reasonably well positioned to develop their economies by taking advantage of their abundance of natural resources and low labour costs. At that time, the EA countries were still extremely poor and entirely dependent on very primitive agricultural production, primarily for home consumption. No economic indicator or econometric model could envisage at that time their rapid growth and accelerated industrialization two decades later.

Since the mid-1970s, the SSA countries endured a series of military coups and violent ethnic conflicts that ruined their economies and impoverished their populations. Inept leadership and political crises led to repeated economic crises, prolonged stagnation, deepening poverty and policies that prevented their integration into the global economy. There were, however, considerable differences between countries, and several of them were politically more stable and had more robust growth. But even countries like Ghana, Uganda and Senegal, which represent the greatest hope of the African continent for accelerated growth, remained poor and unstable, and did not manage to secure the conditions that would lure foreign investors and allow them to accelerate industrialization. Their agricultural sector remained the dominant sector, employing the majority of the population, but they did not manage to introduce the use of advanced technologies and extensive production of more profitable crops for exports. The poor state of their agricultural sector and the dire situation of their rural population is due to several factors, among them the very rudimentary production methods and the poor infrastructure that essentially prevent trade, the gross inequalities in the distribution of land, and the monopolistic, often state-owned supply chains.

The SSA countries implemented major structural adjustment programmes since the early 1980s, but these programmes did not succeed in accelerating their growth and reducing poverty (Fig. 2.6). During the last two decades, most African countries remained stagnant, and in many of them income per capita declined and poverty increased. Although the majority of the world's poor still concentrate in the Asian countries, primarily in India, in the 1990s SSA became the centre of gravity of the world's poverty problem. The rise in income inequality also contributed to widen the income gap between the urban and the rural populations and to deepen poverty in the rural areas.

Table 2.4 highlights other dimensions of the changes in poverty in SSA during the 1990s:

Most of the difficulties encountered by the African countries in reforming their economies were due to internal political instability, often reflected in violent confrontations and wars, and extremely erratic economic policies. External reasons for their difficulties were the changes in the world economy and the structure of world production and trade that were associated with

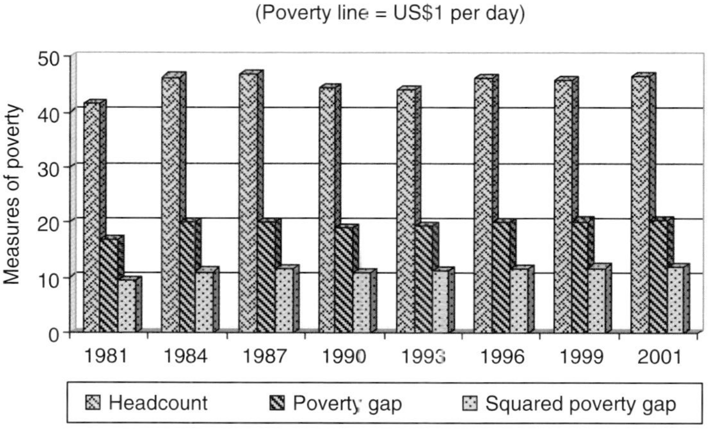

Fig. 2.6. Measures of poverty in sub-Saharan Africa. (From World Bank Report, 2000/2001.)

Table 2.4. Main welfare indicators in SSA subregions. (From World Bank, 2000, 2001.)

	Life expectancy (years)		Infant mortality (per 1000 live births		Primary school enrollment (in % of age group)	
	1990	2000	1990	2000	1990	2000
Western Africa[a]	48	51	123	105	43	61
Eastern Africa[b]	49	46	102	89	36	55
Ethiopia	45	42	131	116	20	44
Kenya	57	47	63	77	67	68
Southern Africa[c]	58	43	64	78	88	80
South Africa	62	48	45	50	99	90

[a]From Xavier Sala-i-Martin (2004).
[b]These figures are based primarily on the household survey and they are therefore somewhat different from the figures noted earlier that use also the income surveys.
[c]There is a third subgroup between these two that comprises many low-income developing countries, but most of them also grew very slowly in the last two decades.

globalization and created many obstacles for their potential to expand their trade and become integrated into the global economy. As a result, the African countries have been increasingly marginalized, and despite the goodwill of the international community to help these countries to significantly reduce their poverty – reflected in the UN declaration of the MDGs – most predictions for the coming two to three decades remain very pessimistic.

With the stalemate in the negotiations for an agreement on agricultural trade at the Doha Round, the supply of plentiful and cheap imported food to the African countries is expected to continue, and African farmers find it increasingly more difficult to compete with imports from the developed countries even in their own urban markets. Household surveys show that a growing number of African rural households are forced either to abandon farming altogether and earn their income from sources other than farming, or to produce mostly for self-consumption.

However, several developments of the recent years may change this negative trend. An important and promising development for the agricultural sector is the sharp rise in energy prices in the world markets that made the production of ethanol from sugar and maize economically profitable and led to a sharp rise in their price. The price of sugar, which is being diverted to ethanol production for automotive fuel, more than doubled from late 2004 until early 2006; the price of natural rubber (a substitute for synthetics produced from petroleum products) rose 60% in the first half of 2006; the price of maize, which is used as feedstock for ethanol production, has also risen.[4]

[4] If oil prices continue to rise, agricultural prices could also be strengthened because biofuel will then be more economically viable. Already, 20% of the US maize crop and 50% of the Brazilian sugarcane crop are used to produce ethanol. If this trend continues, demand for other commodities, especially grains, may also increase to substitute for crops used for biofuels.

On the other hand, high energy costs are raising the prices of fertilizers and prompting an increase in the cost of production of energy and fertilizer-intensive crops such as grains, thus contributing to further increase agricultural prices and reduce yields because farmers are likely to use less fertilizer. A large rise in the prices of certain agricultural goods will raise the incomes of farmers in some agro-climatic regions, but reduce farmers' incomes in other areas; these developments are also likely to result in changes in the composition of farmers' production. However, the rise in grain prices will have strong adverse effects on most consumers.

Since the beginning of the 21st century, the prices of minerals and other natural resources started to rise as an effect of the global growth that concentrated in China and India and increased the strong global demand. Although future improvements in technology and new discoveries are expected to moderate that demand, commodity prices are expected to remain high at least through the end of the decade. The sharp rise in oil prices has received the bulk of media attention, but in fact the rise in the price of metals and minerals during 2004–2006 has been equally strong or even stronger. The rise in the prices of natural resources renewed the growth of many SSA countries, and since 2000–2006, their average rate of growth was 4–5%, while the oil-exporting economies grew at an annual rate of 6–7%. Some countries accumulated huge wealth with the rise in oil and mineral prices that contributed to raise the standard of living of their top income groups.

Figure 2.7 shows the sharp changes in commodity prices since 2001. Despite the rise in agricultural prices by 35% in real terms since their cyclical low level in 2001, these prices trailed considerably behind the prices of oil, minerals and metals.

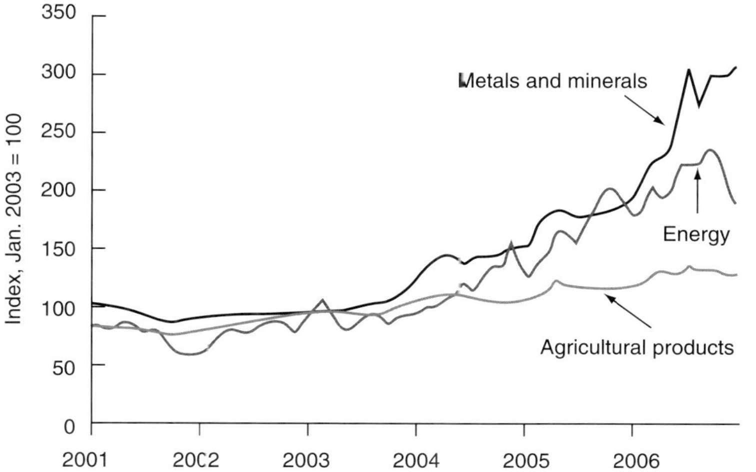

Source: World Bank

Fig. 2.7. The trends in commodity prices, 2001–2006.

Small rural producers in Africa also have difficulties to compete with the larger farmers that have access to advanced technology, high returns to scale in production, greater flexibility to change their product mix, better access to credit and better connections to the local and international supply chains. Limited profit opportunities in the production of food grains and the collapse of the cotton sector with the sharp reduction in its price deterred many farmers in Francophone Africa and severely reduced their profits. The growing population in the rural areas and the continued division of the land between a larger number of households, and the expansion of production to marginal lands, often by ruining the local forests, have reduced the productivity of land and labour. In many developing countries, it was the farmers who suffered the most from the structural adjustments that were implemented. Farmers also had income losses as an effect of the reforms that are required for countries joining the WTO; these reforms include the removal of subsidies on agricultural inputs and all subsidies to agricultural exports.

Agriculture is still the main source of livelihood and work for the majority of the African population, but its contribution to GDP has decreased since the beginning of the 21st century. Non-tariff barriers (TTBs), including high food safety standards in the developed countries, inhibit agricultural exports from Africa, and most countries, with the exception of South Africa, do not have the capacity to enforce these standards and most farmers do not have the capacity to comply.

The manufacturing sector in these countries produces largely for the domestic market, with part of the production exported to regional markets. It continues to be relatively small and is based mainly on the processing of agricultural goods. The development of local industry has been hampered by high production costs despite the low labour costs, by limited access to financing, and the low quality of the products which is due to outmoded production methods. Some African countries have implemented wide-ranging privatization programmes in order to promote their industrial production, but they have not managed to attract foreign investors.

5.2 Asia

SA is the region with the largest concentration of poor people in the world and by 2005, the region was still home to around 400 million poor people. In the 1980s, most SA countries managed to accelerate growth in the rural areas by successfully adopting more productive crop varieties with the implementation of the Green Revolution. As a result, in quite a few of these countries, most notably in India, income inequality declined during that decade. Estimates of consumption-based poverty measures show that India had considerable success in reducing deprivation.[5] The changes in rural areas in

[5] The estimates of poverty declined from 51.3% in 1977/78 to 36% in 1993/94 and 26.1% in 1999/2000.

India included a rapid growth of the dairy sector that spurred significant gains in alleviating poverty and improving nutrition.

However, among the bottom 80% of the rural households and the bottom 40% of the urban households, the share of food expenditure is still over 60%. Despite the country's rapid growth since the early 1990s, hunger still persists in many areas.[6] Cereal consumption and average calorie intake of many rural households has actually declined, and about a quarter of the Indian rural population is still energy deficient.[7] Nevertheless, infant and child mortality rates have been reduced and most urban and many rural households enjoy better and more diversified nutrition.

Poverty remains high in India, and in some states it even increased during most of the 1990s despite rapid growth, and the poor were the first to suffer during crisis periods. Datt and Ravallion (1996) found that although the urban growth reduced urban poverty, it had a marginal effect on poverty in the rural areas. The sharp rise in employment in the urban sector following the economic reforms of the early 1990s and the significant increase in employment of low-skilled workers in the labour-intensive textile and apparel industries had a significant impact on poverty at the national level. However, the majority of the population in India still lives in rural areas, despite the flow of migrants to the urban centres, and the rural population has so far not benefited much from the country's rapid growth that concentrated in the high echelon of the labour market.

Most other SA countries also enjoyed rapid growth at an annual rate of 5–6% and even to 7%, resulting in a decline in poverty in the region. However, Pakistan, Sri Lanka, Bangladesh and Nepal remain susceptible to political instability and internal civil wars with separatist groups. In most SA countries, most notably in India, the decline in poverty was accompanied by a rise in income inequality (Fig 2.8).

India, by far the region's largest country, experienced the most rapid growth and the steepest reduction in poverty. Although the growth reached many rural areas, it was distributed very unevenly between states: the southern states grew at high rates of 6–8% or more; however, many of the northern states grew at a much slower pace and their high rates of poverty barely changed. There has also been a considerable rise in income inequality within the urban and the rural areas. In the urban areas, many were hurt by the privatization of state-owned enterprises, and in the rural areas, many were hurt by the opening of the economy to agricultural trade and to global competition that brought a reduction in many of their product prices and in their incomes.

In most EA countries, with the exception of Japan and the 'four tigers' – South Korea, Taiwan, Singapore and Hong Kong – growth rates were very

[6] Government of India (2001): *Reported Adequacy of Food Intake in India 1999–2000*, National Sample Survey Organisation, Ministry of Statistics and Programme Implementation, New Delhi, India.

[7] The calorie norms are 2400 per capita per day for the rural and 2100 for the urban populations. (GoI, 1993).

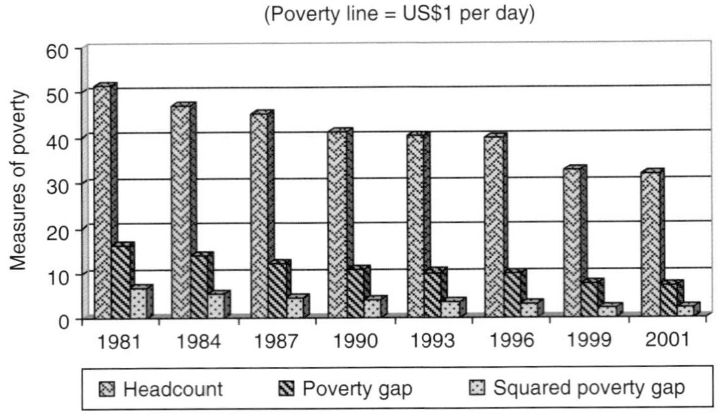

Fig. 2.8. Measures of poverty in SA. (From World Bank Report, 2000/2001.)

low during the 1960s and most of the 1970s, when these countries were still primarily rural and controlled by strong and very authoritarian central regimes that stifled their development. China was just emerging from the grip of the Cultural Revolution, its agricultural lands were still collectivized and productivity was very low. Trade relations of these countries with other regions and other Asian countries were minimal, their economic policies were strictly inward-looking and their infrastructure was in shambles.

The gradual rise in the income gap between the urban and the rural populations in China since the mid-1980s (Fig. 2.9) reflects the rise in the incomes of the urban population with the country's industrialization. The rise in that gap was mitigated by the flow of migrants from the rural areas to the urban centres and by the flow of remittances in the opposite direction, but the gap persists even today. The growing inequality in China also reflects the income

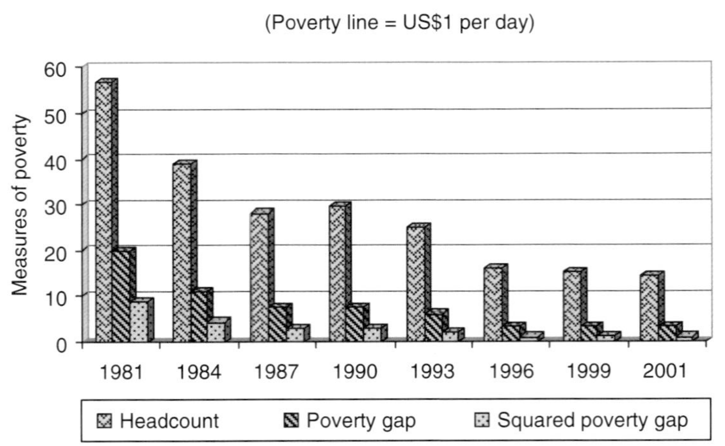

Fig. 2.9. Measures of poverty in East Asia.

gap between different regions and the difference between the rapid rise in the incomes in the eastern provinces and the near stagnation of many western provinces.

In recent years, the rural population has become increasingly more vocal in its protests against the widening rural–urban gap, against arbitrary confiscations of fertile lands for industrial enterprises or development projects, and against the pro-urban and pro-industry bias of government investments.

The policies advocated by the international development organizations to combat income inequality and poverty are based on rational economic principles of the market economy, but they did not take into account the intensity of the frustrations during the transition period when income inequalities tend to rise. The general approach that guides these policies is 'positive-sum' growth strategies that are based on the notion that free markets and free trade will raise over time all incomes, but will raise the incomes of the poor at higher rates. However, these policy guidelines became highly controversial among political parties and policy makers in developing countries, primarily because intense feelings of unjust deprivation could not be compensated by the promise of long-run gains. Violent political confrontations in a number of these countries that erupted during the financial crisis in 1997/98 concentrated on demands that the government take more radical measures to redistribute the national income and increase the incomes of the poor even if others may lose. For the more radical parties, a policy that increases the income of the poor by one 'rupee' is worthwhile, even if it costs the rest of society more than one 'rupee'. The intense feeling of relative deprivation is 'the poverty of watching others whiz ahead while you merely crawl' (*The Economist*, January 2007).

In principle, even the strongest advocates of pro-poor growth and proponents of more radical income redistribution measures must weigh the 'opportunity costs' of these measures and the sacrifices that will be required if these measures are implemented and the incomes of the poor are increased, while the country's growth is slowed down and its potential to reduce poverty in the future is reduced. That rational cost-benefit analysis did not mitigate the political confrontations that still continue in some countries and had a negative impact on their political regimes, sometimes even toppling the government. Although all the EA countries resumed their rapid growth after 2 or 3 years, their regimes remained more fragile, and some of the confidence that foreign investors had in the past when they made their investments in these countries may have been lost. Even in China, foreign investors feel less secure and the current mantra is to spread investments in several baskets 'however large the Chinese basket may seem' (IHT, 2006).

5.3 Latin America

Protection against competing imports has long been an important policy tool used by the LA countries to protect their industries and the weaker segments of their population. During the 1960s and the 1970s, the LA countries implemented nationalist economic policies that combined import substitution, extreme protectionism and export promotion measures, together with subregional integration arrangements (the first being the Central American Integration that was established already in the 1950s). This 'mixed model', rather than import substitution alone, as it is often claimed, was the most common arrangement in the region (ECLAC, 1998). Expansive social programmes resulted in growing fiscal deficits and a growing balance of payments deficit. These policies led to the fall in per capita income levels, a sharp rise in poverty, a wholesale flight of capital and rising debts and inflation, which culminated in hyperinflation at the end of the 1980s.

During the late 1980s and early 1990s, the LA countries implemented, on the insistent advice of the international development organizations, sweeping macroeconomic reforms that worked well in EA: balanced budgets, strong controls to prevent inflation, the opening of the economy to trade, the privatization of state-owned industries and the promotion of export-led growth. The neoliberal, market-oriented and pro-trade policies that most LA countries implemented faithfully following the recipe of the Washington-based institutions in an effort to modernize their economies, strengthen their industrialization, expand their exports and attract foreign investors to accelerate growth, were, particularly in the large countries, quite successful in the first half of the 1990s. In Argentina, President Menem implemented sweeping structural reforms and price stabilization and carried out a broad privatization programme that eventually extended to gas and electric utilities and distributors, telecommunication, railway, airline and oil companies, and a deregulation programme that covered a broad gamut of activities, including foreign investment, capital markets, national markets for goods and services and government regulatory agencies. Regulations governing interest rates, foreign exchange controls, prices and wages were also removed.

As part of these reforms and in an effort to control the fiscal budget and increase the flexibility of their labour markets, many governments reduced their support for the labour unions, permitted dismissals and cuts of the minimum wage, reduced the budgets of many social programmes and privatized many government companies. However, the initial response of their economies to these reforms was rather disappointing and many countries continued to underperform with slow growth rates, not only compared to countries in other parts of the world, particularly in Asia, but also compared to their own performance in the 1960s and 1970s. In the larger economies, most notably in Argentina and Chile, the reforms and restructuring programmes were pursued with considerable success in the early 1990s despite the Mexican financial crisis in 1994. The low inflation in Argentina helped to attract foreign capital, increase investments and increase its per capita income rather rapidly (Table 2.5).

In most other LA countries, low investment rates and large differences in per capita output between countries and, in the large LA countries, between regions, increased income inequality and the income gap between regions and between the urban and the rural populations (particularly in Brazil, Peru and Chile). Income inequality is high in most LA countries, and in 2002 the

Table 2.5. Trade integration: 1960–2000.

	Trade Integration	
	1960–90	1990–2000
Latin America	27.6	39.2
Argentina	13.8	19.0
Brazil	15.8	18.5
Chile	40.2	59.7
Colombia	28.2	36.2
Mexico	22.5	50.5
Venezuela	45.0	51.1

Note: The ratio of total external trade to GDP.
Source: IMF

Gini coefficient ranged between 0.613 in Brazil, 0.55 in Chile and 0.527 in Mexico and Argentina.

In the early 1990s, the larger LA countries adopted more active trade policies that included strong support for export industries and active intervention in the foreign exchange markets to support a stable exchange rate as well as regional initiatives through MERCOSUR and other regional agreements. The rapid growth that followed was stimulated by the expansion of exports, the large inflow of foreign capital and the growth of local industries with the aid of FDIs. It also increased the political power of right-wing parties and politicians that supported more open trade and encouraged large investments of multinational corporations in their economies.

The urban population in these countries reaped most of the gains from the growth, but even in the urban areas, the benefits were distributed very unequally and a large share was accumulated by the narrow elite. In the rural areas, the large farmers and large traders also benefited from the growing exports in agricultural commodities, while the majority of the rural population was increasingly impoverished. In the late 1990s, the rise in income inequality became a polarizing political issue that deeply divided many LA countries, severely affecting political parties and population groups.

In Argentina, even during the roaring 1990s, more than 3.5 million people fell into poverty, and the income gap between the rich and the poor increased sharply. Not only unskilled workers, but also middle class and skilled workers were victims of the Argentinean government's conservative pro-industrialization, pro-trade and fiscal policies. The opening of local markets to competition, the removal of trade barriers and the policy of pegging the peso to the US dollar despite the more rapid local inflation, flooded the local market with cheap goods from abroad that led to the collapse of many small and middle-sized businesses. In Central America, a large percentage of the population remained in rural areas, and these were the region's poorest countries (Table 2.6). The debt crisis during the 1980s considerably slowed down these countries' growth and deepened their poverty. The recovery of the larger LA countries did not affect them very much and they did not

Table 2.6. The share of agriculture in Central America. (From CIA Fact Book, 2004.)

Country	Agricultural sector, % of GDP	% Labour force employed in agriculture
Nicaragua	32.6	42
Guatemala	22.6	50
Honduras	13.7	34
El Salvador	9.5	30
Costa Rica	9.1	20

manage to build a solid industrial base. However, the sharp rise in the world prices of natural gas and many minerals has now helped to increase the financial resources of many of these countries.

The deep financial and economic crisis in the early 21st century led most countries to impose protection measures against free imports and increased the power of interest groups and political parties that opposed trade liberalization. The early signs of the crisis were already noticeable since 1998 with the rise in the trade deficit and the sharp slow down in foreign investments. Despite robust material gains of the leading LA countries during the 1990s, in part due to their successful integration into the global economy, support for trade liberalization waned when many of these countries entered into a deep economic crisis in the early 21st century that impoverished many segments of their population and exposed the weakness of their governments and the corruption of their political institutions.

The deepening crisis led to a strong shift of many regimes and political parties in LA to the left and changed the sentiments of the local population towards globalization by bringing to power political parties that opposed trade liberalization and instead supported the more limited RTAs. The left wing parties and large segments of the population vehemently opposed globalization and contended that their countries' trade was dominated by multinational corporations and corrupt local elites. This shift to the left also reflected lingering resentments in the local population against the very restrictive fiscal policies implemented during the 1980s and part of the 1990s within the framework of the structural adjustment programmes and the conditions required by the IMF and the World Bank. With the sharp rise in unemployment from around 6% in 1990 to nearly 11% in 2002, opposition to the sitting governments came not only from the poor, but also from the middle classes that became weary of the endless economic and political crises, the inefficiency of institutions, the high unemployment, the corruption of the ruling class and the unequal distribution of the national income.

Figure 2.10 shows the high incidence of poverty in the LA countries following the financial crisis of 2001/02. Among the region's large countries, Argentina was most severely affected by the crisis and poverty rose sharply by around 20% and the number of extreme poor nearly doubled.

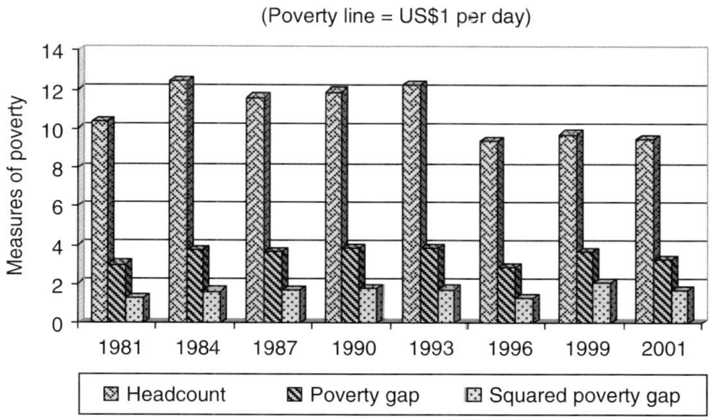

Fig. 2.10. Measures of poverty in Latin America and the Caribbean. (From World Bank, 2001.)

The opposition to globalization and the populist 'anti-Americanism' reflected the perceptions that the multinational trade agreements and the WTO effectively restrict their national sovereignty and open their markets to cheap imports that damaged the local industries. The shared nemeses of most of these social groups were American capitalism and the local oligarchy. Nevertheless, most newly elected leftist regimes did not turn their back on the global economy, and despite the strong rhetoric they generally implemented rather moderate policies and continued their efforts to attract foreign investments. The most prominent example of the search for a middle ground is Brazil's President Lula da Silva, who was elected in 2002, and despite his promises during the election, did not sever Brazil's relations with the IMF and with foreign investors. Argentina, that suffered the most severe depression during the years 1998–2002 and defaulted on its debt in December 2001, managed to recover very impressively since then and had an annual growth of 8% in the following 3 years, pulling more than 8 million people – more than one-fifth of the country's total population – out of poverty.

In the early 21st century, a new leader of *la revolución social* emerged in LA with the election of the Venezuelan president Hugo Chávez. Despite his controversial leadership and inflammatory rhetoric, his economic policies worked quite well. The stunning jump in the country's GDP was largely due to the high oil prices and came after nationalizing the country's oil companies, despite fierce US-backed resistance. In the following 2 years, Venezuela had an annual growth rate of 10%, and large resources from the oil revenues enabled Chávez to implement a wide range of social programmes and a sweeping social agenda. Chávez also provided credit to Argentina, Bolivia and Ecuador that allowed them to free themselves from the 'conditionalities' of the Washington institutions and from the harsh restrictions they had to follow in order to receive credit from the IMF, the Inter-American Development Bank and the G7 countries. Despite this 'freedom' and despite the resources of oil and natural gas in Venezuela and Bolivia, these countries have so far pursued a rather moderate fiscal policy to secure their economic stability.

Concerns about trade liberalization have also increased the political power and influence of RTAs, most notably MERCOSUR, that made considerable efforts to establish regional common trade policies, in part to exercise greater influence in the multinational trade negotiations and in negotiations with other RTAs, particularly NAFTA and the EU. MERCOSUR, based on Brazil, Argentina, Paraguay, Uruguay and Venezuela, remained relatively protectionist, however, and very cautious in bilateral trade deals with the USA, and they have resisted thus far Washington's proposal to establish a 'Free Trade Area of the Americas' in favour of regional economic integration.

6 Concluding Remarks

The impact of globalization on poor countries and on poor people within all countries is the central issue in the debate about the question whether globalization has been good for the poor. The conclusions that have been drawn in very elaborate empirical studies on the basis of statistical averages are statistically correct and significant, but not always very helpful. It is an undisputed fact that globalization brought unparalleled growth and affluence to the majority of the world's population, but it is also painfully clear that it has deepened the divide between developed and underdeveloped countries and between poor and rich people within countries. For a variety of reasons, globalization has not lifted all boats, and some of the boats are sailing faster in this ostensibly flat ocean.

For the majority of the developing countries in Africa, Central America, the Caribbean and Central Asia, the last two decades have been the decades of stagnation and decline. With few exceptions, most notably Vietnam and Bangladesh, the LDCs were left far behind, their economies remained stagnant and their people stayed impoverished. Despite the great affluence that globalization has brought, there are 800 million undernourished people in these developing countries, life expectancy in many of them has declined and child mortality has increased. In a considerable number of these countries, there have been signs of recovery in the early 21st century, thanks to the sharp rise in the world prices of their natural resources, but the outlook for the future is not very optimistic. A recent World Bank report entitled *Global Economic Prospects 2007: Managing the Next Wave of Globalization* predicts that Africa is likely to fall further behind.

At the same time, the majority of the population in the developing countries, particularly the populations in the EA countries, has benefited from greater prosperity and greater hopes thanks to globalization. Within three decades, China has risen from the bottom of the world's income distribution to become one of the world's leading economic powers. In India, great hopes for a splendid future have become almost a national obsession. Not all people in China and India have shared the benefits of their countries' rapid growth, and India still has the largest number of the world's poor. Moreover, globalization has very sharply increased the income inequalities in these two most populous countries. Nevertheless, for millions of people in these two countries, and for millions of people in many other developing countries, globalization has brought an unprecedented rise in the standard of living and very optimistic prospects for the future.

In most developing countries that drifted to the bottom of the world's income distribution and did not manage to integrate into the global economy and share the benefits of the growing global prosperity, their own self-serving and often corrupt leadership, political instability and erratic economic policies were the main reasons for their stagnation. Unfortunately, these countries benefited very little from the advice and guidance they received from the international development organization on the economic policies and the structural adjustments that they implemented. In these countries, the great advantages of globalization accentuated their comparative disadvantages which deterred trade and investments, inhibited their access to and adoption of advanced technologies and discouraged collaboration not only within the country, but also beyond the borders, which is a necessary condition for growth.

In the developing countries that did benefit from the rapid global growth, the rise in income inequality is the greatest danger for their continued prosperity, even though the low-income groups have also benefited from their countries' growth. As noted in this survey, there have been considerable similarities in the reactions of the populations in EA, LA and even in many developed countries, to the rise in income inequality and the accumulation of huge wealth by narrow elites. There is growing resentment to central authorities and to the 'capitalists' who are being blamed for the rise in income inequality, as well as to the international institutions, particularly the WTO and the IMF, that are seen as the agents of globalization, enforcing strict rules of free trade that limit national sovereignty. In some countries, that resentment has led to open hostility between political parties and ethnic groups; in other countries, it has created pressures to reduce the multinational collaboration in trade and restrict trade despite the clear benefits it has brought. The rising inequalities between and within countries are the greatest risk for the continued collaboration between countries, particularly since this collaboration is the necessary condition and the basic principle for globalization and the major driving force for the prosperity that it has brought.

References

Arrighi, G., Silver, B.J. and Brewer, B.D. (2003) Industrial convergence, globalization, and the persistence of the north–south divide. *Studies in Comparative International Development* 38(1), 3–31.

Bhalla, S. (2002) *Imagine There is No Country: Poverty, Inequality, and Growth in the Era of Globalization*. Institute for International Economics, Washington, DC

Chen, S. and Ravallion, M. (2001) *How Did the World's Poorest Fare in the 1990s?* Global Poverty Monitoring Database, World Bank, Washington, DC.

Chen, S. and Ravallion, M. (2004) *How Have the World's Poorest Fared since the 1980s?* World Bank Policy Research Working Paper WPS3341.

CIA World Factbook (2004) *The World Factbook 2004*, Potomac Books 2005.

Collier, P. and Dollar, D. (2002) *Globalization, Growth, and Poverty: Building an Inclusive World Economy,* Oxford University Press, USA.

Datt, G. and Ravallion, M. (1996) Macroeconomic crises and poverty monitoring: a case study for India. World Bank Policy Research Working Paper WPS1685.

Dollar, D. and Kraay, A. (2001) Trade, Growth, and Poverty. *Finance and Development* 38(3).

Dollar, D. and Kraay, A. (2002) Growth is Good for the Poor. *Journal of Economic Growth* 7(3), 195–225.

Economic Commission for LAC (ECLAC) (1998) *Prairie Hills Books*, 1998.

Government of India (2001), Reported Adequacy of Food Intake in India 1999–2000, National Sample Survey Organisation, Ministry of Statistics & Programme Implementation, New Delhi, India.

International Herald Tribune (IHT), 2006.

IMF (2001) *World Economic Outlook 2001*, IMF, 2001.

Ravallion, M. (2000) Do National Accounts Provide Unbiased Estimates of Survey-Based Measures of Living Standards? Development Research Group, World Bank, Washington, DC.

Ravallion, M. (2004) Pro-poor growth: a primer. *Policy Research Working Paper Series 3242*, World Bank, Washington, DC.

Sala-i-Martin, X. (2004) *Why Are the Critics so Convinced that Globalization Is Bad for the Poor? A Comment*. Columbia University and NBER.

Sala-i-Martin, X. (2006) The world distribution of income: falling poverty and convergence, period. *Quarterly Journal of Economics* 121(2), 351–397.

World Bank (1984) *World Tables 1984*, The Johns Hopkins University Press, 1984.

World Bank (2001) *World Development Indicators*. Oxford University Press, USA.

World Development Report 2000/2001, *Attacking Poverty*, World Bank Group, Washington, DC.

World Bank (2005) *World Development Report 2005*, Fairford GLO, 2004.

3 Has Globalization Been 'Pro-poor'?

1 Introduction

The impact of globalization on the world's poor countries and poor people is one of the most controversial subjects in the economic literature, and many theoretical and empirical studies have examined the various aspects that are essential to understand how the globalization process affects the world's poor. These studies concentrated on three issues. First, they sought to establish clear criteria for evaluating globalization's impact on the poor. Second, the studies applied these criteria to the large volume of empirical data that were collected in all countries to evaluate that impact. Third, the results obtained in these studies and the experience of a wide range of countries since the early 1980s, when the globalization process gained momentum, were used to determine which lessons could be drawn for the policies that developing countries in general and the LDCs in particular should implement in the coming years in order to accelerate their growth and reduce poverty.

There is wide agreement among researchers and policy makers that economic growth is the key to reduce poverty, and policies that promote economic growth are therefore the most effective strategy to achieve that goal. However, the formulation of policies to promote growth often involves difficult choices, because economic growth affects the income distribution between and within countries: not all people benefit from their country's growth at the same rate, and in many countries, there are population groups or regions that may be left behind. As a result, the larger share of the gains from economic growth may be accumulated by a small segment of the population that becomes increasingly more affluent, while many other segments of the population may gain very little, and quite a few may even remain poor.

The experience of most – but not all – developing countries during the last two or three decades shows that, over time, the impact of economic growth spreads across more regions and population groups and eventually the majority of the people will share and gain from their country's growth. However, in the short and medium run growth is accompanied by, and often takes place as, a result of structural changes that also change the income distribution by widening the income gap between subgroups of the population that work in sectors and live in regions that benefit heftily from their country's growth and other subgroups that work in sectors and live in regions that benefit very little and remain poor.

The implications for a country's growth policies are still controversial. Rodrik (2000) argued that the debate over growth versus poverty reduction that became embroiled in broader political controversies on globalization is a hollow debate since experience shows that growth and poverty reduction go largely hand in hand. In his view, the poor generally benefit from growth; he argues that the absolute number of people living in poverty dropped in all of the developing countries that had sustained growth over the last two decades. Moreover, although in theory a country could enjoy a high average growth rate without any benefit to its poorest households if income disparities grew significantly, this outcome is rare because the income distribution within countries tends to be stable over time.

Moreover, the question in this debate is not merely about the impact of growth on poverty, but about the impact of globalization on poverty. Specifically, the relevant questions are: Is there a natural process inherent in globalization that brings about a reduction in poverty? Are the spread of growth and the transfer of the production of labour-intensive industries to labour-abundant countries a process that is built-in in the global market system and brings a reduction in poverty? A rather similar question was considered by Kuznets who concluded that rising inequality may be a natural outcome of the early stages of development, as people begin the shift from low productivity subsistence agriculture to high-productivity sectors.

The natural process is an outcome of the rational decision-making process that takes place when markets are free and there is no 'human intervention' in the market through policy decisions in either the developed or the developing countries. However, in practice this is not the case. Policy decisions are made in both the developed and the developing countries; they are also made by the transnational corporations that have a monopolistic power in many markets and their decisions are often influenced by non-economic considerations and moreover, policy decisions are also made by regional and multilateral trade organizations. All these decisions affect prices, incomes, income distribution and poverty. The change in poverty is therefore not inherent in the globalization process or a natural outcome of that process, but it is strongly influenced by the decisions and the policies of all the concerned interests. The answer to whether globalization is 'pro-poor' cannot therefore be dissociated from these policies and policies are changing over time. An answer to whether globalization is pro-poor that is based on a regression analysis of data from one time period may not be correct and may even be misleading when applied in another time period.

However, the evidence of the last decade does not support several of the conclusions suggested in the work of Rodrik mentioned earlier. First, in many developing countries that enjoyed rapid growth, from China to India to most LA countries, income inequality rose very significantly. In India, nearly two-thirds of the population were barely affected by the country's rapid growth in the last decade; in China, the Western regions remained cut off from the flourishing Eastern regions due to poor infrastructure. Although over time the absolute number of people living in poverty indeed declined in nearly all the developing countries that had a sustained rapid growth, the conclusion that 'growth and poverty reduction go largely hand in hand' takes a very narrow approach to the definition of pro-poor growth, as we see below.

The objective of this chapter is to review the debate on the impact of globalization on poverty, examine the definitions that have been suggested in the economic literature for pro-poor growth and assess the experience of the developing countries during the last two and a half decades in light of these definitions. The review of the country experiences focuses on the LDCs, primarily in SSA, but it will also discuss the experience of other developing countries as reference.

2 Estimating the Impact of Economic Growth on Poverty

In most theoretical and empirical studies, the main statistical criterion used to evaluate the impact of growth on poverty is the *poverty elasticity of economic growth*. The objective of this measure is to evaluate the impact of economic growth on the *change* in poverty and its formal definition is as follows: 'The poverty elasticity of economic growth is the ratio between the proportionate change in the mean income of the poor and the proportionate change in the mean income of the general population' (see Ravallion and Chen, 1997; Dollar and Kraay, 2001; Ravallion, 2001).

If that ratio remains unchanged, the mean income of the poor rises with growth at the same rate as the mean of the general population, and this suggests that the income distribution remains unchanged.

Empirical studies on the share of the poor in their country's economic growth started in the 1970s, when a larger number of country studies on income distribution became available. The early papers focused on the relationship between growth and income inequality, and their main focal point was Kuznets' hypothesis that inequality is rising during the initial phase of development, and declines over time at a later stage of the development process. Some of the early empirical studies were concerned more narrowly with the effects of growth on the poor. For example, Adelman and Morris (1973), Ahluwalia (1976) and Ahluwalia *et al.* (1979) focused on whether there is a systematic relationship between economic growth and the income share of the bottom quintile. They concluded that this share tends to decline in the early stages of development, but increases in the long run.

The growth–inequality relationship took centre stage during the 1980s, but recent studies concentrated on whether the benefits from growth are distributed proportionally and whether the poor receive their fair share. These studies took two different approaches: The first used a relative concept of poverty along the line of the definition of the *poverty elasticity of economic growth* and estimated the rate of growth and the poverty elasticity of the mean income per capita of individuals at the lowest quintile of the income distribution. Some of these studies, including, for example, the well-known study of Dollar and Kraay (2002a,b), concluded that the elasticity is practically one, indicating that, on average and in most developing countries, the growth in the country's average income led to an equal proportionate increase in the average income of the poor.

However, the estimates of the poverty elasticity of economic growth varied widely between studies and between countries and time periods. In a later study, Dollar and Kraay (2001) estimated the poverty elasticity on the basis of income distribution data of 137 countries during the period from 1965 through 1995, and concluded that after a significant increase in poverty and income inequality from 1960 to 1975, the trend was reversed from 1975 through 1995 and during that time period poverty also declined. To strengthen their conclusions, they estimated the poverty elasticity by comparing the percentage changes in the incomes of the bottom quintile of the income distribution with the percentage change in the mean per capita income in 80 countries over a period of four decades, and they conducted their analysis after dividing these data into 236 'episodes' of 5-year periods or more. They found that the average income of the world's poorest quintile moved almost one-for-one with the average income overall, implying that, on average, '[the] percentage changes in incomes of the poor are equal on average to the percentage changes in average incomes' (Dollar, 2002, p. 16).

2.1 The growth elasticity of poverty

The growth elasticity of poverty characterizes the growth process in terms of its impact on poverty by measuring the change in poverty as an effect of growth (see Box 3.1). Ravallion (2001, 2004) and Ravallion and Chen (1997, 2001, 2004) estimated in their worldwide and key country studies that the elasticity of the headcount ratio was typically higher than 2; in other words, for every 1% increase in the mean income, the proportion of the population living below US$1 per day (at 1993 PPP) falls by an average of over 2%. In a study in 47 developing countries, Ravallion (2001) found that during the 1980s and 1990s the growth elasticity of poverty, defined in terms of the headcount measure, was equal to −2.50, implying that for every 1% increase in the countries' mean income, the proportion of the population living in poverty declined, on average, by 2.5%. On these grounds, he concluded: 'The poor typically do share in the benefits of rising affluence, and they typically do suffer from economic contraction.'

Dollar and Kraay (2002a,b) found that the average income of the poorest quintile moved almost one-for-one with average income of the general population. Moreover, across countries, the log mean income of the poorest quintile

Box 3.1. The growth elasticity of poverty.

An aggregate estimate of the impact of economic growth on poverty has been a key criterion for evaluating the impact of development policies and for establishing priorities between them that reflect the social values of the different goals of economic growth. The *growth elasticity of poverty* measures the percent change in poverty as an effect of a 1% change in income. This measure shows the reduction in poverty that can be achieved with different development policies for the same rate of economic growth. There is a clear trade-off in using this criterion to compare one strategy that generates a rate of growth of, say, 4% and a growth elasticity of poverty of –2 with another strategy that generates a rate of growth of, say, 6% but a much lower growth elasticity of poverty of –1. This trade-off represents the choice between the well-being of the poor and the well-being of the other income groups.

With the headcount measure, for example, a growth elasticity of poverty of –2 and income growth of 4% means that the reduction in the headcount measure would therefore be 8%. With a growth rate of 6% and a growth elasticity of poverty of –1, the reduction in the headcount measure would be only 6%. Even if the main focus of the policy is on the poor, this evaluation would not be sufficient to establish clear priorities between the two options for two reasons: First, it does not take into account the rise in the incomes of the poor that remain poor, and thus ignores the reduction in their poverty gap; second, it also ignores the gains from the increase in the incomes of the non-poor, even though they have a lower priority.

The estimate of the aggregate elasticity is based on estimates of the percentage change in poverty and the percentage change in the per capita private consumption (or income) during 'spells', i.e. time periods, between successive household expenditures or income surveys in a given country. This estimate is usually derived from a large number of 'spells' and countries, and the elasticity is estimated in a regression analysis that determines the average rate for the entire sample of spells and countries over an extended period of time. These estimates may vary from one another, depending on the poverty lines and poverty measures that are used in the estimation.

Ravallion *et al.* (1991) estimated that during the 1980s the growth elasticity of poverty in the developing countries was around –2.2, while in SSA it was only around –1.5. The estimates of the elasticity in different countries varied widely between studies, and this was the reason for the large differences in the evaluations of the effects of growth on the poor between countries and between time periods.

In empirical studies, the estimates of the aggregate value of the poverty elasticity for a group of countries is calculated as a weighted average of the values of the elasticities in the individual countries, weighted by the size of their population. This estimate is dominated therefore by the value of the elasticity in the larger country (or countries). Estimates for the entire group of developing countries are therefore dominated by the elasticities in China and India and these estimates do not reflect the slow growth and the rise in poverty in most SSA countries. For this reason, the poverty elasticity of economic growth during the 1990s, when both China and India had rapid growth and a steep reduction in poverty, does not reflect the changes in SSA. Moreover, this elasticity may also bias the estimate of the impact of growth a decade later, when a much larger share of the world's poor are concentrated in the African countries.

changed almost one-to-one with the overall log income per capita and the share of the poorest quintile was generally uncorrelated with log income per capita. In other words, economic growth did not have an impact on income inequality and on the share of the poor in the national income. These findings have been corroborated by a number of other studies which showed that, on average, there is little or no correlation between the growth in the average household income per person in a country and the change in income inequality. The correlation coefficient between the annualized change in the log Gini index and the annualized change in the log mean income was in the order of magnitude of –0.09 (Ravallion and Chen, 1997; Bhalla, 2002; Ravallion, 2004).[1] However, Ravallion (2003, p. 742) emphasized that to suggest on the basis of these findings that 'if all incomes double, to say that the poor share fully in the gains from growth is clearly a stretch of the language'. In India, the income gains of the richest decile were four times higher than the gains of the poorest decile; in Brazil and South Africa, they were 15–20 times higher.

Although the growth elasticity of poverty is the key measure used to measure pro-poor growth, there are disagreements about the exact meaning of the concept of pro-poor growth and about the economic developments it characterizes.[2] It is not obvious, for example, what is the value of that elasticity at which growth should be considered pro-poor, or why should a growth elasticity that leaves the income inequality unchanged be regarded as pro-poor even though most of the gains from growth are accumulated by the rich. How much, in other words, should the poor benefit from growth in order to justify the classification of a country's growth as pro-poor? There are also several other measures that have been proposed in the economic literature to characterize pro-poor growth and they differ from one another in the value judgement that they represent with respect to growth, poverty reduction and income inequality, as we shall see in Section 2.2.

2.2 Alternative measures of pro-poor growth

Several definitions have been suggested to determine the characteristics of pro-poor growth. The first definition considers growth to be pro-poor *if and only if* the share of the poor in the national income is increasing. This definition is thus based on a *relative* concept of inequality and growth is pro-poor if the income growth rate of the poor is larger than the average growth rate of the general population (White and Anderson, 2001).

[1] These studies found that the growth process has generally been distribution neutral: the growth incidence curve is thus practically flat and growth tends to raise the incomes of people in all percentiles at much the same rate without having much impact on income inequality.

[2] For example, the findings of Dollar and Kraay (2002a,b) that average incomes of the poorest quintile moved almost one-for-one with average incomes overall do *not* mean, as *The Economist* concluded, that '[g]rowth really does help the poor: in fact it raises their incomes by about as much as it raises the incomes of every body else. . . . In short, globalization raises incomes, and the poor participate fully' (*The Economist*, 27 May 2000, p. 94).

Kakwani and Pernia (2000) suggested a definition that compares the changes in poverty due to growth alone (i.e. holding inequality constant) with the actual change in poverty that includes changes in inequality. According to them, growth is pro-poor when this ratio is larger than 1, and inequality then falls. Under this definition, a decline in inequality that is due to progressive income transfers is pro-poor even if there is no growth.

According to Ravallion and Chen (2003), growth is pro-poor if it brings about a reduction in poverty, regardless of its impact on income inequality.[3]

The relative definition of pro-poor growth may raise a similar dilemma to the one mentioned earlier. Consider the following example:

> Under this definition, an average income growth of 2 percent, that is combined with a growth of 3 percent in the average income of poor would be better than an average growth rate of 6 percent that is combined with a growth of only 4 percent in the incomes of poor – even though both poor and non-poor households are better off with the second outcome.

The relative definition may therefore favour interventions that reduce income inequality even if they slow down the rate of growth.

The absolute definition does not give any weight to the impact of growth in the income distribution and to the reduction in income inequality as long as it does not have an impact on poverty. It should be noted, though, that this definition is the one underlying the goal of poverty reduction that has been specified as one of the key MDGs. In their analysis on the changes in poverty in India, Datt and Ravallion (2002) concluded that poverty reduction during the 1990s was not biased in favour of the poor and growth, measured in relative terms, was not pro-poor. In China, Chen and Ravallion (2001) concluded that the actual decline in the incidence of poverty in rural areas between 1981 and 2001 to 12.5% was much smaller than the decline that would have taken place had the Lorenz curve remained unchanged, and had income inequality remained unchanged, the incidence of poverty in rural areas would have fallen to 2.04% by 2001; the much smaller decline in poverty was therefore due to a rise in income inequality and the income growth has not been distribution neutral.

2.3 Pro-poor growth, income inequality and relative deprivation

The objective of this section is to extend the analysis of these measures of pro-poor growth and highlight the value judgements they represent with respect to the characteristics of the growth process. In this analysis, the implications of the impact of growth on poverty and on income inequality are emphasized by presenting the measures both analytically and diagrammatically.

This analysis highlights the different social priorities that are underlying each of the measures. The focus of this illustration is on the trade-off between

[3] A more advanced version of this definition is related to the Watt's index (see below), which adds the dimension of income inequality.

the three main goals of development policy: economic growth, poverty reduction and income equality. The aim is to show the trade-offs between these goals, and the opportunity costs and inevitable sacrifice that may have to be made in terms of one or two of these goals in order to achieve the goal that has the highest priority.

This diagrammatic illustration not only demonstrates the impact of pro-poor growth on income inequality, but also adds another dimension to the perception of inequality. In addition to the two definitions of pro-poor growth in Box 3.2, this illustration also shows the impact on the perception of 'relative deprivation' when the *additional income* is distributed unequally. What is the impact of growth when the share of the rich in the additional income is larger than the share of the poor? When this additional dimension is introduced, the relative definition of pro-poor growth has to be extended. Accordingly, the following two definitions will be used to characterize the relative pro-poor growth:

- Definition 1: Pro-poor growth reduces poverty at a higher rate than the rate of growth of the general economy and income inequality is declining. The additional incomes can still be distributed unequally in favour of the rich – thus creating a perception of relative deprivation. In this case, the absolute increase in the income of the non-poor can be larger than that of the poor and the income gap between the poor and the non-poor can rise.
- Definition 2: With more progressive pro-poor growth, the additional income is distributed more equally. As a result, income inequality is declining and the income gap between the poor and the non-poor is also declining, thus reducing the perception of relative deprivation.

The second, absolute definition of pro-poor growth remains the same. With this definition, income growth is reducing poverty by raising the incomes of the poor, but income inequality may rise since the share of the poor in the addi-

Box 3.2. Two definitions of pro-poor growth.

Pro-poor growth has been broadly defined as growth that leads to significant reductions in poverty (OECD, 2001; UNDP, 2000).

Two different definitions have been used in order to specify the analytical meaning of this definition.

1. *The relative definition: Growth is pro-poor if it leads to an increase in the income share of the poor.*

With this pro-poor growth, *the reduction in poverty is combined also with a reduction in income inequality.* (White and Anderson, 2001; Kakwani and Pernia, 2000).

2. *The absolute definition: Growth is pro-poor if it increases the poor's income and thus brings about a reduction in the poverty gap.* (Ravallion and Chen, 2003; Bourguignon, 2003; Ravallion, 2004).

Income inequality can either rise or decline as an effect of this pro-poor growth.

This definition is focused on the reduction in poverty regardless of the impact on income inequality.

tional income is smaller than their current share in the national income, thus increasing the income gap between the poor and the non-poor and the perception of 'relative deprivation'.

Under the relative measure of the pro-poor growth in definition 1, the actual change in poverty is compared with a distributionally neutral change that leaves the income share of the poor unchanged. Symbolically: Let $Y^{\phi}_t = \{(y_t): \forall y_t \leq z\}$ be the vector of incomes of all the poor individuals at time t, where z is the poverty line income. The growth rate of the income of the pth percentile is given by

$$\hat{y}_t(p) = \text{dLn}(y_t(p)) = \{[y_{t+1}(p)/y_t(p)] - 1\} \tag{3.1}$$

Let $P(Y^{\phi}_t; z)$ denote the general measure of poverty at time t, and let the head-count measure be denoted as: $P_0(Y^{\phi}_t; z) = H_t$. The condition for 'pro-poor' growth according to definition 1 is then given by

$$P(Y^{\phi}_{t+1}; z) < P((1 + \hat{\mu}_t) \bullet Y^{\phi}_t; z) \tag{3.2}$$

where $\hat{\mu}_t = \text{dLn}(\mu_t) = \{[\mu_{t+1}/\mu_t] - 1\}$ is the growth rate of the mean income of the general population during that time. This condition can be determined for a *specific* poverty line or it can be determined for *all* poverty lines; in the latter case:

$$P(Y_t + 1; z) < P((1 + \hat{\mu}_t) \bullet Y_t; z) : \textit{for all } z \tag{3.3}$$

Ravallion and Chen (2001) suggested the latter, stronger criterion, since there is no commonly agreed level of the poverty line income. When the analysis is made for all poverty lines (and thus also *all* percentiles), it yields a family of growth indicators ('Pen's parade') that Ravallion and Chen used in order to construct the 'growth incidence curve'. That curve shows how these growth indicators vary across the population percentiles that are ranked by income.[4] From that series of growth rates, the country's rate of pro-poor growth is calculated as the mean growth rate of the poor across all poverty lines.[5]

Using the ranked income percentiles, the mean growth rate of the poor's income in a given percentile is given by

$$\hat{h}(\varphi_i) = (1/\varphi_i) \bullet \Sigma_{p \leq \varphi(i)} \hat{y}_t(p) : 0 < \varphi_i \leq 100 \tag{3.4}$$

Dollar and Kraay (2001) calculated the rate of pro-poor growth by comparing the mean growth rate $\hat{h}_t(\varphi)$ of the income of the poorest quintile with the

[4] The measure of pro-poor growth for a *specific* percentile is therefore derived from that curve by comparing the rate for that percentile to $\hat{\mu}$.

[5] That rate is different from the growth rate in the mean income of the poor for a given poverty line.

mean growth rate in the income of the general population, μ_t. That comparison can be extended by calculating the rate $\hat{h}(\varphi_i)$ for several or all income percentiles. With *relative* pro-poor growth, the mean growth rates of the income of the poorest φ% of the population would be *decreasing* functions. In other words, $\hat{h}(\varphi_i) \geq \hat{h}(\varphi_i + \zeta)$ for all $\zeta \geq 0$; or $\hat{h}(\varphi_i) \geq \hat{h}(\varphi_j)$ for $j \geq i$. The *relative* rate of pro-poor growth for a given percentile would be the average pro-poor growth rate over *all* levels of the poverty line income. This criterion avoids the calculation of the pro-poor growth rate for an arbitrary poverty line and it gives a larger weight to income inequality. This measure may therefore be taken as a better representation of these two objectives.

2.4 Raising the absolute level of the poor's income

In Ravallion's second definition of pro-poor growth, a reduction in the *absolute* level of poverty is indicated by the rise in the incomes of the poor regardless of the rise in the incomes of others and this is a necessary and sufficient condition for pro-poor growth (Ravallion and Chen, 2003; OECD, 2004; Klasen, 2005). Any increase in the mean income of the poor that is smaller than the increase in the mean income of the general population raises both the income inequality and the income gap between the poor and the non-poor, but, according to this definition, it is still pro-poor.

To highlight the ethical dilemma that this definition raises, Fig. 3.1 shows the rates of growth of the mean income of the poor in a sample of countries that have been calculated by Chen and Ravallion (2001), (see also Ravallion, 1995). In China, the rapid growth during the 1980s and 1990s led to a large increase in the income of the poor population, but increased the incomes of the non-poor urban population, particularly in the eastern regions, at much higher

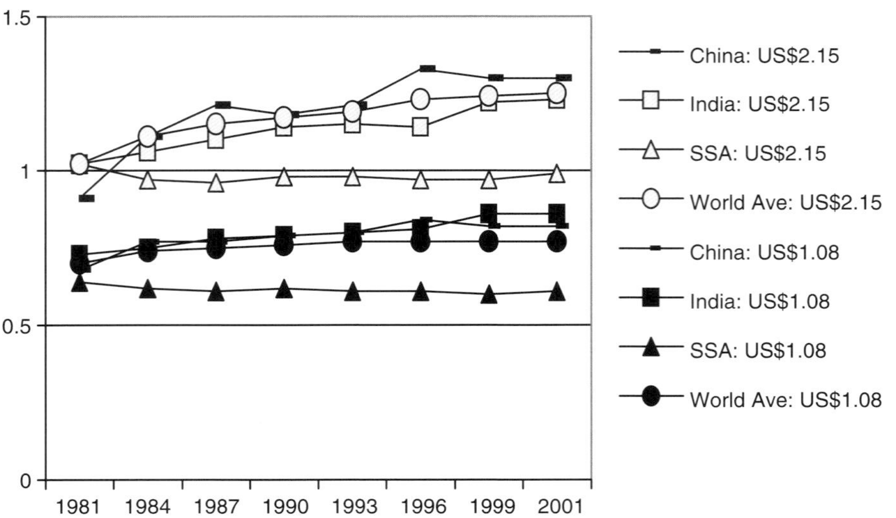

Fig. 3.1. Mean income of the poor – in US$ per day (extreme poverty line US$1.08, moderate poverty line US$2.15). (From Chen and Ravallion, 2001.)

rates. If the poor and/or the policy makers are concerned only with the rise in the incomes of the poor regardless of the change in income inequality, then the absolute measure according to definition 2 is the correct measure of pro-poor growth. However, this view is strongly debated and other writers argue that both the poor and the policy makers are equally concerned with income inequality and with the peoples' relative position vis-à-vis the others and the gap between their income and the incomes of the more affluent segments of the population. This concern has also driven the anti-globalization movement in many countries where protests focused on the rise in income inequality.

A rise in the incomes of the poor that reduces the poverty gap and the poverty measure $P_1(Y^{\phi}_t; z)$ need not lead to a reduction in the headcount measure $P_0(Y^{\phi}_t; z)$. In this case, the impact on income inequality *among* the poor is not certain, but it is likely to decline. A decline in the headcount measure $P_0(Y^{\phi}_t; z)$ need not be associated with a decline in the poverty gap $P_1(Y^{\phi}_t; z)$. In this case, the impact on income inequality *among* the poor is also not certain, but it is likely to rise. The statement of Rodrik (2000) that the number of people living in poverty declined in all developing countries that had sustained economic growth corresponds to the second definition of pro-poor growth, but it may also be the case that the rise in the incomes of the poor was not sufficient to enable them to cross the poverty line income. In principle, that depends on the measure of the poverty line income, but the empirical evidence generally confirms this observation. In quite a few developing countries, the rise in the income of the poor was smaller than the rise in the incomes of the more affluent segments of the population; in these countries, growth has not been pro-poor according to definition 1 and the poverty elasticity of economic growth has been smaller than –1.

3 Criteria for Evaluating Development Strategies

The choice of a development strategy that is the most suitable option for a country, its long-term growth prospects and its efforts to reduce poverty has both economic and political dimensions. From an economic point of view, even the more narrow choice of a development strategy that is pro-poor depends on a wide range of factors that include the existing level and the expected changes in income inequality, the actual rate of growth relative to the rate at which the incomes of the poor are rising, the gains of the poor from growth relative to the gains or losses of other segments of society, the impact of a pro-poor strategy on the country's long-term growth prospects, etc.

3.1 Political and economic priorities between growth, poverty and inequality

From a political point of view, the selection of specific development policies reflects different priorities and a political struggle between parties that represent these priorities and mobilize the support of voters; another consideration concerns the prospects of the policies to achieve their goals. The priorities that are chosen often reflect more intensely concerns about income inequality than about growth. In many developing countries, the widening income gap between the majority of the population and the top elite is a sensitive issue that heats up the political debate and leads to intense confrontation between elected officials and demands from 'the street' for measures to reduce the income gap between rich and poor.

Indeed, the political debate in many low- and middle-income countries has focused in recent years on the highly skewed income distribution and on demands to take active measures to reduce income inequality even at the expense of slowing down the rate of economic growth. However, many (but not all) empirical studies concluded that the changes in income inequality were, in fact, uncorrelated with changes in the level of income (Deininger and Squire, 1996; Chen and Ravallion, 1997; Easterly, 1999; Fields, 2000; Deaton 2001, 2004; Ravallion, 2001; Kanbur 2001; Dollar and Kraay, 2002a,b; Lundberg and Squire, 2003); in other words, there was no *causal* relationship between economic growth and the changes in income inequality. In some countries, economic growth led to an increase in income inequality, while in others economic growth led to a reduction in income inequality. Ravallion (2003) found that among the more rapidly growing economies, inequality tended to fall as often as it tended to rise, and on average, growth was distribution neutral. In an analysis across 117 spells of data between successive household surveys for 47 developing countries, he found a correlation coefficient of only 0.06 between the annualized changes in the Gini index and the annualized rates of growth in mean household income or consumption as estimated from the same surveys.

However, these studies do not make clear whether there is a systemic explanation for these findings; in other words: is there a reason that is embedded in the structure of the economy or related to economic policy that could explain why in some countries economic growth led to an increase in inequality while in others economic growth had the opposite effect. Individual country studies found significant differences between countries: growth was accompanied by greater equality in Bangladesh and Egypt, but by greater inequality in Chile, Brazil, India, China and Poland and, perhaps more surprisingly, in many developed countries ranging from the USA to Israel.

A World Bank (2002) study in 14 countries during the 1990s found that 7 out of the 12 countries that had positive economic growth also experienced an increase in income inequality, and in 2 countries where income declined there was also a decline in income inequality. In all the countries included in that sample, with the exception of Romania, the rise in income was accompanied by a reduction in poverty. Nearly all other country studies also found a clear positive correlation between economic growth and poverty reduction, and there is general agreement that growth is a key factor for the reduction of poverty. Measures to equalize the income distribution that slow down growth may therefore take a heavy toll by slowing down the reduction in poverty.

Country studies show that economic policies that were aimed at stimulating growth by freeing the market, reducing government interventions and abolishing or reducing restrictions on trade were often detrimental to small-scale farmers and unskilled workers in urban areas. According to these studies, policies that were aimed at liberalizing the market and promoting

Table 3.1. Indicators of changes in poverty and income inequality in selected countries. (From IFPRI, 2002; Chen and Ravallion, 2004, World Bank, 2004.)

Country	Income of the lowest 30%	Income of the top 30%	Poverty: headcount	Inequality Gini
	Average annual percentage change			
Vietnam	+3.0	+6.3	−7.8	+2.4
Bangladesh	+0.8	+2.3	−2.8	+1.4
Uganda	+2.0	+3.5	−3.9	+1.8
China	+2.3	+5.4	−11.2	+1.9
India	+1.8	+2.2	−3.8	+0.6

competition in fact often gave more power to monopolistic corporations and enabled them to use their economic and political power as leverage to distort policies or corrupt government officials, thereby augmenting income inequality.

Country studies also found that in the 1990s, there was a strong tendency of rising income inequality *within* countries in most regions with the exception of the Middle East and North Africa; that rise in inequality was particularly rapid in the fast growing EA countries (Table 3.1). In Vietnam, for example, the Gini coefficient grew by an annual average rate of 2.3% between 1993 and 2002, and in China it grew by nearly 2.0% a year between 1990 and 2001.

3.2 The changes in the global economy and growth in the African countries

The African countries found it much more difficult to accelerate their growth in both agriculture and industry. Some of their difficulties were due to their own policies or country-specific economic conditions; others were due to changes in the conditions in the world economy and in the structure of world production and trade that are associated with globalization. These factors will be discussed in more detail in the subsequent chapters, but a brief summary of these factors can highlight the reason for the negative impact on the African countries (Collier and Gunning, 1999):

- With the failure to reach an agreement on agricultural commodities at the Doha Round, the supply of plentiful and cheap food is expected to continue and despite the 'Special and Differential Treatment' given to the exports of the LDCs in the EU and the USA, the African producers find it difficult to compete with imports from the developed countries even in their own urban markets. Household surveys show that rural households are increasingly forced to abandon farming altogether, produce mostly for self-consumption, or earn most of their income from sources other that farming their own lands.
- Major innovations and rapid technological advancements on the one hand and the growing demand for tropical fruits on the other hand

increase the demand and raise the price of these products in the markets of the developed economies. However, small farmers in most African countries are not able to compete with agricultural producers in LA and Asia that successfully diversified their production to high-value commodities such as fruits, flowers, vegetables and livestock due to their countries' poor infrastructure that cuts them off from the global supply chains, and due to their limited capacity and lack of know-how that prevents them from adopting advanced production methods and diversifying their production. These farmers are limited in their capacity to meet the market demand for high quality and food safety which apply particularly to the higher value products.

- Small rural producers are unable to compete with larger farmers that have high returns to scale and better connection to local and international supply chains. Limited profit opportunities in the production of food grains and the collapse of the cotton sector with the sharp reduction in its price deterred the private sector and sharply reduced the access of small farmers to the credit market.
- The growing population in rural areas and the continued division of the land between the larger numbers of households or the expansion of production to marginal lands, often by ruining the local forests, is reducing the productivity of the land.
- In many developing countries, farmers suffered the most from their countries' structural adjustment or their countries' entry to the WTO that required the removal of subsidies for their inputs and the support of the price of their products.

3.4 The lessons from the multi-country studies

The inconclusive findings of country studies and the large differences between geographical regions raise obvious questions about the lessons that can be drawn from the results of the multi-country studies (Box 3.3), because it is clear that, on average, there has been no significant correlation between economic growth and income inequality. The findings raise questions about the expected impact of globalization in the coming years, particularly in light of the large changes in the composition of the poor and the countries in which they concentrate.

Moreover, the country studies did not provide conclusive answers as to whether countries should opt for rapid economic reforms that remove inefficient government controls and free the market in order to achieve higher growth rates, or whether governments should take a more cautious approach with careful planning and economic reforms that take into account the impact on income inequality and establish safety nets in order to mitigate the impact of structural reforms (see also UNCTAD, 2002).

The reason for the conflict between policies that are aimed at accelerating growth and policies that focus on the need to reduce poverty is that growth can be maximized by concentrating production in sectors and regions where

Box 3.3. Data sources of multi-country studies.

The vast majority of the multi-country studies used either the database that was laboriously assembled by Deininger and Squire (DS) or, to a lesser extent, the UNU-WIDER [UW] data set. Both sources include data of a reasonably wide sample of countries and a long time period, but coverage of the LDCs, particularly in SSA, is very thin. The data are often outdated, and the most recent observations are from 1993 to 1995 or earlier. This makes an analysis of the effects of globalization particularly difficult. Income data in the developing countries are notoriously incomplete and the upper deciles tend to under-report their incomes, primarily their income gains from their capital; data on the value of consumption expenditures from households' own production are very difficult to evaluate.

Another data set of multi-country studies is from the University of Texas-Austin Inequality Project (UTIP). This data set is based on annual time series of industrial earnings statistics. Like the DS/UW data, coverage of developing countries, particularly the African countries, is limited and time series are generally outdated, ending in 1993. The UT data set is, however, an annual time series and the focus on industrial earnings makes it particularly useful for analysing the change in the earning gaps, such as the gap between the wages of skilled and unskilled workers. On the other hand, these data focus on inter-sectoral wage differences, and ignore differences in inequality within sectors between different types of skills or education.

Multi-country studies that are concerned with the measurement of poverty had to specify a single poverty line for all countries in order to make comparisons between countries. The usual approach is to define the poverty line as the income that is necessary to maintain a minimum standard of living that is necessary for bare survival. That income includes not only expenditures on food, but also some expenses on health, clothing and other necessities. In most studies, the poverty line for 'extreme poverty' is defined as US$1 per day. In order to allow comparisons between countries and over time, the following adjustments are made: the poverty line is defined at that level for the same year, usually taken to be 1993. Adjustments between countries are made according to the purchasing power of that income in different countries. Adjustments over time are made according to changes in the countries' exchange rate and in their price level (see also Deininger and Squire, 1998).

it is most effective, making investments where capital is most productive and using the more efficient workers. The productivity of the high-skilled workers is higher, and they are generally the more educated non-poor workers. By increasing the wages of poor workers, in contrast, the share of the more productive non-poor workers in the national income is reduced and their work incentives are reduced as well, thus reducing the economy's growth.

During the last decade, for example, growth in India was higher than at any time in the last 50 years, but that growth was not pro-poor and it increased income inequality; moreover, the rural sector remained essentially stagnant and the majority of its population remained poor. In China, the stunning accumulation of wealth in the industrial centres in the country's eastern provinces brought about a sharp reduction in poverty, but only a fraction of that wealth filtered into the rural areas in the western provinces that, by and large, remained poor and undeveloped, and the minor improvements in their standard of living were due to remittances from family members that

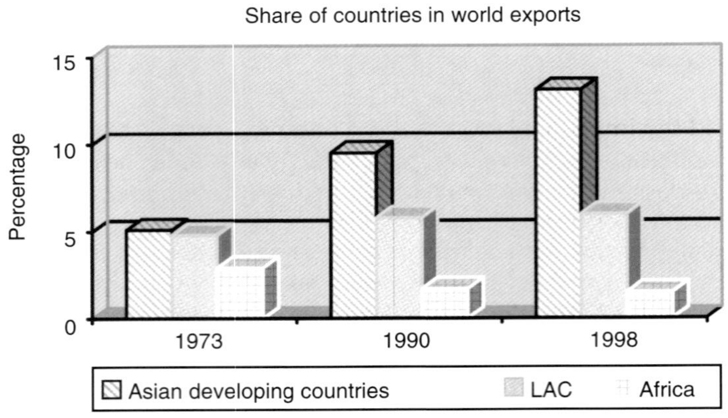

Fig. 3.2. Share of countries in world exports. (From OECD, 2001.)

migrated to the urban centres or abroad (World Bank, 1993). Government investments in infrastructure also concentrated in the urban centres in an effort to attract new industries, and only in the past few years, more public investments were made in rural infrastructure and in the education and health of the rural population (Fig. 3.2).

The following examples illustrate the dilemma that governments in developing countries are facing as they weigh their policy options:

- Should the government maintain price controls over local food crops in order to help the urban poor despite the distortions that these controls create in resource allocations, and despite their negative impact on agricultural growth and on the rural poor?
- Should the government pursue land reforms even though reforms that distribute the lands of the large estates are likely to reduce productivity and exports, or should the government consolidate the lands as well as the rural supply chains in order to raise productivity, even though this may force many small farmers to leave their lands? In the rural sector, land reform is often the main policy instrument to reduce income inequality and to provide a source of income to the landless rural population.
- In many countries, developed and developing, a significant component of agricultural policy included targeted subsidies for specific crops that, in part, were politically motivated. These subsidies often increased the concentration of land in the hands of fewer and larger farmers or agricultural corporations.
- A thorough reorganization of the entire agricultural sector is often part and parcel of a country's integration into the global market and requires thorough structural changes, technological advancements and production diversification. In some countries, this reorganization increased inequality in the distribution of both land and income, sometimes impoverishing many rural households and even leading to active resistance and violence.
- When a country seeks entry to the WTO, it is required to make significant policy changes that include the removal of subsidies and price supports

for selected sectors, thus leading to significant structural changes that are often detrimental to the country's poor.

In practically all countries, the transition period during the structural changes and the implementation of the economic reforms created large difficulties for population groups that were negatively affected. These groups included middle-class, white-collar workers who were not able to adjust to the new structure of the economy and lost their jobs or saw their salaries eroding. In many countries, the groups that were negatively affected coalesced under a common political banner that opposed the reforms and, being significantly empowered by the process of democratization, staged strong opposition to the government's policies and often also to globalization. Against the background of these conflicts, Amy Chua (2003) argued in her book *World on Fire: How Exporting Free Market Democracy Breeds Ethnic Hatred and Global Instability* that in many developing countries, a rapid transition to free-market democracy benefited certain income groups and certain ethnic groups that were able to tighten their control over the economy and thus also on the emerging political system. These controls often led to sectarian conflicts and civil wars when the majority but poorer groups felt economically deprived or even used. The book provides a stimulating discussion on the sobering experience of many developing countries around the world in their combined pursuit of free markets and democratization. The relevant lesson in the context of this chapter is the significance of the two dimensions of income inequality: the inequality in the distribution of the national income and the inequality in the distribution of the gains from growth. The diagrammatic illustration in Section 4 shows the differences between these two dimensions and discusses their policy implications.

4 Pro-poor Growth: Diagrammatic Illustration

The objective of this section is to exhibit the different definitions of pro-poor growth on a diagram that highlights the differences between them. In order to provide the context for this illustration, the section identifies the three factors that jointly determine the impact of growth on poverty and on income inequality: the first is the rate of economic growth; the second is the initial level of income inequality in the economy; and the third is the *change* in income inequality as an effect of the growth process.

4.1 The impact of the income distribution on pro-poor growth

To show the different effects of growth, we use the following notations: Let $F_t(y)$ denote the cumulative distribution function (CDF) of income at time t. The percentage of the population with income equal to or lower than y is denoted by $F(y) = p$. The country's Lorenz curve $L_t(p_Q)$ shows the cumulative share of the Q per cent of the poorest individuals (referred to as the poorest '*quantile*' – quartile, quintile, etc.) of the population. The mean income per capita of the individuals in the Qth *quantile*, denoted by $y_t(p_Q)$, is given by

$$y_t(p_Q) = F_t^{-1}(p_Q) = L_t'(p_Q) \bullet \mu_t \tag{3.5}$$

where μ_t is the mean income of the general population and $L'_t(p_Q)$ is the first derivative of the Lorenz function at p_Q. In the ordered income distribution along the Lorenz curve, the p_Qth individual is the least poor individual in the Qth *quantile*, and $[L'_t(p_Q) \bullet \mu_t]$ is the income of that individual (Gastwirth, 1971; Kakwani, 1980). Thus, for example, with the CDF of income $F_t(y)$ and the poverty line z, the headcount measure of poverty is given by $F(z) = p_Z$, and the poverty line income, $y_t(p_Z)$, is given by

$$y_t(p_Z) = z = F_{t\text{-}1}(p_Z) = L_t'(p_Z) \bullet \mu_t \tag{3.6}$$

By taking the logarithmic differential of Eq. (3.6) at any level of income $y(p_Q)$, the growth rate of the income of the p_Q-th individual in the Qth *quantile* is therefore given by

$$\text{dLn}(y(p_Q)) = \text{dLn}(L'y(p_Q)) + \text{dLn}(\mu_t) \tag{3.7}$$

From Eq. (3.7) we can calculate the change in the income distribution for a given growth rate of the mean income $\text{dLn}(\mu_t)$ and with alternative distribution strategies. Following Gastwirth (1971), this growth rate is given by the ratio:

$$\begin{aligned} \text{dLn}(y_t(p)) &= \{[y_t(p)/y_{t-1}(p)] - 1\} = \\ &= \{L_t'\, y_t(p)/L_{t-1}'\, y_t(p)\}(\text{dLn}(\mu_t) + 1) - 1 \end{aligned} \tag{3.8}$$

This equation can also be written as:

$$\hat{y}_t(p) = \{[y_t(p)/y_{t-1}(p)] - 1\} = \{L_t'(p)/L_{t-1}'(p)\}(\hat{\mu}_t + 1) - 1 \tag{3.9}$$

where a hat over a variable denotes its growth rate. This equation emphasizes that the rate of growth of the poor's mean income $\hat{y}_t(p)$ depends not only on the economy's rate of growth given by $\hat{\mu}_t$, but also on the change in the income distribution given by the ratio $\{L'_t(p)/L'_{t-1}(p)\}$. The larger the 'convergence' that takes place during the growth process, the more pro-poor is that growth.

The growth process can have different effects on the income distribution and thus on the growth rate of the mean income of the poor. The different distributions of a given level of income between the poor and the non-poor are given in Fig. 3.3 by different points along the negative 45° line that crosses through D and K. According to the *relative* definition:

1. A–D: Income inequality is rising and the ratio $[y_t(p)/\mu_t]$ is falling. The mean income per capita of the poor is rising, however, and their poverty gap is falling, thus reducing the level of absolute poverty. The impact on the incidence of poverty depends on the level of the poverty line income. The rise in the mean income of the general population is given by the horizontal distance from B_1 to H. The increase in the poor's mean income (given by the horizontal distance from B_1 to D) is therefore smaller than the increase in the mean income of the general population. However, the mean income of the poor is still rising even though the income inequality is also rising and $\hat{y}_t(p) < \hat{\mu}_t$. With this transition, growth is 'pro-rich' according to two definitions of relative pro-poor growth, 1 and 2, but it is pro-poor according to the 'absolute' definition 2.

A-G: Income inequality at point G is equal to the income distribution at point A and the income distribution and the ratio $[y_t(p)/\mu_t]$ therefore remain unchanged. However, the *increase* in the mean income of the poor is smaller than the increase in the mean income of the general population and the share of the poor in the additional income is therefore smaller than the share of the non-poor. Hence, although the increase in income is distribution neutral, that increase is 'pro-rich' according to the definition 2. Since income inequality remains unchanged, the growth rate of the poor's mean income is equal to that of the general population, $\hat{y}_t(p) = \hat{\mu}_t$.

An indicator of whether income is distributed unequally is the ratio $[y_t(p)/\mu_t]$. If this ratio is smaller than 1, income is distributed unequally; but the common measure of income inequality is the ratio between the incomes of the poor and the non-poor, or the ratio between the highest and the lowest 'quantiles' (deciles or quintiles). The change in inequality (and the 'convergence' or the 'divergence' between the rich and the poor) is given by the ratio between the rate of change in the income of the highest 'quantile' and the rate of change in the income of the lowest 'quantile'. These are common definitions of the change in income inequality that are used to determine whether growth has been 'pro-poor' (Dollar and Kraay, 2002a,b). With this growth, the share of the poor in the increase in income is still smaller than that of the non-poor and that growth is therefore still 'pro-rich' according to definition 2.

2. A–B_δ: Income inequality is declining and growth is pro-poor according to the relative definition 1. However, the additional income is still distributed unequally and the share of the poor in the additional income is smaller than the share of the non-poor. This growth is therefore still 'pro-rich' according to the relative definition 2. The difference between the actual reduction in poverty and the reduction that leaves income inequality unchanged is given by:

$$\hat{\mu}_t - \hat{y}_t(p) = (1 + \hat{\mu}_t)\{[L_t'(p)/L_{t-1}'(p)] - 1\} \quad (3.10)$$

When income inequality is declining, $[L'_t(p)/L'_{t-1}(p)] < 1$

3. A–H: Income inequality is declining and the additional income is distributed equally between the poor and the non-poor. In this distribution as well as in distribution to the right of point H, such as the distribution at K, growth is pro-poor according to definition 2 as well.

4.2 Alternative patterns of pro-poor growth: diagrammatic illustration

The diagrammatic illustration in this section shows the alternative definitions of pro-poor growth in order to highlight the differences between them and demonstrates some related concepts. This section shows these definitions in a closed economy; later, the illustration is extended to a multi-country multisector model in which the definitions of pro-poor growth are extended to take into account the effects of growth on the income distribution between and within countries. The diagram combines the various definitions of 'pro-poor' growth that have been presented above.

Figure 3.3 shows the impact of alternative growth and distribution scenarios on the incomes of the poor and the non-poor segments of the population. The initial income distribution is given at point A; the mean income is given by $\mu(y_A)$, and the poverty line is given by z. The initial level of income of the non-poor is given by $y_A(R)$ and the income of the poor is given by $y_A(P) < z$. The initial income inequality is thus given by the ratio $y_A(P)/y_A(R) = t_g(\alpha) < 1$, where $\alpha < 45°$.

With the rise in the mean income from $\mu(y_A)$ to $\mu(y_B)$ the points along the negative 45° line that crosses through G describe different distributions of the same additional income between the two individuals. In the transition from A_1 to G, the income distribution remains unchanged, the income of the poor rises from $y_A(P)$ to $y_B(P)$ and the increase in their income is given by $\Delta y(p)$. The income of the rich rises from $Y_A(R)$ to $Y_B(R)$. $\Delta y(p)/\Delta y(\mu)$ is the ratio

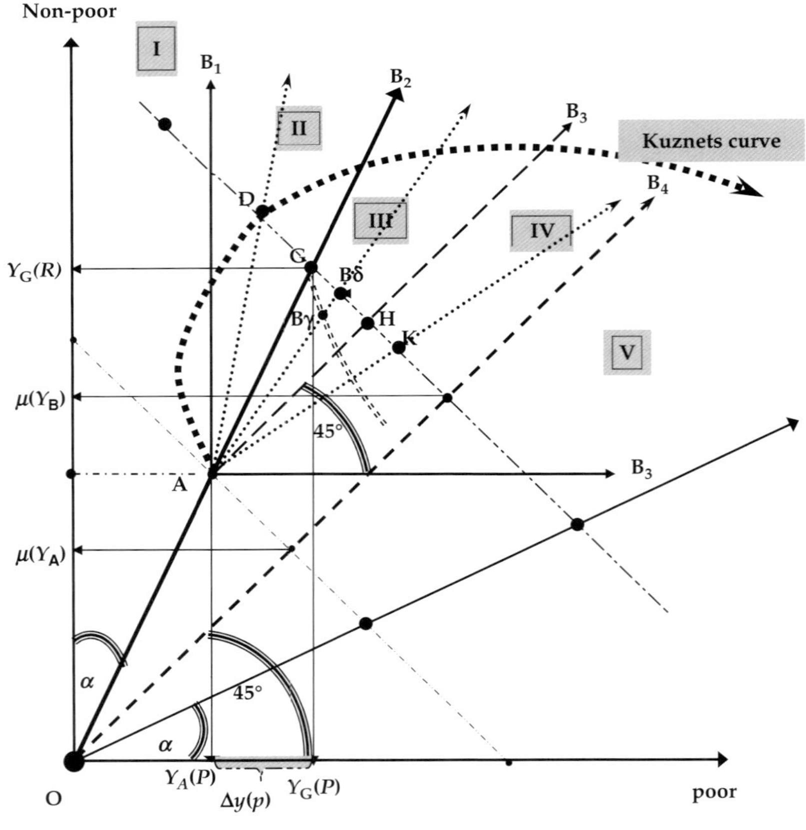

Fig. 3.3. Alternative trajectories of *pro-poor* growth and the Kuznets curve.

between the increase in the income of the poor and the increase in the mean income of the general population. If the income distribution remains unchanged, then $dLn(y_t(p)) = dLn(\mu_t)$ and the ratio $\Delta y(p)/\Delta y(\mu) = t_g(\alpha) < 1$.

The trajectories that determine the distribution of the additional income in one of the ranges I–V show different options for the distributions of the additional income between the poor and the non-poor. Their impact on poverty, income inequality and growth is as follows:

I – The distribution of the additional income along a trajectory in this range is regressive, the share of the poor is declining and the poverty gap is rising. The impact of regressive income transfers on growth is discussed below.

II – The increase of the additional income of the poor in trajectories in this range is lower than the increase in the additional income of the non-poor and income inequality therefore rises. Nevertheless, with that increase in the mean income of the poor, the poverty gap is reduced and the number of the poor is also likely to decline. An increase in the mean income in this range is therefore pro-poor according to the second definition since it reduces the absolute level of poverty, but it is not pro-poor according to the first definition since the share of the poor in the national income is declining and income inequality is rising. In Fig. 3.3 this is shown by the fact that a trajectory in this range is to the left of the trajectory OG and the national income becomes more regressive. In terms of the above notations:

$0 < dLn(y_t(p)) < dLn(\mu_t)$, the ratio $[y_t(p)/\mu_t]$ is therefore falling, and income inequality is rising. Although the distributional shift along this trajectory is 'pro-rich', the poor have *absolute* gains.

Rodrik (2000) argued that growth and poverty reduction go hand in hand; the impact of growth on the reduction in poverty is clearly shown in the diagram, but it is unclear whether Rodrik would argue that the reduction in poverty itself also accelerates growth (and thus they go 'hand in hand'), and how the reduction in poverty is affected by the changes in the income distribution. In any event, economic growth is driven not only by rational decisions of people in the free market, but also by government policies that affect the market prices and the response of the various agents that operate in the market – consumers and producers – to the incentives of the price system that may include direct incentives (e.g. through the tax system) given to them by the public sector. Growth is also affected by the introduction of more advanced technologies and more capital goods in production. People respond to all these incentives and to incentives like better working conditions, by increasing their productivity and possibly their working time and the resulting increase in their output, as well as their response to other changes in their workplace, in turn, affect the economy's overall output and the national income (see also Alesina, 1997).

OG – As noted earlier, growth along this trajectory leaves the income distribution unchanged and $dLn(y_t(p)) = dLn(\mu_t)$. This is, in fact, the pattern of growth found in the empirical analysis of Dollar and Kraay (2002a) according to which, on average, a rise of 1% in the world's mean income raises the mean income of the world's poor population by 1%.

III – Trajectories in this range, like the one that crosses through the points B_γ and B_δ indicate distributions of the additional income that increase both the absolute income of the poor and their share in the national income. With these distributions of the additional income, the distribution of the national income is therefore becoming more progressive. In the figure, this is shown by the fact that the trajectory AB_δ is located to the right of the trajectory OG. The distribution of the additional income along this trajectory is therefore pro-poor according to both definitions of the term.

It should be noted, though, that since a trajectory in this range is located to the left of the trajectory AB_2, along which the additional income is distributed equally between the two groups, the distribution of the additional income along this trajectory is still regressive and the share of the rich in the additional income is still larger than the share of the poor. Hence, growth along this trajectory still raises the *absolute* income gap between the poor and the non-poor.

None the less, the distribution of the additional income along this trajectory is less regressive than the original income distribution and this growth therefore contributes to reduce the overall income inequality. In other words, the new income distribution is less regressive than the initial distribution at G.

However, the growth rate along this trajectory may also be lower. One reason why progressive transfers of income to the poor may lower the economy's rate of growth is the much lower productivity of the poor. Progressive income transfers increase the incentives of the poor to work more but reduce the incentives of the more productive non-poor. Another reason is because income transfers from the rich, whose marginal propensity to save is higher, to the poor, whose savings are much smaller, would reduce savings and investments in the economy and thereby reduce the potential of the economy to grow. A third reason is the impact of progressive transfers on the profitability of investments. By reducing, for example, the profitability of investments in local industries, higher taxes on high incomes or on capital gains may lower investments and slow down growth of the industrial sector and thus the growth of the entire economy. In extreme cases, progressive transfers may slow down the economy's growth and the rise in incomes to the extent that the rise in output and income may be so meager that it may lead to a decline in the income of the poor and a rise in the poverty gap despite the economy's growth.

In the more general case, the incomes of the poor are rising, income inequality is declining, but the economy's growth rate is reduced. This illustration may clarify the somewhat enigmatic definition of Ravallion and of Kakwani and Pernia quoted earlier:

> Although with growth in this range the share of the poor in the *additional* income is still smaller than the share of the non-poor, the distribution of the additional income is less regressive than the original income distribution, and the overall income inequality is therefore declining.

Policy makers have a dilemma with progressive transfers of incomes when they have a negative impact on growth: in the figure, this dilemma is illustrated by comparing the rise in the mean income with the growth from A to G with the rise in the mean income with the growth from A to B_γ. With the growth from A to B_{γ}, income inequality is declining, the poverty gap is falling, but the economy's rise in income is lower than its rise with the growth from A to G. Is the economy then better off (with the decline in poverty and in income inequality) or is it worse off (with the decline in the economy's growth)? This trade-off between a higher rise in the incomes of the poor and lower growth presents a dilemma that is inherent in this policy.

When income inequality is measured by the ratio between the poor's mean per capita income relative to mean income of the non-poor, then income inequality along this trajectory is equal to:

$$\alpha\{(1 + d\gamma/\gamma)\}/\{1 + [(d\rho/\rho)]\} < \alpha \tag{3.11}$$

where α is the initial income inequality and $d\gamma/\gamma$ and $d\rho/\rho$ are the growth rates of the mean incomes of the poor and the non-poor, respectively. When income inequality is measured by the ratio $[y_t(p)/\mu_t]$ then the change in income inequality is equal to:

$$\alpha\{(1 + d\gamma/\gamma)\}/\{1 + [s_p(d\gamma/\gamma) + (1 - s_p)(d\rho/\rho)]\} < \alpha \tag{3.12}$$

where s_p is the share of the poor in the national income. These are the common definitions of pro-poor growth (Datt and Ravallion, 1992, 2002; World Bank 2000; Chen and Ravallion, 2001, 2004; Dollar and Kraay, 2001).

AH – Distributions along this trajectory divide the additional income equally between the two income groups, so that $\Delta y(p) = \Delta y(\mu)$. In other words, the income gains of the poor are equal to the income gains of the rich and income inequality is declining. This, however, is not the meaning of the statement that 'the poor share fully in the gains from growth', and this is therefore a wrong interpretation of the findings of Dollar and Kraay – in which case income inequality remains unchanged. However, even in this case, the overall income distribution remains unequal.

IV – The distribution of the additional income along this trajectory is clearly progressive with the allocation of a higher share of the additional income to the poor. However, the entire income distribution can still remain unequal. Distributions in this range reduce both income inequality and the absolute income gap between rich and poor. It has sometimes been argued that only with distribution policies in this range, in which the *absolute* income gains of the poor are larger than those of the non-poor, the growth process is pro–poor (See Atkinson and Brandolini, 2001). That pattern of growth and distribution is, obviously, by far more pro-poor than that in definition 2. The question is what would be the impact on the economy's growth of this radical pro-poor distribution of the additional income.

V – That trajectory of economic growth can be described as truly Marxist since it revolutionizes the country's income distribution by giving the poor a larger share of the country's resources. This is obviously an extreme reference point in the rhetoric of the recent years, but it has been brought up in the platforms of quite a few leftist parties and even by some of the new leaders in LA, most notably Hugo Chávez, the newly (re)elected Daniel Ortega and the veteran warrior Fidel Castro (and this is only a partial list). The attraction of the Marxist revolution that will redistribute all wealth to all people is still in the minds of many ordinary people and political parties.

4.3 The opportunity costs of equity and the Kuznets curve

The different trajectories represent different income distributions and rates at which poverty can be reduced. Each trajectory expresses different weights that are given to the three goals of the country's development strategy. The different strategies represent different priorities with respect to the levels of growth, poverty reduction and income inequality, but, in themselves, they do not suggest any criteria how to determine the priorities and make the choice between these strategies.

A possible way for making this choice is to quantify the opportunity costs of an increase in income inequality in terms of the reduction in the mean income that may be required, i.e. the opportunity costs of raising equity in terms of the reduction in the economy's growth rate that may result from the progressive income transfers. Likewise, the opportunity cost of reducing poverty can also be expressed in terms of the reduction in the economy's growth. These trade-offs can be presented in Fig. 3.3 by the 'growth distribution curve' which shows the decline – or increase – in the rate at which the economy's mean income is rising with growth as an effect of the impact of the policy aimed at reducing income inequality and/or poverty.

In the figure, these costs can be determined as follows:

- If the income distribution and the gap between the mean income of the rich and the poor remain unchanged, the growth in the incomes of the poor is given by $\Delta y(p)$, the rise in the mean income is given by $\Delta y(\mu)$ and $\Delta y(p)/\Delta y(\mu) = t_g(\alpha) < 1$.
- Consider the case in which, with progressive income transfers, the change in income is manifested by the transition from G to B_γ. The reduction in the income gap between the rich and the poor is determined by the impact of the progressive transfers on accelerating the incomes of the poor, given by $d\mathrm{Ln}(y_t(p))$ and on the rate at which they slow down the rate of growth of the economy's mean income, given by $d\mathrm{Ln}(\mu_t)$, and thus also the incomes of the rich. The decline in income inequality is measured by the ratio $[d\mathrm{Ln}(\mu_t)/d\mathrm{Ln}(y_t(p))] < 1$.
- The reduction in the income gap of the poor below the poverty line is determined by comparing that gap at G and at B_γ. The increase in the income of the poor at B_γ, denoted by $\Delta y(p_\gamma)$ is usually larger than $\Delta y(p)$ and the poverty gap is therefore declining. All this can be easily shown in the diagram itself, but that would require an additional diagram to prevent the entire illustration from being almost incomprehensible.
- The reduction in the mean income as a result of the progressive income transfers is given by the difference between the mean income at G (which

is equal to the mean income at B_δ) and the mean income in B_γ. In other words, without any change in the income distribution, the mean income would have increased by $\Delta y(\mu)$ and that increase is larger than the increase when growth is combined with progressive income transfer, and the increase in income is then given by $\Delta y(\mu_\gamma)$.

- The 'growth distribution curve' is that which crosses through G and B_γ and determines for each trajectory from A the gains due to the reduction in the income gaps and the opportunity costs in terms of the reduction in the mean income.
- These quantitative measures of the reduction in the rate of increase in income and the reduction in poverty thus add another dimension to the discussion whether growth has been pro-poor. Even when we can verify that the growth has been pro-poor, reduced the poverty gap and increased the share of the poor in the national income, there may still be a question about the opportunity costs of this pro-poor growth. Only when we know these costs can we provide a complete answer to whether that pattern of growth and that strategy of economic development have been desirable.

In a similar way and using essentially the same measures, we can evaluate the opportunity costs of the gains in income – or in growth – as an effect of a regressive transfer of income with, for example, regressive, 'pro-rich' taxes, in terms of the resulting increase in income inequality and the increase in the poverty gap.

The opportunity costs are only quantitative measures of the changes in the relevant performance indicators, but the choice of the specific strategy on the basis of these criteria depends on the weights the policy makers attach to these indicators and the cost they or their constituents are willing to pay in terms of the slow down in the growth rate in order to achieve a more equal distribution. Obviously, once the costs can be quantified in these terms, the choice between the alternatives can be made in more rational terms.

The relations between economic growth, the poverty gap and income inequality are marked in the figure by a curve that is referred to as the *Kuznets curve*. According to this curve, the initial phase of development involves a transfer of resources from the rural to the urban sector in order to develop an industrial base and make the necessary investments in infrastructure. That phase involves regressive income transfers because considerable resources are required for these investments, which only the government can make, by mobilizing these resources through taxing the incomes of the rural population.

Simon Kuznets (1955, 1965), in his well-known analysis, describes the relations between growth and income inequality and concluded that it has an inverted-U pattern: as per capita income rises, inequality first worsens and then improves. The major driving force for this change is presumed to be the structural changes that occur because labour shifts from the poor and less productive traditional sector to the more productive and differentiated modern sector. The hypothesis was supported by a number of studies – including

Kravis (1960), Ahluwalia (1974, 1976) and Robinson (1976). Later on, with the accumulation of better and more data, the Kuznets' inverted-U in individual countries was challenged and sharply criticized (Ahluwalia *et al.*, 1979, 1980; Anand and Kanbur, 1993; Deininger and Squire, 1996).

According to Kuznets, the initial stage of industrialization has a negative effect on rural incomes due to the migration of the younger and more productive rural workers to the urban areas in search for higher income. In the urban area, the industrial sector is still in its infancy, but the more skilled workers already receive higher wages due to their higher productivity and the rise in profits of the local industries, largely thanks to government investments in infrastructure and pro-industry tax policies.

Over time, the income distribution gradually becomes more equal for several reasons. First, more and more rural migrants find jobs with higher income as the industrial sector is expanding. Second, as the rural migrants get higher incomes, they increase their remittances to their families that remained at home in the rural areas (Cáceres and Saca, 2006). Third, as more unemployed workers find work in the expanding industries, they also acquire more skills and receive higher wages as the local industries increase their profits with more productive workers and their exports to foreign markets where their product prices are higher. Fourth, the rising incomes in the urban sector increase the urbanites' demand for agricultural products and the price of these products is gradually rising. Over time, the income gap between industrial workers is declining with the rise in their skills, and the income gap between the urban and the rural sectors is also shrinking with the rise in agricultural prices, the increase and diversification in agricultural trade and the improvements in agricultural technologies that raise the productivity in the rural sector.

4.4 Other goals of economic development

The main concern with the focus on pro-poor growth as the principal yardstick for measuring the performance of the economy is that other criteria may be disregarded or relegated, even though they may be important. One point to consider is that these other criteria may represent constituents that are also important, not least because their support may be essential to implement the pro-poor policies. Another consideration is that pro-poor policies can be too myopic since they may come at the expense of higher growth rates and thus lower the capacity to help the poor in the future.

International development organizations and aid donors focus primarily on the needs of the poor 'here and now'. The main performance criterion of aid to the LDCs is the success in reaching the poor and alleviating their poverty; growth is 'good' if (and only if) it is 'good for the poor'; aid is effective if (and only if) it helps the poor.

The criteria that are commonly used to identify the poor and to evaluate the performance of pro-poor policies are usually not confined to income or consumption expenditures alone, but include education, health and other indicators that define human development. These criteria are used most extensively in the UNDP's Human Development Report (UNCTAD, 1999). Pro-poor policies, in contrast, usually concentrate first and foremost on measures of income or expenditures and assign much less importance to the other indicators. Obviously, income measures are only partial indicators of well-being, but they are widely available and easily comparable between countries and over time. Other indicators like infant mortality, children's education or life expectancy are also widely available, but the creation of a single indicator combining all these different measures is very complex and such a combined index is bound to be very controversial.

Moreover, a country's indicators of economic progress should include not only indicators of the individuals' well-being, but also indicators that manifest the country's long-term growth prospects and its capacity to reduce poverty in the future. These indicators would include measures that represent relevant public goods such as public investments in infrastructure, improvements in governance, effectiveness of institutions, etc.

The focus on income measures of poverty and on pro-poor growth as the main indicators of a successful development strategy also ignores other goals that the country and/or the policy makers may have and irons out the dilemma that these policy makers often face in determining priorities between alternative and perhaps even conflicting goals. This has sometimes been a source of dispute between the international development organizations and the donor countries on the one hand, and the governments in the developing countries on the other hand.

In many developing countries, concerns over the rising income inequality have assumed higher priority in the government social agenda than concerns over poverty. Even in countries where poverty remains high – India, Vietnam, most countries in Central Asia and Central America and even quite a few SSA countries – the sharp rise in income inequality and the accumulation of huge wealth by the top percentiles of the population became an explosive political issue. The rising inequalities created strong resistance and opposition to the policies of the old parties and the sitting governments, particularly since they were often tainted by corruption, and in many countries the opposition forces brought new leaders and political parties to power. At the same time, the poor populations, particularly in the rural areas, are not engaged very actively in their countries' politics, either because of their distance from the urban centres or because of their preoccupation with mere survival.

The opposition therefore concentrates in the urban areas and its main driving forces are the middle-income classes, particularly after many of them were impoverished by the policies of their governments. On these grounds there is also strong opposition in many countries to globalization and to the FTAs advanced by the WTO and promoted by the 'Washington Consensus' institutions, sometimes as part of the conditionalities for aid. When countries opened their markets to imports of cheap products, primarily from EA, many industries were bankrupted and many workers were left unemployed. In addition, jobs were lost to outsourcing to countries where wages are lower or business conditions are better.

The middle-income classes are politically more powerful than the poor and governments cannot ignore their plight, their demands and their pressures. The focus on pro-poor growth often fails to give enough attention to these political issues which governments must address and which may involve conflicting demands. These genuine needs and concerns of many developing countries must also be taken into account in an open dialogue between their leaders and the international development organizations and donor countries that give them aid or other assistance.

4.5 Can growth, poverty reduction and equity be conflicting goals?

The dilemma between growth, poverty reduction and equity is sharpened when the time dimension in the policy decision making is taken into account and the choice between growth and poverty reduction is presented as a choice between poverty reduction today and poverty reduction tomorrow (Cornia, 2005). Kaldor highlights the dilemma between more equal distribution of income that reduces the incomes of the 'capitalists' who tend to save more and thus reduces local investments and therefore also prospects of longer term growth, and between less egalitarian but more 'pro-investment' and 'pro-growth' policies.

This dilemma dominated the debate on the merits of the tax policies that are often promoted by conservative, right-wing political parties and the policies advocated by left-wing socialist parties that favour much more aggressive income redistribution measures and wider social safety nets. In practice, these tax policies reflect not only economic considerations and political affiliations, but also the pressures of local politics that often dominate the debate, but the conflict between 'pro-growth' and 'pro-poor' does exist and cannot be swept under the carpet.

The opposition to 'pro-growth' and less progressive tax policies tends to be particularly strong because these policies are often perceived as 'pro-rich' policies. However, the economic justification for one position or the other, leaving aside local politics, is not that straightforward in today's global economy: innovations in information and communication technology (ICT), for example, created many rich people in both developed and developing countries whose initial wealth was only their skills. 'Pro-investment' and 'pro-growth' policies increased their incentives to study and acquire the necessary skills and thus contributed to accelerate growth, but inevitably they widened the gap between these nouveau (very) riche and the poor.

In Kaldor's world, the division of people into 'capitalists' and 'workers' is essentially the same as in the world of Karl Marx. People are born capitalists or workers and this is their fate and the fate of their children, unless they rise up to change it. In today's world, this is no longer a preordained fate and people rise on the income ladder to the heaven of great richness or fall to the hell of poverty with their own skills and with the whims of the stock market. The key to this ladder is their education. Rich people can indeed give their children better education, but in the developed countries, and increasingly in the developing countries, the state takes greater responsibility and makes larger investments in education that enable more people to gain education and thus contribute to the growth of the entire economy.

In this world, a politically less confrontational and more accommodating development strategy that is concerned primarily with reducing the income gap between rich and poor or between the urban and the rural populations by reducing the incentives of the 'capitalists' is therefore a double-edged sword. On the one hand, it is regarded by many as ethically and often politically correct. On the other hand, this policy may have high costs in terms of the prospects of more rapid growth by reducing government investments and by lowering the incentives of entrepreneurs and of their employees to develop the needed skills and use them in their work in ways that ultimately will be to the benefit of all.

Moreover, in today's global economy, higher tax on higher incomes or on capital gains will not reduce the riches of the 'capitalists', but rather chase away the talented people to other countries and give the rich incentives to transfer their capital to countries where the taxes on capital are lower. In other words, the policy options that governments now have, regardless of their social affinity to the 'left' or to the 'right', are very different from the options Kaldor had in mind.

5 Pro-poor Growth in a Multi-country Framework

The extension of the analysis to a multi-country (or multisector) framework is necessary for analysing the effects of growth on the income distribution between and within countries and eventual effects on income inequality between individuals. At the country level, a similar analysis can be conducted in order to distinguish between the effects of growth on different regions or sectors and between the effects on different income groups within regions. Diagrammatic illustrations of these effects are necessarily constrained to two dimensions and thus also to two countries (or two groups of countries) and to two income groups within each of the countries.

To distinguish the measures of income inequality between and within the two countries, Fig. 3.4 is divided into four quartiles: the upper right quartile shows the shares of the two countries (labelled 'developed' and 'developing' in the diagram) in the world's income and in the world's population. Their relative shares in the global population are given by the slope of the line OA. The slope of that line is assumed to be 45°, indicating that the two countries are assumed, for simplicity, to have equal shares in the world's populations. The line OA* shows the distribution of the world's income between the two countries, and the slope of that line, denoted by α, shows their relative shares in the global income. In the figure, $\alpha < 45°$, indicating that the share of the developing countries is smaller than that of the developed countries; the slope of that line thus measures the degree of income inequality between these two countries.

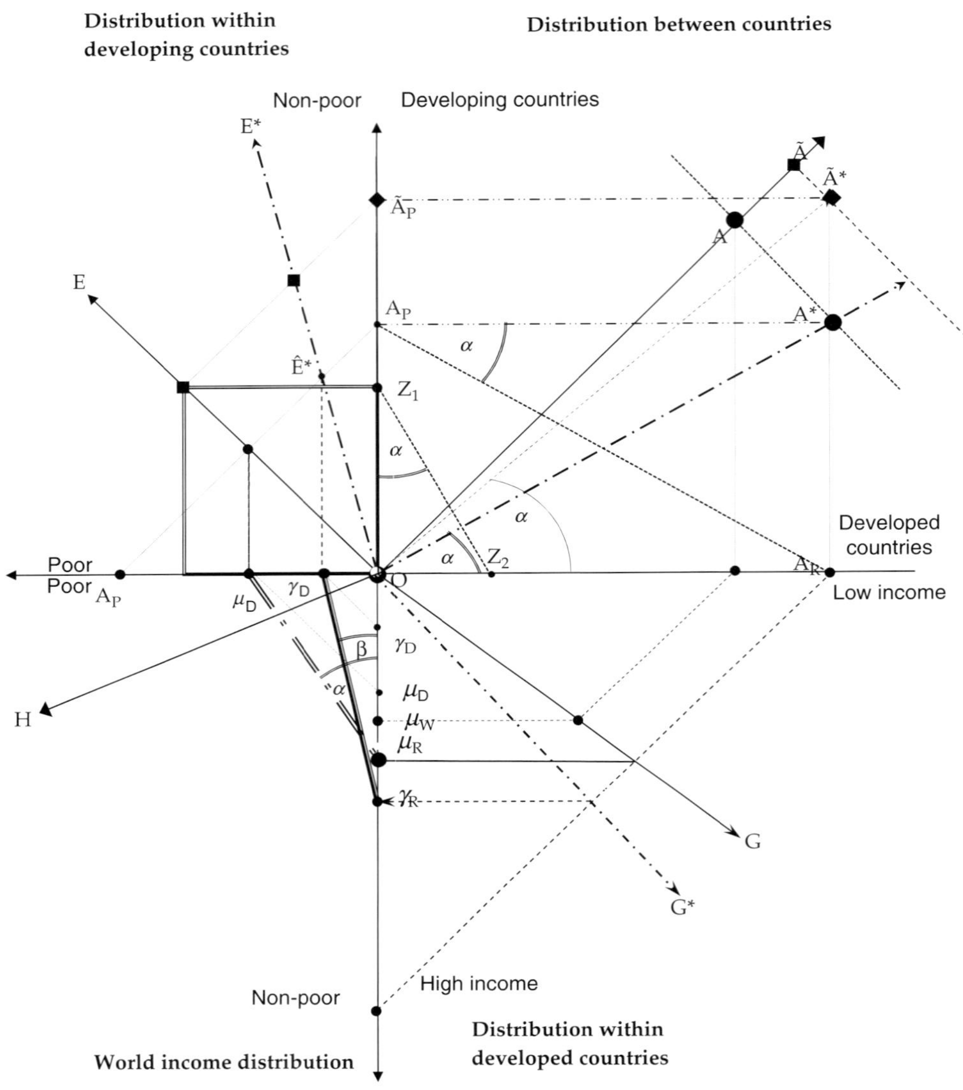

Fig. 3.4. Global poverty and income inequality.

The upper left quartile shows the income distribution *within* the developing country. The line OE shows the shares of the two income groups in the country's population and has a slope of 45°, indicating that the two groups are assumed to have equal shares in the population of the developing country. The intersection of the negatively sloped line from A_P with a slope of 45° with the positively sloped line OE determines the mean income per capita in the developing countries, denoted by μ_D. The line OE* shows the income distribution within the developing country between the two income groups, and the slope of that line measures the country's income inequality. The intersection of the positively sloped line OE* and the negatively sloped 45° line from A_P at point $\hat{E}$* determines the shares of each of the two income groups in the country's total income.[6] For the given poverty line Z_1 the per capita income of the individuals in the group denoted as 'poor' and given by γ_D falls below the poverty line and they are therefore poor in both relative and absolute terms. The income of the 'non-poor' is higher than the poverty line income, and the incidence of poverty in the country is therefore determined by the share of the poor in the developing country's population, assumed to be 50%. The poverty gap of the poor is the gap between their income γ_D and the poverty line Z_1.

The lower right quartile in Fig. 3.4 shows the income distribution *within* the developed country. The line OG shows the shares of the two income groups in the country's population and has a slope of 45°, indicating that the two groups are assumed to have equal shares in the population of the developed country. The intersection of the negatively sloped line from A_R with the positively sloped line OG determines the mean income per capita in the developing countries, denoted by μ_R. The line OG* shows the income distribution within the developed country between the two income groups, and the slope of that line measures the country's income inequality. The intersection of the positively sloped line OG* and the negatively sloped 45° line from A_R at point $\hat{G}$* determines the shares of each of the two income groups in the country's total income.[7] For the given poverty line Z_2 the income of both income groups is above that poverty line so that the individuals in the low-income group are poor in relative terms but not in absolute terms.

It should be noted that, since the poverty line income is determined in an absolute value (say, US$1 per person per day), its share in the mean per capita income in the developed country is therefore smaller than its share in the mean per capita income in the developing country, and the ratio Z_2/Z_1 is equal to the ratio $\mu_D/\mu_R < 1$.

The lower left quartile in the figure shows the income distribution between individuals in the world. All individuals were divided into two subgroups of poor and non-poor, and the line OH shows the shares of the two groups in the world's population. According to the data of the two countries, only 25% of the world's population is poor and this is shown by the slope of the line OH. There are two measures of income inequality: one is income inequality between countries and the other is the gap between the incomes of the rich and the poor individuals. In that quartile, the measure of inequality between countries is determined by the ratio μ_D/μ_R, and is given by α. The inequality between the individuals in the poor and the rich quartiles, given by the ratio of their incomes, is determined by the ratio $\gamma_D/\gamma_R = \beta < \alpha = \mu_D/\mu_R$. Hence, with these income distributions between and within countries,

[6] More precisely, it determines the share of each of the two groups in the country's total share in the world's income.

[7] That is, the share of each of the two groups in the country's total share in the world's income.

income inequality between individuals is considerably larger than income inequality between countries.

Finally, and as if the diagram is not complex enough as it is, I wanted to give guiding points on how the diagram can be used to analyse the impact of *changes* in income or income inequality between and within countries: Suppose that the income of the developed countries (only) is rising from A* to Ã*. The mean income in the world is then rising from A to Ã. In the developing countries, the square points on the negative 45° line from $\tilde{A}_P$ show how the higher income will be distributed between the two income groups. If there is no change in the income distribution within countries, then income inequality between countries will decline from α to $\acute{\alpha}$.

The impact on income inequality between individuals depends on the rise in the incomes of the non-poor in the developing countries. If the incomes of the non-poor in the developing countries will still be lower than that of the non-poor in the developed countries, then income inequality between individuals will still be the ratio between the incomes of the poor in the developing countries and the incomes of the non-poor in the developed countries. If, however, the incomes of the non-poor in the developing countries rise above the incomes of those in the developed countries, then the global income inequality will be equal to the internal inequality within the developing countries. These transitions are the reason for the rather abrupt changes in the measures of global inequality, as we shall see below.

Before concluding this section, I would like to show how this diagrammatic illustration can help in explaining or at least suggesting a different interpretation to the very dramatic development in global inequality of the last three decades. Sala-i-Martin's thorough study on the changes in the global income inequality since 1970 was discussed already, in a different context, in Chapter 2. In Chapter 4, I will discuss in more detail some of his main findings, but I would like to illustrate in this section his main summary findings on the changes in global income inequality in order to examine how the methodology developed in this section can assist in explaining these changes.

Figure 3.5 shows the measures of global income inequality during the period from 1970 to 2000. Although the graph that shows the changes during these years looks like a roller coaster, the changes in the levels of the Gini coefficient were rather small, and in a section entitled 'Global Income Inequality: Convergence, Period!' Sala-i-Martin concluded that after reaching its highest level in 1979, the Gini measure declined by almost 4%. The other measures of inequality that he calculated essentially supported this conclusion.

His main explanation for the short-term reversals in this trend is the following: 'The central point is that business cycles in the largest countries or groups of countries are associated with short-term reversals in the trend of world inequality' (Sala-i-Martin, 2002, p. 385). The diagram in this section suggests another explanation: the higher growth rates in China and India from the mid-1980s and the stagnation in SSA changed the composition of the low percentiles of the global income distribution with the large rise in the share of the poor from the SSA countries. In 1984, the headcount measures of extreme poverty (below US$1 per person per day) were 38.9% in China,

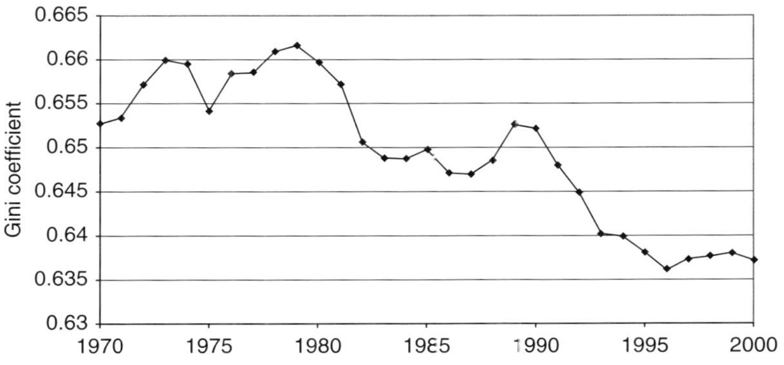

Fig. 3.5. The Gini coefficient of global income inequality. (From Sala-i-Martin, 2006.)

49.8% in India and 46.3% in SSA; in 1993, the headcount incidence changed to 24.9% in China, 42.3% in India and 44.0% in SSA. The rise in income inequality during these years reflects the stagnation or even decline in mean income of the people at the bottom percentiles as a result of these changes in the composition of the poorest income groups, which widened their income gap vis-à-vis the population in the developed countries, rather than the effects of business cycles.

5.1 How did countries evaluate their development priorities?

The idiosyncratic and widely different priorities chosen by governments in different countries for their development policies – focusing either on growth, poverty reduction or equity – were highlighted in a global survey that was conducted in 2005 by several regional development banks.[8] The survey rated the countries' performance on the basis of various economic policies, and a performance-based allocation (PBA) system was designed to reflect the specific institutional mandates, the country's policies and other characteristics of its governance structures. In their rating of social policies (that accounted for about 30% of all the policies under consideration), one of the main observations was that all the institutions included growth, poverty reduction and equity as their central goals, and did not make a significant distinction between them.[9] In practice, however, the weight given to equity in the evaluation of

[8] Their report was presented at the first Multilateral Development Bank (MDB) Technical Meeting of the leading economic institutions in developing countries that took place in Manila in January 2005. The subject of the report was PBA methods among alternative economic policies. The meeting was jointly planned by the African Development Bank (AfDB), the Asian Development Bank (ADB) and the International Development Association of the World Bank Group (IDA).

[9] They also noted the following goals: gender equality and social inclusion, pro-poor targeting programmes, enhancing the human capital of the poor, reducing the rural–urban gap, equality in the allocation of resources, social protection of labour and social safety nets.

their *actual* policies was much lower and their most important performance measure was economic growth. Even the goal of poverty reduction had a much lower rating.

The focus on pro-poor growth and poverty reduction may therefore reflect not so much the actual priorities of the developing countries themselves. The focus of the World Bank since the early 1990s on poverty reduction as the main performance measure was marked at the time by a formal declaration of the then President of the World Bank, Lewis T. Preston, that, henceforth, poverty reduction would be the overarching objective of the World Bank. The need and the way to recognize other objectives and establish priorities between them is discussed in more detail and illustrated in subsequent chapters.

6 Concluding Remarks

The measures that are commonly used to quantify and evaluate the level of poverty, the performance of pro-poor growth policies and their success in alleviating poverty have well-recognized limitations. They are determined on the basis of a preset poverty line that divides the low-income population into subgroups that are recognized as poor because their income or consumption expenditures fall below the poverty line, and subgroups with income or expenditures just above the poverty line. The pro-poor policies concentrate on the first subgroups and relegate the impact on the others to a secondary rank.

In many developing countries, researchers overcame this deficiency by specifying two poverty lines or even conducting the analysis for all poverty lines along the Pen's parade, and by adding other indicators of poverty to define a human development index. In the developed countries, the poverty line is usually determined relative to the mean or the median income of the general population, thus taking into account the relative deprivation of the poor vis-à-vis the general population. *Perceptions* of relative deprivation and of being unjustly shortchanged are highly subjective; but during periods of rapid growth, when most incomes are growing and large wealth is accumulated rapidly by a small minority at the top, they tend to be particularly strong.

The dividing lines between those who succeeded to climb up the income ladder and accumulate huge wealth and those who trailed behind and benefited very little from that growth are often closely associated with the dividing lines between population subgroups along geographical, ethnic, religious or national lines. In these cases, the deepening gap between these groups heightens the tensions between them, which then also take on ethnic or religious overtones.

Paradoxically but characteristically, the poor themselves are rarely in the front line or lead the protests against the rich. Indeed, the struggle for greater equality has rarely been 'a struggle of the starvelings that rise from their slumbers', as portrayed in the *International*. Instead, this struggle was, in most cases, led by the intellectual elites and driven by groups that were not at the bottom of the income distribution, but politically better organized and united by strong feelings that they were unjustly denied their fair share in their country's growing wealth. The poor themselves seldom have the political power to 'unite and come rally'.

The globalization process has deepened the inequalities between and within countries. On the one hand, economic growth has greatly accelerated and spread the rise in the standard of living across more countries and population groups than at any time in the past. On the other hand, the distribution of the new wealth was highly skewed in favour of a small number of countries and a relatively small affluent minority within these countries, while most others gained much less and quite a few even became worse off.

The process progressed in that direction with the collapse of entire economic sectors that could not survive the global competition, the rapid technological advancements that made many products redundant or too expensive, leaving their producers bankrupt and their workers unemployed, while new entrepreneurs and 'start ups' became rich almost overnight. The growing monopolization of production and of large segments of the wholesale and retail supply chain by multinational corporations was another factor that forced large and rapid changes in the structure of world production and trade and widened the gap between those who became better off and those who were left behind.

The goals of pro-poor growth and the debate over the most appropriate strategy to achieve these goals often fail to reflect the perceptions of *relative* deprivation and the intense bitterness of those who managed to remain afloat but felt shortchanged by not receiving the share they felt entitled to in their country's growing wealth. The premise of a social agenda that focuses on the elimination of poverty as the central goal of a country's social and economic policies and the main performance criterion of its policies does not reflect the full dimension of the tensions between the haves and the have-nots, particularly since many of the have-nots are not counted among the poor according to the existing criteria and are left mid-air.

This focus may explain the efforts of policy makers and/or their consultants to prove as much as the data and the statistical measures permit that the growth process driven by globalization has indeed been good for the poor. However, it does not explain the strong opposition of the middle classes to globalization in many developed and developing countries despite its many achievements. Laudable as the goal of poverty reduction is, in practice, economic policies are highly biased by political considerations and pressures that are not solely focused on the poor, and the policy makers are more strongly influenced by these pressures of the urban middle class and other well-organized interest groups, including wealthy corporations and rich farmers. The struggle against the growing income inequalities is primarily a struggle of the middle classes and it is exacerbated by the rapid movements up and down the income ladder due to the rapid economic changes that in many countries take place on a scale and at a pace that have never been experienced before.

Massive structural adjustments are precipitated by the changes in the global economy and are often combined with large changes in employment and income as enterprises and entire sectors are compelled to change their mode of operation, production technologies, managerial structure and often also their geographical location. For the large majority of the population, economic growth has become a roller coaster rather than a stable movement upward, and while some managed to climb on the income ladder to unparalleled heights, many others fell into deep holes as they lost their jobs and their income.

The common measures of poverty do not reflect the subjective perceptions of people that mirror their concerns and uncertainties. Obviously, feelings of relative deprivation and unjust disenfranchisement are necessarily very subjective and differ between people and political parties. The statistical measures of income, poverty and income inequality try to introduce a rational dimension into these subjective perceptions and the debate that follows, but for that very reason, they often fail to explain the intensity of the opposition to globalization in many countries that may seem baffling given its many benefits and its contribution to reduce poverty.

The deep divide between the unqualified support for globalization by the leading international development organizations, that seems to have tapered-off somewhat in the very recent years, and the widespread opposition to globalization that is increasingly intensifying in many developed and developing countries is one of the reasons for the collapse of the Doha Round. Many people, including many intellectuals in the developing countries, have come to see globalization as identical to American hegemony, which gives rise to calls for a struggle between the 'conqueror and the conquered' marked by a growing inequality between the impoverished and 'subjugated' South and the wealthy and 'patronizing' North.[10] This divide remains a subject that economists and policy makers are fully aware of, but it has not been given enough attention to deal with its ramifications.

References

Adelman, I. and Morris, C.T. (1973) *Economic Growth and Social Equity in Developing Countries*. Stanford University Press, Stanford.

Ahluwalia, M.S. (1976) Income distribution and development: some stylized facts. *American Economic Review* 66(2), 128–135.

Ahluwalia, M.S. (1974) Income inequality some dimensions of the problem. *Finance and Development,* 3–7.

Ahluwalia, M.S., Carter, N. and Chenery, H. (1979) Growth and poverty in developing countries. *Journal of Development Economics* 6, 399–341.

[10] The words in quotation mark are taken from articles of a prominent and mainstream Egyptian scholar, Sheikh Yussuf al-Qaradawi, who also wrote: 'This globalization wants to strip us bare and take away our identities. It wants to disseminate among us its commodities and ideas, its cultural conserves containing death and destruction.' (Quoted by Zvi Bar'el, Haaretz 10 November 2006.)

Ahluwalia, M.S., *et al.* (1980) Who benefits from economic development? Comment. *American Economic Review* 70(1), 242–245.

Alesina, A. and Rodrik, D. (1994) Distributive politics and economic growth. *Quarterly Journal of Economics* 108, 465–490.

Alesina, A. (1997) The Political Economy of High and Low Growth. Paper for the Annual World Bank Conference on Development Economics, Washington, DC.

Anand S. and Kanbur, R. (1993) Inequality and development: A critique. *Journal of Development Economics* 41, 19–43.

Atkinson, A.B. and Brandolini, A. (2001) Promises and pitfalls in the use of secondary data-sets: income inequality in OECD countries as a case study. *Journal of Economic Literature* 39, 771–800.

Bhalla, S. (2002) *Imagine There's No Country: Poverty, Inequality and Growth in theEra of Globalization*. Institute for International Economics, Washington, DC.

Bourguignon, F. (2003) The growth elasticity of poverty reduction. In: Eicher, T. and Turnovsky, S. (eds) *Inequality and Growth*. MIT Press, Cambridge.

Cáceres, I. and Saca, N. (2006) *What Do Remittances Do?* Analyzing the Private Remittance Transmission Mechanism in El Salvador. IMF Working Paper, November, Washington, DC.

Chen, S. and Ravallion, M. (2001) How did the world's poor fare in the 1990s? *Review of Income and Wealth* 47(3), 283–300.

Chen, S. and Ravallion, M. (2004) *How Have the World's Poorest Fare Since the Early 1980s?* World Bank Policy Research Working Paper No. 3341, World Bank, New York.

Chua A. (2003) *World on Fire: How Exporting Free Market Democracy Breeds Ethnic Hatred and Global Instability*. Heinemann, London.

Collier, P. and Gunning, J.W. (1999) Explaining Africa's economic performance. *Journal of Economic Literature* 37, 67–111.

Cornia, G.A. (ed.) (2005) *Inequality, Growth, and Poverty in an Era of Liberalization and Globalization*. WIDER Studies in Development Economics. Oxford University Press, Oxford.

Datt, G. and Ravallion, M. (1992) Growth and redistribution components of changes in poverty. *Journal of Development Economics* 47, 403–428.

Datt, G. and Ravallion, M. (2002) Has India's post-reform economic growth left the poor behind? *Journal of Economic Perspectives* 16(3), 89–108.

Deaton, A. (2001) Counting the world's poor: problems and possible solutions. *The World Bank Research Observer* 16(September), 125–147.

Deaton, A. (2004) Measuring Poverty. Working Papers, Woodrow Wilson School of Public and International Affairs. *Research Program in Development Studies*. Princeton University.

Deininger, K. and Squire, L. (1996) A new data set measuring income inequality. *World Bank Economic Review* 10(3).

Deininger, K. and Squire, L. (1998) New ways of looking at old issues: inequality and growth. *Journal of Development Economics* 57(2), 259–287.

Dollar, D. (2002) Global economic integration and global inequality. In: *Globalization, Living Standards and Inequality: Recent Progress and Continuing Challenges*. Reserve Bank of Australia 2002 Conference, Australia.

Dollar, D. and Kraay, A. (2001) *Trade, Growth, and Poverty*. World Bank Policy Research Working Paper 2615 World Bank, Washington, DC.

Dollar, D. and Kraay, A. (2002a) Growth is good for the poor. *Journal of Economic Growth* 7(3), 195–225.

Dollar, D. and Kraay, A. (2002b) Spreading the wealth. *Foreign Affairs* 18(January/February), 120–134.

Easterly, W. (1999) Life during growth. *Journal of Economic Growth* 4(3), 239–276.

Fields, G.S. (2000) *Distribution and Development: A New Look at the Developing World*. Russell Sage Foundation/MIT Press, New York/Cambridge, Massachusetts.

Gastwirth, J.L. (1971) A general definition of the Lorenz curve. *Econometrica* 39(6), 1037–1039.

International Food Policy Research Institute (IFPR), *Data Set*, IFPRI Various years.

Kakwani, N. (1980) *Income Inequality and Poverty: Methods of Estimation and Policy Applications*, World Bank, Oxford University Press, New York.
Kakwani, N. and Pernia, E.M. (2000) What is pro-poor growth? *Asian Development Review* 18(1), 1–16.
Kanbur, R. (2001) Economic policy, distribution and poverty: the nature of disagreements. *World Development* 29(6), 1083–1094.
Klasen, S. (2005) Economic Growth and Poverty Reduction: Measurement and Policy Issues. OECD Development Centre, Working Paper No. 246.
Kravis I.B. (1960) International differences in the distribution of income. *The Review of Economics and Statistics* 42(4), 408–416.
Kuznets, S. (1955) Economic growth and income inequality. *American Economic Review.*
Kuznets, S. (1965) *Economic Growth and Structure: Selected Essays.* Oxford University Press, New Delhi.
Lundberg, M. and Squire, L. (2003) The simultaneous evolution of growth and inequality. *Economic Journal* 113, 326–344.
OECD (2001) *Employment Outlook*, OECD Publishing (June, 2001).
OECD (2004) *Employment Outlook*, OECD Publishing (June, 2004).
Ravallion, M. (1995) Growth and poverty: evidence for developing countries in the 1980s. *Economic Letters* 48, 411–417.
Ravallion, M. (2001). Growth, inequality and poverty: looking beyond averages. *World Development* 29(11), 1803–1815.
Ravallion, M. (2003) The debate on globalization, poverty and inequality: why the measurement matters. *International Affairs* 79(4), 739–753.
Ravallion, M. (2004) Pro-Poor Growth: A Primer. *Development Research Group*, World Bank, 2004.
Ravallion, M. and Chen, S. (1997) What can new survey data tell us about recent changes in distribution and poverty? *World Bank Economic Review* 11(2), 357–382.
Ravallion, M. and Chen, S. (2001) *Measuring Pro-poor Growth*, Development Research Group, World Bank Policy Research Working Paper No. 2666, World Bank, Washington, DC.
Ravallion, M. and Chen, S. (2003) Measuring Pro-Poor Growth. *Economics Letters* 78(1), 93–99.
Ravallion, M. and Chen, S. (2004) China's (Uneven) progress against poverty. *Policy Research Working Paper 3408*, World Bank.
Ravallion, M., Datt, G. and van de Walle, D. (1991) Quantifying absolute poverty in the developing world. *Review of Income and Wealth* 37, 345–361.
Robinson, S. (1976) A note on the U hypothesis relating income inequality and economic development. *The American Economic Review* 66(3), 437–440.
Rodrik, D. (2000) Growth versus poverty reduction: a hollow debate. *Finance and Development* 37(4).
Sala-i-Martin, X. (2002) *'The Disturbing 'Rise' of Global Income Inequality*. NBER Working Papers 8904, National Bureau of Economic Research, Cambridge, Massachusetts.
Sala-i-Martin, X. (2006) The world distribution of income: falling poverty and convergence, period. *Quarterly Journal of Economics* 121(2), 351–397.
UNCTAD (1999) *Human Development Report*. Oxford University Press, New York.
UNCTAD (2002) *Escaping the Poverty Trap: The Least Developing Countries Report*. United Nations Conference on Trade and Development, United Nations/New York, Geneva.
UNDP (2000) *Human Development Report*, Oxford University Press/UNDP 2000.
White, H. and Anerson, E. (2000) *Growth versus Distribution: Does the Pattern of Growth Matter?* Institute of Development Studies, University of Sessex, Mimeo, Brighton, UK.
World Bank (1993) *The East Asian Miracle: Economic Growth and Public Policy*. Oxford University Press, New York.
World Bank (2000) *Attacking Poverty: World Development Report 2000/2001*, Oxford University Press, New York.
World Bank (2002) *Globalization, Growth and Poverty: Building an Inclusive World Economy*, Policy Research Report, World Bank, Washington, DC.
World Bank (2004) *World Development Report 2004*. World Bank 2004.

4 Have the Policies of Economic Development Been 'Pro-poor'?

1 Introduction

While the last two decades brought great affluence to many developing countries, the LDCs still experienced deepening poverty and widespread hunger. The goal of reducing poverty had always been an important objective and the emphasis on growth was an outcome of the accepted hypothesis that growth and poverty reduction are compatible and complementing goals that go hand in hand. This may not be the case in the multi-country, multisector framework when growth rates are different and income inequalities between and within countries are also changing.

The early writings on economic development in the mid-1950s emphasized primarily the impact of economic development on income inequality as an inevitable outcome of the structural changes during the development process that is based on industrialization. The most well-known description of that process is given in Kuznets' path-breaking model on the stages of economic development. In that model, the developing country's industrialization initially leads to a rise in income inequality as an effect of the migration of unskilled workers from the rural areas to the urban centres in their search for work. However, at the initial stages, the country's industry is not developed enough to absorb all of them and therefore, many migrants remain unemployed or must find employment in the informal sector. As a result, the income gap between the skilled workers who are already employed in the industry and the unskilled workers is rising. Only over time, as the country's industry expands to absorb more workers and the rural migrants attain the necessary skills, the income gap is gradually falling.

Kuznets published his paper in the mid-1950s, when a description of the difficulties that await developing countries in the process of industrialization was timely because industrialization was regarded as the main strategy for accelerating growth in order to close the gap between the economies of the developing countries and the economies of the European countries that were still in the process of recovering from the devastation of World War II. However, in the last two decades many developing countries encountered difficulties and experienced a prolonged economic stagnation that prevented the development of a solid industrial base and impoverished the rural population and many unskilled workers in urban areas.

The theory of economic development emphasized the three goals of development: economic growth, poverty reduction and the reduction of income inequality. In the long run, these goals need not be conflicting, but in the short run, there may well be a conflict between these goals, and a country may have to make a choice between them that represents a choice between 'short-term pain and long-term gain'. The short-term pain may be due to restrictive macroeconomic policies that reduce the transfer of resources to the poor and avoid drastic income distribution measures in order to promote growth today and postpone the reduction in poverty to the longer run when the country will have the necessary resources (see Klasen *et al.*, 2003).

The short-term pain may be due to the structural changes that are necessary to accelerate growth, particularly in countries that make the transition from an agricultural-based economy to industrialization. It is also the result of changes when sectors that are no longer profitable close down, when bloated state-owned enterprises are privatized, and when farmers lose their markets due to cheap imports. Many workers may lose their jobs, and many of them may become unemployable since their skills are no longer suitable for the new economy. In the LDCs, the short-term pain is particularly acute because these countries cannot provide even the minimum social safety nets to secure the basic needs of those who lost their entire livelihood.

The objective of this chapter is to survey the policies implemented in developing countries in the last two decades in order to accelerate their growth, combat poverty and make the adjustments in their economies in order to be integrated into the global economy. This process affected different segments of society in different ways and it was partly due to policies that were not related to globalization, and partly due to policies that had to be implemented in order to adjust their economies to the global economy.

The chapter will focus on the role of the international development organizations, primarily the IMF and the World Bank, and the policies and reforms that they have advocated. These policies arguably had a more decisive impact on many developing countries than the process of globalization itself, and it is therefore not appropriate for an analysis that seeks to explain the impact of globalization on growth and poverty to lump all the developing countries together and concentrate on the impact of the globalization process. It is indeed necessary to evaluate the impact of the development policies that the developing countries implemented under the guidance of the Washington Institutions. This analysis must, in particular, seek to answer the following questions: Were these the right policies to accelerate growth and reduce poverty? How beneficial (or detrimental) was their impact? What was the difference between these policies and the policies implemented by the EA countries? And, last but not least, since the reforms designed by the

Washington Institutions can hardly be credited with producing the desired results, what guidelines should be formulated for development policies in the future?

The basic principles of the reforms are summarized by a set of 'commandments' that became known as the 'Washington Consensus', since it reflected views shared by the IMF, the World Bank and the US Treasury about the 'right' policies for developing countries. The reforms were implemented by the developing countries within the framework of the structural adjustment programmes (SAPs) carried out under the guidance of the IMF and the World Bank.

The programmes were highly controversial and there are still heated debates among economists and policy makers whether the approach that was advocated by the SAPs was appropriate to help the developing countries to make the necessary structural changes in their economies. Many books and articles have discussed these programmes, and many studies were conducted to prove or disprove their usefulness. Perhaps the most vehement criticism is Stiglitz's book *Globalization and its Discontents*.

The reason for taking up the discussion of these programmes again in this chapter is because the difficulties faced by most LDCs today in their efforts to be integrated into the global economy may require them to make additional and very thorough structural reforms in the coming years that would be as comprehensive as the SAPs. These reforms should not be specified as a set of mandatory conditions, but as reforms that the countries themselves would have to initiate, design and assume the responsibility for their implementation. The reforms would therefore need to be based on a broad consensus in these countries that will determine their structure and give them legitimacy. The reforms must also reflect the lessons that have to be drawn from the experience these countries had with their previous round of reforms. To be sure, the SAPs have been modified over the years with the accumulation of more experience and better understanding that was gained in an arduous learning process. The focus of the chapter is on those aspects of the SAPs that are likely to be the principal components of future reforms.

2 What Has Remained of the 'Washington Consensus'?

The principles of the macroeconomic policies that have guided development economists and the economic development institutions take their raison d'être from the fundamentals of classical economic theory. These have also been the policies promoted by the leading international development organizations, the World Bank and the IMF.[1] These policies became known as the Washington Consensus (Box 4.1) – a phrase that was initially coined by John Williamson in

[1] These organizations were established in the Bretton Woods Agreement in July 1944 and became known as the 'Bretton Woods Institutions', and since they are both located in Washington, they are also referred to as the Washington Institutions.

Box 4.1. The Washington Consensus.

The Washington Consensus summarizes the shared themes in the policy advice given by Washington-based institutions in the early 1990s, primarily the IMF, the World Bank and the US Treasury Department; the policies advocated were believed to be necessary for economic growth. The concept and name were first presented in 1990 by John Williamson, and referred to LA following the financial crises of the 1980s, but they later came to symbolize, often unjustly, the more general neoliberal blueprint for economic development.

The consensus includes the following reforms:

- Fiscal discipline;
- Allocating a larger share of public expenditures for education, health and infrastructure investment;
- Tax reform to lower existing high income tax rates on high incomes and lower the marginal tax rates; at the same time, raise the tax rates on the low tax brackets (typically below median income);
- Reduce government intervention in the capital market and let the interest rates be market–determined and positive in real terms;
- Reduce government intervention in the foreign exchange market to allow competitive exchange rates;
- Reduce government intervention in foreign trade through comprehensive trade liberalization that will abolish quantitative restrictions and establish low and uniform tariffs;
- Open the capital market to FDI;
- Privatize state enterprises;
- Abolish regulations that impede market entry or restrict competition, unless they are necessary for safety, environmental and consumer protection;
- Maintain law and order and protect property rights.

the early 1990s to describe the principles of the macroeconomic policies that constituted a 'standard' reform package promoted by the Washington Institutions. Subsequently, the term has been widely applied to describe a less precisely defined gamut of policies that were the basic formula for promoting economic growth and introducing free market-oriented economic reforms.

The concept Washington Consensus became a familiar term, though not always with positive connotations, and Williamson himself eventually became doubtful as to whether the phrase that he coined served to advance the cause of rational economic policymaking. These policies can be summarized by the motto 'get the fundamentals right'. The measures included:

- Coherent and consistent macroeconomic policies that secure economic stability;
- Responsible and sustainable fiscal policy that maintains a balanced government budget, and comprehensive tax reform to promote savings and investments and to accelerate growth;
- Improve governance, increase the efficiency of public institutions, combat corruption and strengthen the rule of law and property rights;

- Reduce the monopoly of private or state-owned enterprises, and reduce government influence over the price system through subsidies, taxes or direct controls;
- Open economy with free, unrestricted and competitive trade according to the country's comparative advantage in order to allocate its resources more efficiently;
- Privatize state-owned enterprises.

In the debate about these principles, it has been argued that both proponents and opponents of the policies in the Washington Consensus have used the term far too dogmatically, and the criticism was mainly on the following points. First, the term became associated with a very narrow ideology that 'free markets can handle everything'. Second, the Washington Consensus was taken as a rigid set of policies that should be implemented in full, in all countries and under all circumstances as the basis for economic policies, regardless of the specific economic conditions and their stage of development. The IMF and the World Bank were thus blamed for advocating a 'one-size-fits-all' set of policies, which were also criticized for insensitivity to the needs of the poor. The criticism and accusations levelled against the 'Washington Consensus' were perhaps no less dogmatic than the term itself, but they were not entirely baseless.

Trade liberalization has been a central part of the policies advocated by the Washington Consensus, and it played a major role in promoting growth and reducing poverty. That role has been demonstrated in a series of multi-country studies (Deininger and Squire, 1996; World Bank, 2000, 2001a,b; Chen and Ravallion, 2001; Dollar and Kraay, 2002b; Sala-i-Martin, 2006). These studies found a strong positive impact of pro-trade policies on growth and poverty reduction in a wide range of developing countries, but concluded that growth did not increase income inequality. Deininger and Squire (1996, p. 566) concluded that there was 'no systematic link between growth and inequality, but we find strong positive relationship between growth and poverty reduction'; Dollar and Kraay (2002a) found that the share of income accruing to the bottom quintile tended to rise at the same rate as the rise in the average income, indicating that income inequality remained unchanged; Sala-i-Martin (2006) found that during the years of rapid globalization, there was a reduction in global income inequality. These studies therefore offer a clear recipe for pro-poor growth policies: Get the fundamentals right as fast as possible and open up the economy for trade as widely as possible.

3 How Should LDCs Make the Choice of Growth and Trade Policies Today?

A developing country that implements structural reforms in order to accelerate growth faces a wide range of choices between alternative macroeconomic, trade and monetary policies. There is no single recipe for a set of policies that will be best for achieving these goals and there is no single set of policies that fits all countries. The choice between policies depends on their suitability for a country's specific socio-economic, geographic and agroclimatic conditions and constraints, on the political feasibility of implementing them given the country's own priorities as well as the competing interests of different groups, civil organizations or parties, and on the capability and political power of its leaders.

The choice therefore depends not merely on an accounting balance between the total gains and the total losses for the country, but also on the opportunity costs of achieving these gains for the country and for the different interest groups. In other words, since the decision has to be political and represents some sort of consensus among competing political groups, rather than a decision of a 'benevolent dictator', the consensus also depends on the distribution of the gains and costs among these groups. Without a broad consensus, a democratic government may lose its capacity and its legitimacy to function, make the necessary choices and implement the appropriate policies. A non-democratic government that does not have a broad consensus may be forced to invest some of its resources to secure its control given the opposition to its policies. The following examples show the dilemma that may be involved in these choices.

3.1 Outward- versus inward-oriented growth strategy

The debate over the merits of export promotion versus import substitution was at the centre of the discussion in the economic development literature already in the 1950s. The consensus among economists at the time was in favour of outward-oriented, export-led growth but they did not rule out an import-substitution strategy in the *first stage* of the industrialization process, and it has been recognized that virtually all the successful exporters of manufacturing products began their industrialization process with an inward-oriented strategy that promoted import substitution. In the larger countries, the gains from export-led growth that are due to scale economies in production can also be reaped in import substitution.

In the debate about a growth strategy for a labour-abundant economy, the dominating view was that, for the majority of developing countries, the prospects of growth are closely linked to industrialization. Even in the agricultural SSA countries, their long-term growth potential is linked to increasing the domestic value-added of their exports through processing and raising the ratio of manufactured to primary products in their production. During the 1990s, the discussion on the choice of a trade policy was dominated by the success of the EA countries to accelerate their growth by increasing their exports. The key questions in the discussion were:

- How significant is an outward-oriented strategy for the country's growth and at what stage of its development is it most effective?
- Is industrialization more effective through export promotion or through import substitution?

- What is the role of the government in promoting exports with public investments in infrastructure, education and subsidies to selected industries and to FDIs?

In their highly influential research on trade and industrialization, Little *et al.* (1970) provided a thorough account of the connection between an export-oriented growth strategy and the rate of economic growth. They emphasized the effects of competitive conditions in the world markets which force both the exporting enterprises and the governments to eliminate distortions, thus leading to more rapid growth. Behind the walls of protection, in contrast, efficiency considerations are secondary and (infant) industries have little incentive to select the more cost-effective technologies.

Bhagwati (1982) highlighted another distortion of an inward-oriented trade regime that is due to its heavy reliance on market restrictions and state intervention and is therefore more likely to create an environment that is more congenial to distorting profit seeking. With open, outward-oriented strategies, in contrast, the state is less involved and profit seeking is more likely to be productive. Bhagwati emphasized that an outward-oriented trade strategy helps industrialization and growth. He therefore concluded: 'The question of the wisdom of an outward-oriented (export-promotion) strategy may be considered to have been settled.' (Bhagwati, 1987, p. 257).

In a comprehensive cross-country study, Chenery *et al.* (1986) showed, however, that the growth rate in semi-industrial economies that adopted import-substitution policies was only few tenths of a per cent lower than in countries that adopted export-promotion strategies. Later on Easterly *et al.* (1994) established a clearer causal relationship between enhanced exports performance and more rapid growth in extended cross-section studies.

The drive to promote industrialization was one of the main reasons for an inward-looking trade policy aimed at restricting imports of manufactured goods in order to give incentive to local producers to develop their industries for the domestic market. The 'infant industry' argument was widely accepted as a valid exception even in the case of free trade. Import-substitution policies that promote labour-intensive industries, particularly of unskilled labour, contribute also to increase employment, but these industries were often extended well beyond the original plans and covered many products in which the country does not have a comparative advantage.

An inward-looking policy that is aimed at developing an industrial base through temporary but wide-range protective duties therefore runs the risk of diverting the demand of domestic consumers and producers to products in which the country does not have a comparative advantage, and thus increases their costs and distorts the allocation of the country's resources.

Moreover, during the transition period, domestic production of exportable products would have to be confined mostly to the domestic market, thus preventing the local producers from capturing the returns to scale in production and reducing incentives to improve efficiency. Production costs of exportable goods would be higher since they would have to compete with the producers of importable products on the country's resources and purchase many importable intermediate products or capital goods from domestic producers at a higher price thus raising their market prices. The more limited domestic market can also reduce competition and give monopolistic control to some enterprises in the production of certain goods, thus raising their price for both consumers and other producers.

A series of country studies organized by the World Institute for Development Economics Research (WIDER) in 1987 and 1988 demonstrated that Korea, Taiwan, Turkey and other countries where economic progress was led by the rapid growth of export industries had an initial stage of industrialization based on import substitution. In Turkey, for example, the export-led growth during the first part of the 1980s was built on an industrial base that had been created at an earlier stage of import-substitution industrialization as well as on heavy export subsidies and various administrative measures to promote exports.

In Korea, infant industries were protected by high tariffs during the early stages of industrialization alongside the export drive, and while exports were liberalized, the domestic market remained protected. The monopolistic or oligopolistic conglomerates that dominated many Korean industries were able to practice price discrimination that was considerably augmented by export subsidies. Infant industries matured by acquiring more advanced technologies, and the effective protection rates were only gradually reduced. These industries turned to international trade not due to market forces, but because they were driven by government regulations to export their products (Lee, 1997). However, the concentration of manufacturing production in Korea and in most other EA countries (except Taiwan) in large conglomerates and the power exercised by transnational corporations over exports distorted the competitive conditions of the world markets substantially and reduced the efficiency gains of the export-led growth (Helleiner, 1994).

The success of EA countries in achieving high rates of economic growth made export promotion, spurred by open trade and market liberalization, the preferred strategy. The World Bank study on *The East Asian Miracle* (1993) emphasized the prominent role of export-oriented policies in the economic 'miracle' of the EA countries. The low-skilled labour-intensive clothing industry was the quintessential foundation on which their industrial base was gradually built and diversified into other industries as they accumulated more know-how, trained a more skilled labour force and opened their economies to foreign investment. But the pressures of international competition and the greater role of the market did not mitigate by much their rent-seeking behaviour compared to countries that promoted industrialization through import substitution, as the theory promised, since their exports were highly influenced by government intervention through subsidies, trade restrictions, administrative guidance and credit allocations (Wade, 1990; Amsden, 1991; Rodrik, 1992, 1995).

Country studies of Dollar (1992), Sachs and Warner (1995) and Edwards (1998) showed that most countries that adhered to import-substitution policies had relatively lower growth rates, but quite a few developing countries, Chile being perhaps the most prominent example, that switched to outward-looking policies were forced to abandon these policies at least temporarily either due to domestic political pressures related to rising unemployment or to growing foreign debts. However, these countries were driven to maintain market-oriented and stable exchange rates and to control monopolistic enterprises.

Rodrik and Rodriguez (1999) analysed the empirical studies that estimated the negative relation between trade barriers and economic growth and concluded that the available evidence was not sufficient to support the hypothesis that there is a strong negative relationship between growth and trade policies – where the latter was narrowly defined to cover trade taxes and subsidies and other trade-distorting measures specifically aimed at restricting imports or promoting exports. However, the statistical findings supported the hypothesis that an increase in exports is positively related to increased growth, although the increase in export may be the result rather than the cause of economic growth.

The development strategies of the EA countries were based on the premise that government interventions are necessary at the early stages in order to facilitate the transition from the production of primary products for the domestic market to the production of manufacturing goods for the home market and for exports, and in order to make investments in the non-traditional sector sufficiently attractive. The rise in investments was largely engineered by these governments through direct credit at low or even negative interest rates, and through investment subsidies, direct administrative subsidies and the use of public enterprises. In these countries the weakness of the market system and the absence of the necessary financial and economic institutions made the state the only entity that could offer the necessary incentives and provide a measure of security.

3.2 The gains and constraints of government interventions

Since the mid-1980s, China has epitomized a successful choice of policies that achieved rapid growth and a steep reduction in poverty. However, many of its macroeconomic, trade and monetary policies differed significantly from those advocated by the Washington Consensus and they left little room for the free market. Although the main recommendations of the Washington Consensus were to reduce government intervention in the financial and the foreign exchange markets, open the economy to trade and reduce government intervention with comprehensive trade liberalization, as well as conduct responsible and balanced fiscal and monetary policies with free interest and exchange rates, the Chinese government has been, and still is, heavily involved and actively intervenes in all these markets. Its ongoing active involvement in the foreign exchange market that effectively determines the exchange rate of the Yuan, despite the protests of many countries, is perhaps the most widely publicized example.

China's stunning economic success, despite (or because of) these tight and manifestly undemocratic controls, is certainly a controversial subject that economists and political scientists grapple with. It raises fundamental questions whether, in certain countries, certain circumstances or at a certain stage of development, government controls over the economy are, in fact, better than a free market. However, even if in certain circumstances the answer is positive, there are obvious difficulties to decide when the free market should have a greater role, when the government should release its tight control over the economy and whether it would actually be willing to do so.

Economic theory was fully cognizant of the compromises that democracy requires and that, in some sense, makes it a second-best solution. Ideally, a completely selfless dictator would be approved by the entire community as its decision-making authority. The Communist regime in China had some of the characteristics of that ideal but hypothetical benevolent dictator. In the last two decades it was able to secure relative political stability (the Tienanmien crisis in 1989 and the high level of corruption notwithstanding) and to prevent the chaos that often characterizes new democracies. This stability is particularly essential during the infancy stage of economic growth when a country's institutions and its system of governance are not yet in place and are not effectively operational. In China, the Communist party achieved these controls after the extremely gruesome period of the 'Great Leap Forward' and the 'Cultural Revolution' that cost millions of lives and left deep scars on the entire population.

The first 'test' of the more liberal regime after the death of Mao came in the 1990s when economic growth accelerated as a result of the strong urban bias of the Chinese government's development strategy to promote industrialization. This strategy led to a sharp rise in income inequality and in the rural–urban income gap. At the initial stages, in the mid-1970s and through the 1980s, government interventions to promote industrialization were necessary and the power of the strong central regime to reallocate resources may have been an advantage. However, it is by no means certain that the rural population, that still constitutes the majority of the population in China, agrees to the widening income gap between the rural and the urban sectors, but in retrospect it cannot be denied that this strategy was successful in achieving the desired allocation of resources to promote industrialization.

In recent years, protests over the rising inequalities and the stagnation in many rural areas and many provinces have clearly begun to worry the Chinese central authorities and prompted them to increase public investments in the rural sector in general and in the more remote western provinces in particular. With this and other measures targeted on rural areas, the Chinese government now tries to mitigate the opposition to its heavy involvement in the economy. It is, however, the stunning growth that was the main reason for the rather mute opposition to the rising inequalities, even though they are still deeply divisive and potentially explosive.

3.3 Necessary and sufficient conditions for growth

The experience of the last two decades clarifies that a stable macroeconomic policy, free trade, low inflation, a stable exchange rate and all the other remedies that are usually prescribed by the Washington Consensus are only *necessary* conditions to promote growth, increase trade and attract foreign investments. By themselves, however, these measures are not *sufficient* to achieve these goals. In countries at low levels of development, these necessary conditions may fail to promote trade, accelerate growth and reduce poverty. In EA, government interventions may have been essential at the early stages to transfer resources in order to establish an industrial base, give the incentives to promote exports and create the conditions for growth.

The focus on industrialization created a bias in favour of the urban sector and it did not have broad popular support. In many EA countries, this set off a wave of social unrest despite rapid growth, and that unrest exacerbated the impact of the financial crisis during the late 1990s that brought some countries to the verge of an economic collapse. In nearly all EA countries, that crisis did not last very long and they managed to recover rather rapidly; in 2006, the countries in the region registered their fifth consecutive year of strong growth, backed up by a substantial decline in poverty. In some countries, however, most notably in Indonesia and Thailand, the political tensions spread beyond the economic sphere to ethnic, political and other areas, which lowered consumers' and private investors' confidence.

In many other developing countries, primarily in SSA and LAC, governments also intervened very actively in the economy, and many of the interventions had little popular support. All too often, government policies were biased or even dominated by sectoral and even personal considerations of their leaders. In many countries, leaders made their decisions with the sole objective of securing their control over the country in order to stay in power, and huge amounts were spent on the military forces, or smuggled to foreign banks.

Considerations of political or personal power may be a factor in governmental decisions everywhere, but in established democracies, rules and institutions are designed to ensure that such considerations will at least be within certain limits. In many developing countries, such rules and institutions were not yet in place and their leaders' decisions have often been outright destructive. In Nigeria, the supposedly democratic government of Olusengun Obasanjo has been blamed in the newsmedia for holding billions of US$ in foreign bank accounts, while the majority of the population still lives in poverty.

The goal of building effective institutions and better governance that would prevent these destructive decisions was, in principle, one of the goals of the SAPs that the World Bank and the IMF designed. These programmes were implemented initially during the 1980s in developing countries that were in need of credit to cover their foreign debts and applied to the World Bank to receive concessional credit for that purpose. These programmes will be discussed more extensively in the second part of this chapter, but it must be noted here that during the 1980s and until the mid-1990s, the goal of building effective institutions and better governance took second place to the more central and strictly economic goals that were listed in the Washington Consensus. Over time, with the accumulation of more extensive and wider experience (including many failures), it has been recognized that rule-based policy making and an effective system of governance are essential conditions that must precede all other reforms and are necessary to achieve the economic goals.

3.4 Government interventions in trade and foreign investments

Classical economic theory tells us that by removing the restrictions on trade and allowing the free flow of imports and exports, a country should, in principle, increase its gains from trade, maximize the benefits from its comparative advantage and accelerate its growth. Indeed, open trade was one of the key principles of the Washington Consensus. In the short run, however, opening a country to trade also has disadvantages since it requires massive structural adjustments to promote the sectors in which the country has a comparative advantage. In theory, these structural changes are made when the economy is opened to trade through the incentives given by the price system. At the initial stages, these changes have very negative effects on the economy by forcing many enterprises to go out of business, resulting in job losses, while it takes time and large investments for enterprises in which the country does have a comparative advantage to build up their production capacity and to train the necessary labour force.

Both the IMF and the US Treasury favoured a 'shock therapy' in the implementation of trade liberalization that will open the economy to the world prices and let these prices provide the incentives for restructuring the economy according to its comparative advantage and thereby establish a more efficient allocation of the country's resources. Actual developments in practically all countries that implemented this policy proved the superiority of the gradualist approach. Despite the failures of this 'shock therapy' in the countries that implemented the SAPs during the 1980s, and despite the very different experience of the EA countries, this was still the policy that was advocated and promoted in most ex-centrally planned economies. The result was another failure that could have been foretold since these countries lacked the necessary legal and regulatory framework, private property rights, functioning markets for equity and debt capital, competition policies and competitive and efficient banking institutions.

In the 1980s and the 1990s, the EA countries had large benefits from their export-oriented policies that played a major role in accelerating their growth. However, the process in which these countries promoted their exports was significantly different from the one recommended by the Washington Consensus. Most importantly, these countries did not just open their economies to free trade and allow the world market prices to provide the incentives and the guidelines for reallocating their resources and restructuring their economies. Instead, their governments actively intervened in restructuring the economies in two main ways: First, they provided various incentives to selected sectors in order to make them more competitive in the world market. Second, they continued to protect their local industries and workers against a flow of cheap imports by continuing to maintain high tariffs and various administrative restrictions.

In many countries, particularly in LA and recently in quite a few developed countries, the open trade policy raised strong opposition due to its damaging impact on some of these countries' traditional sectors and the loss of many jobs. The high social costs during the transition period are the main reason for the strong opposition of many political parties to the multilateral trade agreements that are aimed at promoting free trade. As a result of these negative effects and the opposition to free trade, the WTO came to symbolize the 'demon' of the globalization process itself.

The lack of attention in the SAPs to the social costs of an open trade policy during the transition period and the negative impact on large segments of society raised strong opposition in many countries to these programmes. It has only gradually and often belatedly been realized by the Washington Institutions that the social implications of these programmes cannot be ignored even when these are, in theory, 'only' short-term difficulties.

3.5 The different forms of trade liberalization

In many developing countries that opened their economies to trade as part of their reforms and SAPs, the agricultural sector and the rural population suffered the most. The difficulties of the traditional sectors to make the necessary adjustments and the lack of adequate infrastructure and efficient organizations to enable these sectors to use their comparative advantage with trade by changing the composition of their crops inflicted great losses on them. Many farmers were forced to revert to self-sufficiency and became much poorer since the urban markets were flooded with cheaper imports of agricultural products – thanks, in part, to the high subsidies to agriculture that farmers in the USA and the EU receive.

As these developing countries opened their economies to free trade, their farmers found themselves in a catch-22 trap since they were not able to compete with the cheaper foreign imports while their high transport costs and other costs did not allow them to shift to the production of export crops in which they could otherwise have a comparative advantage. Their cheap labour did not suffice to give them the competitive edge in the world market due to the high costs in other segments along the supply chain, the monopolistic power of local traders and their lack of access to advanced technologies, better means of production and credit that left them at a comparative disadvantage in the production of most crops. These farmers needed government support at the initial stages in order to make the necessary adjustment with investments in the rural infrastructure, access to cheap credit and better organization of the supply chain.

Government interventions in developing countries were also necessary to attract foreign investments by reducing corporate income taxes, granting tax holidays and industrial subsidies. However, these measures also had disadvantages, because they reduced government revenues and the resources available for welfare programmes or public investments. In fact, most of these investments contributed very little to local employment of low-skilled labour or to the country's growth. Instead, their main contribution was to enrich a narrow elite.

Measures to liberalize the capital account and open the country to foreign investments were also biased in many countries in favour of the rich in urban areas and contributed to raise income inequality.[2] China and India are the prime example of countries that had a major success in attracting foreign investments: Total FDIs in India increased 60-fold in US dollar terms from 1990/91 to 2001/02, and foreign investors made effective and highly rewarding use of India's skilled workers, but these investments had a rather limited impact on the population in the rural areas or the country's poor, and they did not contribute much to increase employment of unskilled workers.

Stabilizing macroeconomic and monetary policies that were necessary to create the conditions for growth required strict restrictions on credit, stable exchange rates, sharp cuts in government expenditures, tight controls over the financial markets and scaled down social safety nets. In the short run, all these measures were not conducive to promote employment, but forced enterprises to cut wages and reduce their workforce.

The developing EA countries, most notably China, opened their capital markets very gradually and established strict controls over the interest rate in order to avoid these shocks. Foreign investors and foreign banks were also very cautious and during the 1980s, the inflows of capital and foreign investments were very small. In the mid-1990s, foreign investments gradually increased and, despite the financial crisis in 1997/98, the capital inflows to most Asian countries (including Thailand, Indonesia and the Philippines) were rapidly renewed and the accumulation of huge foreign currency reserves reached proportions that required them to establish some controls in order to secure the stability of the exchange rate.

In contrast, the LA countries, most notably Argentina, implemented during the crisis of the early 21st century strict stabilization programmes and emergency adjustment measures that were highly biased against the low-income unskilled workers and brought a steep rise in the country's unemployment and poverty. The need to complement the emergency measures with social programmes to relieve the difficulties of the vulnerable groups that were hurt the most by these measures were realized in the LA countries only when public protests and new elections toppled the governments in a considerable number of them and brought to power new and very radical leftist leaders.

4 The Lessons from the Structural Adjustment Programmes

The SAPs were designed by the World Bank and the IMF as guidelines for economic reforms in the developing countries. Their aim was to assist these countries to make the necessary structural changes in their economies in order to accelerate growth and reduce poverty. The Washington Institutions

[2] On the experience in LA, see Sz'kely (2003).

designed these reforms for countries that applied to these institutions for concessional loans that they needed in order to cover their large balance of payment deficits and debt payments when the market financial institutions refused to give them more credit. In the 1980s, the number of developing countries that accumulated large debts and applied for concessional credit grew so rapidly that this became one of the Bank's main functions, and was undertaken jointly with the IMF.

The Washington Institutions demanded that countries that applied for credit will implement comprehensive SAPs as a condition for receiving this credit. To improve the controls over this credit, the IMF introduced in the 1980s the Structural Adjustment Facility and later the Enhanced Structural Adjustment Facility (ESAF) that monitored the implementation of the economic reforms to ensure that the countries met the conditionalities that they assumed. Before proceeding to discuss the programmes themselves, a brief background of the origins of the debt crisis and the reasons that led so many developing countries to accumulate large debts will provide a better understanding of the reasons for the difficulties that these countries encountered in meeting the conditionalities. The debts have remained an unresolved problem for many developing countries until today.

4.1 The debt crisis and the implementation of the structural adjustment programmes

The debts of most African and LA countries started to pile up at a rapid pace already before the global financial crisis of 1973. The process began with the transition to floating exchange rates and the oil crisis that followed and was exacerbated by expansive government policies. The countries' leaders mismanaged expenditures and implemented various social and economic programmes in order to secure public support; in addition, some also took large loans for their own personal accounts outside their countries. Government expenditures grew without internal or external restrictions and the budget deficits blew up well beyond the capacity to finance them from the countries' own resources; this resulted in the need to borrow heavily in the world capital markets at high interest rates. Huge amounts were also spent on largely wasteful programmes to develop large local industries with ambitions to become global economic powers.

In the LA countries, the weak growth performance was associated with high macroeconomic instability that was due, in part, to volatile trade and monetary and exchange rate policies that were planned to stem the rising foreign debts, since these debts could no longer be financed by borrowing from foreign financial institutions. The high instability, the accumulation of large debts and the large government and balance of payments deficits also reduced very drastically the local savings and investments and drove out the panicked foreign banks, thereby aggravating the crisis. In SSA, unstable regimes, frequent political upheavals, repeated coups and internal conflicts, as well as ineffective and highly erratic economic policies deepened the economic crisis. The instability of governments eroded the discipline of public administrations, led to widespread corruption at all levels of government, and the corrosion of the system of law and order.

Political and economic volatility in these countries deterred foreign investments and endangered the relations with foreign suppliers. Government policies, primarily the mounting budget deficits, undisciplined monetary policies and frequent exchange rate depreciations further aggravated economic instability, often driving inflation out of control, damaging business activities and making investments riskier and untenable. The high inflation was particularly harmful to the poor that were less able to protect themselves against the rising prices and had no savings to rely on. In rural areas trade collapsed, and even the production of traditional agricultural export products fell rapidly as farmers resorted more to production for self-consumption.

In 1982, Mexico told its creditors it could not repay its debts. This pattern was repeated over and over in the following years as other countries found themselves in similar situations. But their debts continued to rise, and new loans added to the burden. The IMF and the World Bank stepped in with new loans at concessional interest rates to help these countries pay off their outstanding debts, but when countries needed assistance, the IMF enforced its requirements of 'conditionality' by paying out its loans in installments – each one dependent on the achievement of scheduled policy reforms. These conditionalities required the countries to implement reforms that were designed to put their economies in order.

Since then these developing countries have been advised by the IMF and the World Bank on strategies to reform their economies as a condition for getting the credit from the international financial institutions to repay their debts. The recommended reforms included practically all aspects of economic policy: liberalization of trade and FDI, privatization of state-owned enterprises, industrial deregulation, strengthening of law and order and securing of property rights, tax reforms, liberalization of prices and interest rates, introduction of a competitive exchange rate and maintaining fiscal discipline.

Trade liberalization and the removal of various trade restrictions were essential ingredients in these programmes, and they had two goals: First, reducing the countries' large trade deficit and restoring a more balanced external trade account; second, facilitating the restructuring of their economies by promoting competition that would secure a more efficient allocation of resources by bolstering the more efficient sectors and reducing production in the inefficient ones in which the countries do not have a comparative advantage. In most countries where a SAP was implemented, these measures were quite radical and initially quite damaging, particularly since the preferred approach was not to implement these measures gradually, but rather in the form of a 'shock therapy'.

During the 1980s and the 1990s, the multinational financial institutions granted structural adjustment loans to many developing countries under the condition that they implement these reforms. The World Bank and the IMF made 1055 separate adjustment loans to 119 poor countries from 1980 to 1999, and 36 poor countries received 10 or more adjustment loans during these years.

The debate over the effectiveness of the SAPs designed by the IMF and the World Bank has been, and continues to be, one of the more controversial, perhaps even confrontational subjects in the literature on economic development. It has been the battleground for many vehement exchanges between economists of the World Bank and the IMF, representing the leading international development organizations in charge of implementing these adjustments, and economists and political scientists from the academia, as well as leading intellectuals from the developing countries themselves, who often opposed many aspects of these programmes and, more fundamentally, their basic approach. Box 4.2 lists a typical set of conditions that were included in the SAPs.

4.2 Privatization

Another part of the SAPs was the privatization of state-owned enterprises. The goal of privatization was to reduce the direct control of the government over the economy and the monopolistic power of these enterprises, to strengthen the private sector and to increase the efficiency in the allocation of the country's resources. In addition to eliminating the wasteful use of resources, privatization was necessary to cut government expenditures and reduce the budget deficit since many of the state-owned enterprises had losses and accumulated large debts. The public sector reforms included

Box 4.2. Common IMF/World Bank conditionality.

Reduce the size of the government: Cut budget expenses by trimming the size of the government payroll and the size of government programmes.

Privatize state-owned enterprises: Further reduce government expenditures and streamline the budget by either privatizing state-owned enterprises or by stopping the operation of losing enterprises.

Cut subsidies and social spending: To reduce government expenditures, eliminate more wasteful social programmes and reduce the size of others.

Increase interest rates: To combat inflation, increase the interest rate charged for credit and award local savings by raising their rate of return.

Promote foreign investments: To attract foreign investments, cancel regulations that restrict or prohibit foreign ownership of local businesses.

Trade liberalization: To promote exports and improve the allocation of resources, reduce import tariffs and eliminate all other restrictions on imports that protect domestic producers and prevent free competition. Equalize all tariffs to reduce distortions in production and in the allocation of domestic resources.

Reorient the economy to promote exports: Give incentives for farmers to produce cash crops (coffee, cotton, etc.) for foreign markets; encourage manufacturing for export markets and eliminate regulations that bias the allocation of resources towards production for the local market.

Other balance of payments reforms: The reforms related to the balance of payments included capital account liberalization and devaluation at the initial stage of the reform in order to compensate for trade liberalization.

divestiture of parastatals and downsizing public work in order to cut government expenditures, even though that meant ignoring the important contribution to provide employment to unskilled workers.

Although privatization was inevitable given the gigantic size of the public sector and its share in the economy, this was a nightmare for the thousands of employees who lost their jobs, many of them in remote areas where no alternative employment was available. The sweeping implementation of the privatization programmes in countries where state-owned enterprises controlled a large part of the industry and were the main source of employment for a large share of the population in many regions left many people without alternative employment, with no other income sources and with little prospects of finding employment because most industries in the region collapsed. Many of the people who lost their jobs became poor practically overnight, since until the reforms were implemented, they had guaranteed employment and guaranteed pensions for life and therefore had very little savings.

The majority of the privatized enterprises had to close down; others needed time to make the necessary adjustments in order to find an appropriate niche for their products that allowed them to exploit their comparative advantage and keep their employees. A shock therapy of the type that was implemented in the SAPs either through privatization or through trade liberalization generally did not give enterprises the time needed for adjustment, and when the failing industries had to close down, it often resulted in great damage to the entire economy.

However, the shock therapy applied as part of the reforms did not achieve the desired goal of increasing economic efficiency. The waste of human resources due to prolonged unemployment and the waste of other resources with the destruction of the machines and the industrial constructions that were left unused and the rapid deterioration of the infrastructure in the affected areas obviously give reason to doubt the efficacy and desirability of this approach. Moreover, the search for greater efficiency cannot and should not be the only guide for government policies and the only criterion in the calculus of costs and benefits. At the very least, these countries should have prepared in advance adequate social security and public health care programmes to support the needy population in the course and in the aftermath of the structural adjustment.

In Bangladesh, a World Bank study that analysed the performance before and after privatization of 13 enterprises that were privatized in the first half of the 1990s showed very mixed results (Box 4.3). Some firms increased their profits after privatization, others continued to sustain and even increase their losses, and some went out of business or were shut down. Privatization in the jute sector showed that the main problem was not in the ownership form but in the policy regime and management. Both private and public jute mills had high losses, which have generally been attributed to the fact that production costs tended to be higher than the export price of jute because of the outdated equipment in the mills.

Industry-wide data also showed that labour productivity does not depend on ownership but on managerial efficiency. The failure to increase

Box 4.3. Views of the people of Bangladesh.

'I was a laborer of AB Biscuit Industries of Tongi. The factory is now closed for the last eight years. I am unemployed. I am not getting any other job and I do not find any self-employment opportunity to carry on. Now I cannot support educational expenses of my children.'

– A female worker of AB Biscuit Factory

'My husband tortured me physically after I had lost my job. He is unemployed. I have one child. I am now living in my brother's house.'

– A female worker of a textile factory of Tongi

'In the competitive market, the entrepreneurs are less careful about the environment of the work place. Workers have no voice in this regard as they are concerned to keep their employment sustained.'

– A labour leader

'Hundreds and thousands of workers are now becoming unemployed due to closing down of both state-owned and private industries, but the government is virtually not doing anything for them. The workers who have lost their jobs are now living a sub-standard life. Many of them have been compelled to withdraw their children from schools. Their children have now become child labor, sex labor. They have become maidservants. They are now breaking the bricks on the streets. Should the government do anything for them?'

– A female worker

'We can now have many foreign goods cheaply. But government should take care of our own industries as well.'

– A rural consumer

efficiency through privatization was attributed to inefficient management, the scarcity and misuse of credit, as well as the virtual absence of technological and marketing support. Many privatized firms had experienced an increase in non-performing loans and defaults. The decline in jute production following the privatization had negative ramifications throughout the entire economy. Many jute factories closed and thousands of workers were laid off.

In El Salvador, the privatization of public services, such as electricity distribution, was part of the third generation of reforms that began in 1996. The privatization of electricity services was aimed at increasing the savings of the state by reducing public spending that was partly due to low rates for low-income users. Two years after the sale of the electricity-distribution companies, and in spite of the gradual upgrading of technology and equipment by four foreign firms, there were no significant improvements in terms of service quality or coverage.

Furthermore, rural areas were seriously affected both directly, by increases in electricity rates, and indirectly, due to price increases in other services that depend on electricity, such as deep-well pumps for potable water that are used in many communities. In addition, users with low levels of consumption, who constituted the majority of the population, were the ones most affected by the increases in electricity rates. Low-end users saw their rates increase by 47% on average, while users with the highest consumption levels experienced an average increase of 24%. Higher electricity bills contributed to the closing of many micro and small enterprises.

Subsidies are still given through social-investment funds to ensure that services are provided to rural communities and to communities in other areas not deemed profitable by the private companies. The creation of a new regulatory entity to cover privatized services also generated new costs. Furthermore, the process of privatization did not succeed in opening the market to free competition and full consumer choice. The four privatized distribution companies compete for only a 20% share of consumers in the market – mostly large industrial, commercial and service companies. The remaining 80% of consumers have no possibility of switching providers.

4.3 Privatization in the ex-centrally planned economies

Similar developments took place a decade or two later in most ex-centrally planned economies where the entire industry was state-owned. More than a decade and a half after the privatization programme was implemented, many areas are still economically destroyed and many of the younger people left to other areas in search of work. The older population, consisting mostly of people who had been in their late forties or older when the privatization took place, is still unemployed and totally impoverished; however, they were forced to stay as they had little prospect of finding alternative employment elsewhere.

Privatization of state-owned enterprises in these countries doomed their economies from the start to prolonged stagnation and decay. The wages in these countries are much higher than in EA and SA, and their enterprises cannot therefore compete with producers from the Asian countries in the production of labour-intensive products such as textile and apparel either for the EU market or even for their own market. Even in agricultural production, the local farmers could survive only with government support and with other measures to protect them against foreign competition. In many of these countries, the transition period after the liberation from the communist regime still continues.

Moreover, privatization in these countries did not bring greater efficiency, but rather a transition from one type of inefficient monopoly to another when ownership was transferred from the state to a private oligarch. The monopolistic position of the new owners made it possible for them to continue their operations and charge high prices for their products, but did not oblige them to take any responsibility for their employees. In the agricultural sector, the privatization of collective farms was often accompanied by disruptions until the land distribution was determined; production became much more difficult and much less profitable, since there were no arrangements for the supply of fertilizers and other inputs and for the purchases of the final products from the farmers. Despite potentially huge advantages of countries like Ukraine in the production of grains, their current productivity is far lower than that of other producers, and they cannot compete in the world market.

The privatization of hundreds of state enterprises within the framework of the adjustment programmes and public sector reforms in many African countries forced many companies to close down and resulted in thousands of retrenched employees who were not paid for many months and were not able to collect any severance package. Women were particularly vulnerable because many of them occupied jobs in state-run institutions and the private sector did not accept them when the state enterprises were privatized.

These adjustments were necessary, however, to eliminate extremely inefficient public enterprises in the face of dwindling government budgets. The public sector reforms involved divestiture of parastatals, downsizing and retrenching public workers. Some countries introduced a Social Development Fund that was administered by the World Bank and was aimed at carrying out various projects in order to absorb many of the unemployed, as part of a more comprehensive Poverty Reduction Strategy. Most of these projects and programmes did not have a sustainable impact on employment or on poverty, either due to the overwhelming rise in the number of unemployed, or because the funds were rapidly exhausted.

In Hungary, the privatization of public utilities was prepared and executed in the first half of the 1990s. By the end of 1995, around 42% of the property of the public-utility sector was under foreign ownership. Hungary is now second only to Great Britain in Europe in terms of the proportion of private ownership in the utility sector, but it is unique in that the country's fundamental utilities have become the property of state utility companies of foreign countries. As a consequence of privatization, the quality of services – except telecommunications – barely changed at all, while prices increased at a much higher rate than did wages. Large segments of society now have severe difficulties to pay their electricity, gas and water bills.

Some ex-centrally planned economies also introduced a Social Development Fund administered by the World Bank to create programmes to provide work for the unemployed. Again, most of the related projects and programmes did not have much impact on employment or poverty, because the large increase in the number of unemployed as a result of the privatization exhausted these funds too rapidly.

However, the proximity of the East European ex-centrally planned economies to Western Europe, the benefits that some of them now have when they join the EU, and their skilled labour force give them significant advantages and good prospects for future growth, particularly since they already have a rather solid, though inefficient, industrial base. In recent years, several large West European industries, including car manufacturers, found enough incentives to transfer their factories to these countries when they became members of the EU.

4.4 Criticism of the SAPs

The criticism of the SAPs echoed in earlier debates on development strategies centered on the argument that the programmes often remained too abstract and generic in that they failed to recognize the substantial differences between countries. In view of these differences, it must be doubted that structural adjustment strategies can be formulated in the abstract or that they can provide a menu of actions and policy recommendations that apply to all countries. Lumping all the 'developing countries' together and advocating a 'one-size-fits-all' development strategy with a blueprint of actions is bound to be oversimplifying and possibly even misleading. For each country, it is necessary to recognize its unique geographic, socio-economic and political conditions, establish priorities that reflect these conditions, and identify the critical areas on which the adjustment programme should focus. Recommendations such as 'promote labour-intensive growth and open trade' may fail to recognize country-specific needs and differences in the underlying conditions.

Another criticism of the neoliberal economic principles and their implementation in the SAPs of the World Bank and the IMF concerned the absence of complementing and supporting measures to reduce poverty and the failure to design an 'adjustment programme with a human face'. SAPs should have included social policies and funds for the protection of vulnerable groups instead of focusing exclusively on policies aimed to spur economic growth.

Contrary to the expectations, the experience with the SAPs shows that the conditions to minimize the role of the government to functional interventions aimed at 'getting the fundamentals right' and otherwise get out of the way, and the demand to open the economy for trade, did not achieve the desired goal of reducing poverty. Therefore, it must be concluded that this was not the appropriate model for countries where poverty is widespread and where institutions are not functioning.

Trade liberalization through sweeping and uniform reduction of all tariffs and the removal of all other trade restrictions proved to bring more damage than gains. Even sectors in which the country had or could have a comparative advantage were unable to rapidly make all the necessary adjustments and transform their production to export products. Built-in structural constraints, the need to train workers and establish connections to the supply chain prevented enterprises from making the necessary adjustments rapidly enough to become competitive in the world markets over a short period of time. Obviously, the inefficient sectors that were protected until the reforms were implemented by high tariffs and trade restrictions had to find out very rapidly that they were unable to continue their production, since they could not compete with the flow of cheap imports.

The main immediate results were therefore massive lay-offs and a wave of bankruptcies that were particularly harmful to the poor who lost their employment; in addition, the trade deficit widened despite the reduction in imports. At the early stages, the only expansion in exports was of the country's natural resources, but these exports contributed very little to local employment since the value-added was very small. A Multi-Country Participatory Assessment of Structural Adjustment (SAPRIN, 2002) concluded: 'Import liberalization has led to the failure of local enterprises and the destruction of domestic productive capacity. This has resulted in an employment loss in important sectors of local economies, thereby reducing the purchasing power of large segments of society and overriding presumed consumer benefits from the opening up of trade.'

The World Bank and the IMF strongly advocated an outward-oriented and liberal trade regime, and the experience of the EA countries was their main reference for demonstrating the benefits of the increase in exports to promote growth and reduce poverty. In practice, the drastic measures of trade liberalization that the SSA and LA countries were required to implement as a condition for the adjustment loans were significantly different from the actual measures that the EA countries implemented. Instead of getting the government 'out of the way', the EA reforms involved active government interventions. Instead of open market and free trade, their reforms were based on government support to selected industries in order to promote their exports and continued protection on the local industries and local employment by means of high tariffs.

Other evaluations of the SAPs that focused primarily on the countries of SSA concluded that these programmes did not deliver sustained and sustainable growth. The focus on the export industries to promote growth could not be sustained due to the lack of funds for the investments needed to build an industrial base. The very low domestic savings rates were not sufficient to provide funding for investments from local resources, and foreign investment was far too low to provide adequate funding. As a result, the SAPs could not promote industrialization over a short period of time and exports continued to concentrate on local natural resources and the traditional agricultural goods.

In the low-income countries of SSA and Central America, the policy of rapid structural adjustments focused on industrialization and universal, uniform and across-the-board trade liberalization irrespective of the countries' levels of development, clearly did not produce the desired outcomes. Instead of bringing about an expansion of exports and a diversification of their agriculture and manufacturing sectors, the reforms led to a damaging process of de-industrialization and slowed down the growth of their exports. The main expansion in their exports was in resource-based industries, while the majority of local industries was at an infancy stage and could not survive the intense competition in the world market.

It has also been realized that market forces by themselves could not achieve an efficient allocation of a country's resources or an effective diversification of its structure of production in favour of exporting goods in which the country has a comparative advantage. To achieve that result, the economy needs a proper infrastructure and working institutions that many of these countries do not yet have.

In a thorough report that followed a detailed 13-countries comparative study, the UNDP (2002) criticized the following points in the IMF–World Bank SAPs:

- These macroeconomic policies generally overemphasize macroeconomic stabilization which involves a drastic retrenchment of aggregate demand.

Public expenditure is subjected to huge reduction and credit expansion is severely restricted.

- These policies have a doctrinaire antipathy to public investment, on the grounds that public investment crowds out more efficient and productive private investment. UNDP – in contrast – found that many public investments had a 'crowding in' effect: These investments concentrated in sectors providing broad externalities, thus making private investment more profitable.
- Trade reform has often been implemented without adequate precaution to enable the existing structure of traditional industries to make an orderly adjustment to a more open trade regime. They found that 'shock therapy' that implemented a wide range of trade and price reforms had a high social toll.
- Drastic curtailment of public expenditure also reduced the ability of the government to provide direct assistance to the poor and its ability to fund public works programmes and publicly funded programmes.

4.5 Why did the SAPs fail?

It is now widely recognized, even by the Washington Institutions themselves, that the SAPs failed on a number of counts:

- The SAPs did not have any material effect on poverty. In fact, poverty increased in Africa, the Middle East and North Africa, and in the transition economies of Europe and Central Asia; in LA poverty remained unchanged. Only in EA and SA the incidence of poverty declined, but this was primarily due to these countries' own policies and rapid growth.
- Over time, the dismal results of the SAPs became evident despite the efforts of the World Bank and the IMF to improve the results of their policies by increasing the number of conditions for new loans and debt rescheduling, focusing more on poverty reduction.
- The SAPs did not take into consideration the varying and differing circumstances in the countries; nor did they pay much attention to the priorities of the debtor countries themselves or what they regarded as their own self-interests.
- Stiglitz (2002) noted that since the poor countries have poorly developed financial institutions, forcing these countries to 'liberalize' those institutions by opening their economies to world financial flows can only undermine the existing fragile credit institutions and shift rules on credit away from those whom local businesses have come to rely on in the past.
- The SAPs were short-term and expected results within 3 years – which was the standard duration of a SAP. Structural adjustments are bound to take much longer and must be geared in the long run.
- The number of structural demands imposed by the IMF grew from 2 in 1987 to 17 in 1997, even though the number of IMF programmes that could be realized as planned declined: by 1999, only 16% of IMF programmes were carried out as planned.

- The World Bank also stipulated a great number of conditions for its lending, on average 36 requirements for each SAP, but only one-third of these demands were categorized by World Bank staff as 'very relevant'.
- IMF-conditionality consists of *quantitative performance criteria* (e.g. the size of the currency reserve, the increase of the money supply, budget outcome, borrowing needs) and *structural performance criteria* (e.g. privatization, deregulation of financial markets, the reform of social security). The quantitative criteria often required restrictive fiscal policy even when the country had negative growth. The performance conditionalities were tailored according to neoliberal criteria and prohibited more proactive regulations.
- The primary concern of the SAPs was the countries' economic growth; poverty eradication was assumed to follow as an outcome of growth.
- Both the IMF and the World Bank, however, recognized later on that it is highly uncertain whether their policies would indeed lead to economic growth.
- It has also been recognized that these institutions themselves were quite uncertain which of their policies would actually work and which would not. Their detailed conditionalities and interventions were therefore barely helpful. The elements of the adjustment programmes that contributed most to positive outcomes were the removal of gross macroeconomic imbalances, which were evident in very high rates of inflation and exchange rate misalignment. But the structural reforms which were undertaken did not have any positive effects on growth.
- The failure of the SAPs was evident from the fact that even programmes that were carried out fully had no material impact on growth and development.
- In Stiglitz's analysis (2002) the conclusion is clear: The first step that is needed in development practice is an end to free-market ideology, and willingness to see individual circumstances in their complexity and distinctiveness. Instead of blind ideology, 'the most fundamental change that is required to make globalization work in the way that it should is a change in **governance**' (p. 226, emphasis added).

4.6 Governance

The requirement to improve governance has become a central and often repeated mantra in many recent reports on the developing countries. It has not always been clear what the goals of this requirement are and what specific measures should be taken to improve governance. In order to clarify this requirement, Box 4.4 lists the various definitions of governance that have been used.

Kaufmann *et al.* (2004) suggested the following indicators to measure governance:

- Voice and accountability;
- Political stability/absence of violence;
- Government effectiveness;

Box 4.4. Governance.

The term *governance* deals with the processes and systems by which an organization or society operates. Frequently a government is established to administer these processes and systems.

The word derives from Latin origins that suggest the notion of 'steering' a society in contrast with the 'top-down' approach of government.

The World Bank has used the following alternative definitions:

- The exercise of political authority and the use of institutional resources to manage society's problems and affairs.
- The use of institutions, structures of authority and even collaboration to allocate resources and coordinate or control activity in society or the economy.
- Governance consists of the traditions and institutions by which authority in a country is exercised. This includes:

the process by which governments are selected, monitored and replaced,
the capacity of the government to effectively formulate and implement sound policies,
the respect of citizens and the state for the institutions that govern economic and social interactions among them.

UNDP: 'Governance is defined as the rules of the political system to solve conflicts between actors and adopt decision (legality). It has also been used to describe the 'proper functioning of institutions and their acceptance by the public' (i.e. their legitimacy). And it has been used to invoke the efficacy of government and the achievement of consensus by democratic means (participation)'.

- Regulatory quality;
- Rule of law;
- Control of corruption.

This list of indicators does not make clear what it would take to change and improve governance, as Stiglitz (2002) and many others have noted: Does it require a better organization of the various public institutions? Does it require better training for the staff members of these institutions? Does it require different rules according to which these institutions and their staff members make the decisions on how to 'manage society's problems and affairs'? Or should it determine the process by which a legitimate government is selected?

The World Bank definition of governance suggested by Kaufmann, Kraay and Mastruzzi in Box 4.4 emphasizes the process in which a legitimate government is determined; on the other hand, their list of indicators focuses on the mode of operation, the effectiveness and the trustworthiness of the public institutions. Indeed, public institutions can function effectively only if the government itself has legitimacy and can thus give them the authority to function, and can define the rules according to which these institutions should make the decisions on how to manage the country's affairs.

In the other definitions as well as in the list of indicators that Kaufmann, Kraay and Mastruzzi are using, governance is defined as the tool with which the government is governing. This tool consists of the various public institutions, their staff and the rules according to which the staff members make their decisions. In a democracy, however, governments are changing, usually according to specific rules and a predetermined schedule, whereas the core of the rules by which the institutions are operating remain unchanged – protected by the country's constitution. There is, of course, a gradual process by which some of the rules may change over time, and this must be approved by the parliament and by the majority of its members, but the core rules remain unchanged. In contrast, in a non-democratic regime, the government itself determines the rules.

Obviously, this subject deserves a much more elaborate discussion. I found it necessary to bring up this topic, however briefly, because in many discussions and research papers, including the World Bank's *World Development Report* (WDR; 2004), 'improving governance' is regarded as the panacea for all the economic malaise of the developing countries. With the lack of a precise definition of what that concept exactly means and what the specific actions are that need to be taken in order to 'improve governance', this somewhat vague term may seem like searching for the coin under the light, or, perhaps, blaming the usual suspects.

The first step to improve governance is, therefore, to examine whether the necessary institutions are in place and are given the correct rules and the authority to implement these rules. That step should include an analysis of the way the government itself is working, but its focus is on the process by which the rules are determined and on the effectiveness and consistency. The institutions essentially represent the division of labour and responsibilities between and within the government ministries that are in charge of implementing these rules.

The second step is an evaluation whether the structure of the institutions and the division of labour and responsibilities between and within these institutions is adequate and secures an efficient execution of these rules. However, the authority to change the structure of these institutions – should that be necessary to improve their work – can be given only by the government.

In many African countries, the governments themselves did not function in a way that made it possible to change or improve governance. In quite a few countries this is still a major problem. Later on in this chapter, there is a small collection of recent news reports from Somalia that describe how people related to the unfolding events in their country. One of these events was the selection of a new coalition government, but the report on this event makes clear that the government did not have legitimacy in the perception of the people. In part this was because the government was very fragile and lacked the authority to enforce its rulings; in part this was also because the government did not even control large parts of the country. Under these circumstances, it is obviously futile to try to improve governance.

Somalia is obviously an extreme case, but it is by no means entirely atypical. As we enter the 21st century, only a handful of African countries have governments that are strong enough and have the legitimacy and the authority to use the institutions of governance, and determine clear rules that define the responsibilities and authority of these institutions. During the 24 years in which Daniel arap Moi was in power in Kenya, it was equally futile to discuss improvements in governance in that country; nevertheless, many efforts were made to do just that. Only after a new president was elected in Kenya in 2002, this has become possible and more sensible. This is just one among many examples that in a non-democratic regime, government and governance are interlinked even though in theory at least they are not the same thing.

Even corruption in these countries is not the same thing as corruption in a democratic country. In a non-democratic country, corrupt decisions of staff members in one of the institutions are not acts that violate the rules by which they are supposed to work or the responsibility and the authority that they are given, since the rules themselves are not clear and they are not anchored in the country's constitution. In the absence of clear rules and responsibilities, these institutions' staff members, at all ranks, have much more leeway to define the rules by themselves or give them their own interpretation according to which they make their decisions.

Curiously, though, in an extensive study of Radelet (1999), who used the set of World Bank governance indicators suggested by Kaufmann, Kraay and Mastruzzi, the conclusion was that many African countries are well governed and there is no evidence that Africa's governance, on average, is worse than elsewhere. Sachs *et al.* (2004) made a similar analysis using the Corruption Perceptions Index of Transparency International and also found that all but one of the African countries in their sample received scores of 'good' (i.e. low corruption) or 'average'. Nevertheless, many of the relatively well-governed African countries have been unable to increase the material well-being of their populations. The main conclusion of the analysis was that 'Africa's crisis requires a better explanation than governance alone'. More policy or governance reforms, by themselves, will not be sufficient to overcome their poverty trap (p. 121).

The analysis of the role of governance must emphasize that markets can function only under the umbrella and the supervision of a wide range of non-market institutions that have the authority and the responsibility to maintain the rules of the game in the market and in the society at large. Their role in what we usually (and somewhat arbitrarily) identify as the economy arena is to protect property rights and enforce contracts through a system of law and order that determines regulations, supervision and arbitration (Rodrik, 2002).

4.7 The impact of the SAPs

Trade liberalization can clearly succeed only if the necessary institutions are in place. Trade liberalization can boost economic growth only if the basic rules are maintained and the institutions guarantee that the markets are indeed free and competitive. All too often, econometric studies sought to establish the statistical relations and the causal link between trade liberalization and economic growth, but disregarded these necessary conditions and concentrated directly on the statistical measures themselves.

Rodriguez and Rodrik (2000) noted that the priority given in these studies to the role of trade reforms might create expectations that these reforms by themselves are the panacea for growth and poverty reduction in the developing countries. Greater openness to trade is an important ingredient of the package of policy reforms that are necessary to achieve this goal, and adequate macroeconomic policies are also important components of these reforms, but effective governance and well-functioning institutions are necessary to make the reforms possible.

In an article on 'Institution Rule', Rodrik *et al.* (2002) demonstrated in a sample of 140 countries that the role of institutions, particularly the role of property rights and the rule of law, is far more important in determining a country's growth than the role of other factors and policies that were regarded in the past as essential to drive the process of economic growth. In the wisdom of hindsight, these results should not be surprising: Without the safety of property rights and the rule of law, production, investments and trade cannot take place.

Openness by itself cannot attract private investors as long as their investments are not protected, and comparative advantage in today's world depends much less on the relative costs of labour and far more on the relative risk of investments and the quality of the country's infrastructure and institutions. The limited experience or sheer incompetence of many of the private and public institutions in the LDCs is their main handicap in building up their economy and attracting foreign investments (see also Dollar and Kraay, 2002a).

However, the institutional environment and the legal regime in the countries of SSA, the Middle East or LA may be quite different from the Anglo-American or the EA ones. Clearly, the systems of social and economic institutions that were developed over many decades and centuries in different environments differ widely from one another. Loopholes in the system are revealed when the structure of the economy is changing and the existing arrangements no longer function.

The reform of institutions encounters huge difficulties that become most apparent during the transition period when old arrangements and institutions no longer work but new and stable ones are not yet in place. The breakdown of law and order and rampant corruption are just two of the more common manifestations of the difficulties that many developing countries experience with the breakdown of their system of law and order. In these countries, globalization, far from bringing people closer together, introduces a deep divide between them.

An essential part of the structural adjustment is to help these countries to reform their institutions and restructure their economies, since they do not have at present the necessary resources, know-how and experience to guide them in carrying out this undertaking. The obstacles faced by these countries now are much larger and much more complex than those that the EA countries had to overcome when they started their reforms.

The IMF's own studies concluded that the impact of the SAPs on growth was barely discernible; UNCTAD (2002) concluded that 'even when well implemented, past adjustment programmes have not delivered sustainable growth rates sufficient to make a significant dent in poverty in most LDCs' (p. 176). In fact, they may have actually increased the rate of poverty, or at least, they 'are not associated with sustained reductions in the incidence of poverty in the LDCs even when they are well implemented' (p. 176).

The conditionalities imposed by the IMF and the World Bank on the LDCs in order to implement the SAPs were therefore a complete failure: The LDCs which put these programmes into operation experienced a fall in their GDP per capita and an increase in poverty. In countries that complied with these conditionalities, nearly 50% of the poor population suffered further income losses during the adjustment period, and poverty continued to rise even during the post-adjustment period.

The weakness of the SAPs was that they were not nationally owned, and thus not well implemented. The local authorities did not have to pay much attention to the constraints established by these conditionalities on their economic policies since funds were highly fungible. The conditionalities that were targeted mainly on macroeconomic policies did not establish effective constraints on their capacity to allocate funds according to their own priorities. Moreover, in many areas – including agriculture, trade, finance, public enterprise, deregulation and privatization – the policies established by these conditionalities were totally unsuitable for the LDCs and did not contribute to promote growth and reduce poverty.

The social, economic and political conditions in these countries required different SAPs and different policies in many other areas to enable the countries to break out of the vicious circle of stagnation and poverty. Very often, the programmes that they were required to implement restricted them in developing their productive capacity, their markets and their entrepreneurial potential. Only at later stages, with the growing criticism on SAPs, the conditions were somewhat relaxed and programmes that were more attuned to the country's social needs and were aimed at mitigating the negative social effects of the structural adjustments were introduced.

Over time, the SAPs were gradually changed and greater emphasis was put on building up and improving local institutions and governance. These reforms emphasized the restrictions on the role of the central government in the economy and the desirability of expanding the role of local non-governmental organizations at the district or even at the village level. However, trade liberalization and the establishment of a free market remained the backbone of the reforms and were perceived as the main means of reducing the wasteful use of the countries' resources. It has been recognized, though, that these are long-term goals that may take a long-term adjustment process, and the capacity to achieve them varies widely between countries.

4.8 The East Asian model and its constraints for the LDCs

The model for a successful implementation of the SAPs was the 'East Asian miracle', and outward-oriented trade policies were regarded as essential for the success of these reforms. In fact, the actual trade policies of the EA countries did not follow a rigid doctrine of open trade free from all government interventions. Instead, the EA governments intervened very actively with various incentives and subsidies that were targeted on selected sectors in order to build these industries and promote their exports; during the transition period and even until today, they retained provisions that protected the other sectors that produce for the local market and thereby protected also their employees.

In many World Bank and IMF studies, the EA miracle was attributed first and foremost to openness to international trade and to a market-friendly approach. On these grounds, the Washington Institutions advocated reforms that adopt a similar market-friendly approach, which came to be regarded as the basic pillar of the Washington Consensus and a necessary condition to gain from trade. By cutting the tariffs on their imports, removing all other trade barriers, opening their economies to a free flow of imports and exports, and promoting export-led growth, the reforming countries should expect to reap large benefits from trade. The EA countries, particularly those who were regarded as the 'new globalizers', set the example for the expected benefits from these policies.

There were, however, significant differences between the principles of the Washington Consensus and the actual policies that the EA countries implemented and continue to implement even today:

- Trade liberalization led to an increase in income inequality. In some countries this was because the competition of cheap imports depressed wages in many sectors. As noted earlier, even farmers' incomes in many developing countries were depressed by the competition of cheap imports from the developed countries that benefited from high subsidies to agricultural producers and exporters, particularly in the EU and the USA. The EA countries tried to mitigate the rise in income inequality, even though their pro-industry, pro-urban policies still led to an increase in inequality.
- A report by UNCTAD (1997) found that trade liberalization in Latin America widened the wage gaps, depressed real wages of unskilled workers (who account for more than 90% of the labour force), and raised unemployment. In the manufacturing sector, this was due to the competition of cheaper imports of labour-intensive products from the Asian countries. Earlier it was noted that even in SSA, that competition handicapped industrialization despite the low wages of workers.
- The World Bank reports concluded that since 1980 there was a 'serious . . . increase in within-country inequality in industrialized countries reversing the trend of the period 1950–1980' (World Bank, 2001a, p. 46). Data from the Luxembourg Income Study (2001) show that, among 24 developed countries in their sample, 18 experienced increasing wage inequality.
- Critics also challenge the view that trade liberalization is equally beneficial to all countries and eventually to all sectors and all income groups. This conclusion is based on the neoclassical Heckscher–Ohlin trade

model that suggests that over time the prices of all inputs and outputs are likely to converge as an effect of trade. The accumulated empirical evidence shows that, for various reasons, globalization has not provided these benefits and instead widened the gap. There has been a large increase in the gap between the LDCs and the emerging economies and the developed countries that led to the marginalization of many of the LDCs. There has also been an increase in the income gap between regions and segments of society within the developing countries. In China, the gap between the average per capita income in the coastal eastern provinces and the western provinces reached a ratio of 10:1.

- Despite the steep decline in the incidence of poverty in China, income inequality increased. Moreover, although China has gradually liberalized its markets and gradually opened its market to trade, particularly during and following its negotiations to join the WTO, it has not loosened the central controls over trade prices or removed all trade restrictions. In addition, inter-provincial commerce is also still quite restricted and local markets have been liberalized only very gradually. In general, the Chinese authorities were careful not to implement 'big bang' reforms in order to avoid the collapse of many industries and the political and economic chaos that is likely to follow.
- In India, total merchandise trade (exports plus imports) amounted to about 20% of India's gross domestic product in 1998 or 10% points less than in China and only about one-fifth of the level of export-oriented countries like Korea (IMF, 2001). Although both India and China have relaxed their control over trade and cut many tariffs, it cannot be concluded that they are open to trade. The share of their trade in the GDP is still relatively small and a large part of their imports is of raw materials (including oil) for use in their industrial production, rather than of final consumer goods.
- Both India and China continue to protect many of their local industries, partly in order to protect their workers. The demand of the developed countries that these two countries and other emerging economies will relax their controls and their restrictions on imports of industrial goods was one of the main demands in the Doha Round of the WTO.
- The World Bank report on *Globalization, Growth and Poverty* (2001a, p. 36) recognized that there need not be causal connections between opening up the economy to trade and faster growth. In most of its earlier reports and studies, the Bank took the view that trade liberalization is a key element in unleashing growth and reducing poverty.
- In an overview of the trends in poverty that is summarized in the Bank's own 'Global Poverty Monitoring database', a bank staff study (Chen and Ravallion, 2001) concluded that 'in the aggregate and for some large regions', all measures suggest that the 1990s did not see much progress against consumption poverty in the developing world. The IMF 2000 survey also concluded that 'progress in raising real incomes and alleviating poverty has been disappointingly slow in many developing countries'. These conclusions are strongly influenced by the findings on the

relatively small increase in the incomes of the poor population in India that widened the urban–rural income gap and on the rise in income inequality even within the rural sector.

The main conclusion from the failure of the SAPs is that unleashing the market forces in the structural reforms that is one of the main pillars of the Washington Consensus, is not a panacea for rapid economic growth and poverty reduction. Under the social, economic and political conditions in the developing countries, these neoliberal reforms may, in fact, achieve the opposite results.

In light of this experience, the LDCs must consider whether their alternative strategy is to emulate the policies of the EA countries and use selective interventions to promote growth of sectors in which they have a comparative advantage and whether they are likely to achieve similar results. The main concern with this strategy of targeted interventions is that, in the LDCs, the effectiveness of this strategy is highly dependent on the effectiveness of the public administrations and on the extent to which their decisions would be unbiased and guided first and foremost by economic considerations. With weak public institutions, widespread corruption, unstable central government and erratic economic policies, there are high risks in these countries that the selection of the sectors for targeting the incentives might be strongly influenced by political or other pressures or by corruption, that the industries would fail to meet the conditions that they assume when they receive the credit, and that public investments in infrastructure, in labour training, in research and in other areas will not create the conditions that are necessary for these industries to become competitive.

4.9 The neoclassical model and its constraints

Although the EA model is by no means based on the principles of the neoclassical model of the gains from trade and efficient trade policies, the SAPs are largely based on these principles. It is worthwhile therefore to highlight potential risks and pitfalls even in that seemingly straightforward set of prescriptions for the adjustment process. A textbook description of the neoclassical free-market model that shows the gains from trade liberalization is based on the assumption that the market is a more effective mechanism than any public administrators or central planning for selecting the sectors in which the country has a comparative advantage.

According to this model (Table 4.1), as the economy is opened to trade, the price differences between the domestic and the world markets would elicit a speedy and sizeable response in investments, employment and output according to the country's comparative advantage. Competing cheap imports and higher price for exports would drive sectors to specialize in that direction and thereby lead to a more efficient allocation of resources. However, in a more realistic setting, consider some of the obstacles that the country may encounter in this process:

Table 4.1. Global income inequality. (From IMF/World Development Outlook, various years.)

By countries	1980	1990	1999
Ratio of mean incomes	86.2	125.9	148.8
Ratio of median incomes	76.8	119.6	121.8
By individuals			
Ratio of mean incomes	78.9	119.7	117.7
Ratio on median incomes	69.6	121.5	100.8
By individuals (excluding China)			
Ratio of mean incomes	90.3	135.5	154.4
Ratio of median incomes	81.1	131.2	153.2

The ratio of the top 10% to the bottom 10%.

- Critics of the neoclassical approach emphasize that the world trading environment is grossly uneven and characterized by unequal exchange, transfer-pricing practices, manipulation of prices by monopolies, imperfect competition, etc. The perfect-competition neoclassical model is therefore unsuitable to describe the impact of trade liberalization in a world where the conditions assumed in this model do not exist. In our uneven and imperfect world, trade liberalization may, in fact, result in economic losses and greater hardships for the lower income groups.
- As a result of these unequal conditions, trade may actually increase rather than reduce the divergence of prices and incomes and thus increase income inequality. To remedy these effects, redistribution policies may be necessary to ensure more equitable distribution. But these policies may change the impact of trade liberalization by changing the relative factor prices and thus also the country's comparative advantage.
- Unleashing the market forces in a 'shock therapy' with sweeping and simultaneous reforms of the country's entire price and trade regimes, as the textbook model suggests, is certain to overwhelm the existing industries and lead to the collapse of many of them. These industries, their employees, the country's institutions and other segments of the economy that jointly constitute the market forces need time to make the adjustments. The free-market prices are the signals that guide the economy how to allocate its resources to the sectors in which they are most efficient by giving incentives to entrepreneurs and workers to make the transition. But this process takes time until the prices are stabilized, the information on the new prices is spread and the entrepreneurs and the workers make the adjustments.
- Since the adjustment process takes time, it gives an advantage to those entrepreneurs that identify more rapidly the potential to make profits in certain sectors. The more rapid entrepreneurs may use their advantage to take monopolistic control over these sectors and thereby distort the market price system.

- The transition of industries to the production of export goods requires investments and may thus involve considerable costs in the short run. Entrepreneurs that have easier access to cheaper credit will have a considerable advantage in that transition that they may also use to increase their monopolistic power. In fact, financial institutions have an interest to give credit only to a small number of entrepreneurs and thus to increase their profits.
- The adjustment process is not only an economic process that operates in response and according to price signals, but also a political process in which some entrepreneurs in some sectors may apply pressure on the government to intervene and either give them exclusive rights or change the price signals, at least temporarily, in their favour. The textbook description cannot 'assume away' these pressures and assume that the convergence to the new equilibrium is smooth, equal and rapid.
- During that adjustment, employees in sectors that go out of business are losing their jobs and remain unemployed. The search for new employment, the need for training to acquire the necessary skills, and possibly the need to move to other regions in search for work is also a process that takes time. During that time, these workers suffer income losses, and these losses when they are unemployed must be weighted against their future gains when the economy finally makes the transition to the new equilibrium with free trade and full employment.
- Finally, the sectors in which the economy has a comparative advantage may need a very different organization of production and supply, may require different inputs, some of them imported, and may require a different organization of their trade both within the country and for exports. In short, the new general equilibrium that market forces and economic theory promise will certainly be a Pareto improvement, but the road to that equilibrium can be quite arduous.

4.10 The impact of government-coordinated policies

The current trend of the SAPs is to strengthen lower level, local and non-governmental institutions. However, there are a number of reasons why reforms of the macroeconomic policies and government interventions would still be necessary in order to contribute to poverty reduction. One reason is that the donor countries and the international development organizations cannot replace the central government in disbursing the aid to local communities and targeting that aid to specific sectors. The other reason is that macroeconomic policies would be necessary to implement countrywide policies that promote growth and increase employment by supporting infant industries, giving incentives to labour-intensive industries, providing public goods and making public investments in road infrastructure, water supply, electricity, etc., and, most importantly, securing economic stability.

With all these measures, the central government can shorten the time it takes for the impact of growth to trickle down to the low-income groups, and by providing social safety nets, it can reduce the hardships faced by these groups. The central government may also have to intervene to prevent monopolistic controls over certain industries and secure the rules of fair trade.

For all these reasons and many more, the guiding principle of the SAPs that the government should 'get the fundamentals right and then get out' may not be the best formula for the success of the reforms. This simplifying and somewhat simplistic formula has been rejected by practically all the developed countries, even though in these countries the market has a much more dominant role. Moreover, the EA countries made a clear decision to give the central government a key role in their economic reforms, and this worked very well for most of them most of the time, despite some ups and downs, particularly during the financial crisis. In India, for example, the wide-ranging reforms that the government implemented in 1991/92 included improvements in the incentive system, reductions in the barriers to entry, relaxation of administrative constraints and regulations on the private sector, reduction of distortions caused by badly planned customs and excise duty rates, etc.; however, the government remained the key player that determined the pace of the reforms, the opening of the market for trade, the provision of public works, etc (Table 4.2).

In China, India, Malaysia and most other EA countries, and recently, in Bangladesh and Vietnam, the sharp rise in urban incomes deepened the divide between the urban and the rural populations, between the skilled and unskilled workers. This divide is potentially explosive and without government interventions to reduce the gap and raise the incomes of the poor, the market forces by themselves may detonate this time bomb, leading to open and violent conflicts that will be damaging to all. This has been the experience in many countries, particularly the LA countries, where income inequality is particularly high.

For the other developing countries, particularly for the LDCs, the choice between the two strategies, the one represented by the Washington Consensus and the other represented by the strategy of the EA countries, is a major dilemma. On the one hand, the strategy of the Washington Consensus has failed in the SAPs; on the other hand, a strategy that is based on an active role of the central government is an uncharted road. All the gains that an effective leadership of the central government can offer are conditioned on the effectiveness of the government and its institutions, the integrity of its leaders and

Table 4.2. Income and poverty indicators in selected Asian countries: 2002.

	Bangladesh	India	Nepal	China	Indonesia
Population (million, 2002)	136	1049	24	1280	212
Per capita income, 2002 (US$)	380	470	230	960	710
Per capita income (PPP$, 2002)	1770	2650	1370	4520	3070
Per capita GDP growth, 1990–2002	3.0	4.1	2.4	8.6	2.3
Gini ratio	0.32	0.33	0.37	0.45	0.34
Poverty measures: HC	36.0	34.7	37.7	16.6	7.5
Poverty gap	8.1	8.2	9.7	3.9	0.9

their commitment to the country's growth and welfare, the power and control of the rule of law, etc. Obviously, the past experience makes clear that the risks are very high and the choice must be made on a country-by-country basis.

4.11 The Poverty Reduction Strategy Programme

Towards the end of the 1990s, the leading international development organizations, the IMF, the World Bank and the Organization of Economic Co-operation and Development/Development Assistance Committee (OECD/DAC) came to realize that new and entirely different national and international policies are needed to tackle the problems of poor countries. A broad consensus emerged that these countries' external debts are unsustainable and present a major impediment to their growth and their capacity to combat poverty; the Heavily Indebted Poor Countries (HIPC) Initiative was an effort to respond to this challenge. But there was also consensus that the reforms must go far beyond debt relief (Box 4.5).

Box 4.5. Debt relief in the Heavily Indebted Poor Countries (HIPC) Initiative.

The HIPC Initiative was first launched in 1996, with the aim of ensuring that no poor country would have a debt burden that it cannot manage. It was a conditional forgiveness of multilateral loans for 41 poor countries that was designed as coordinated action by the multilateral organizations and the governments in order to reduce to sustainable levels the external debt burdens of the most heavily indebted poor countries. In September 1999, the programme was modified to provide faster, deeper and broader debt relief and to strengthen the links between debt relief, poverty reduction and social policies. In 2005, in the context of the MDG, the HIPC Initiative was supplemented by further debt relief for countries that meet the following conditions: (i) face an unsustainable debt burden, beyond traditionally available debt-relief mechanisms; (ii) establish a track record of reform and sound policies through IMF- and World Bank-supported programmes; and (iii) have developed a PRSP. Debt reduction packages were approved for 29 countries, 25 of them in Africa, providing US$35 billion (net present value terms) in debt-service relief over time.

Debt relief under the current HIPC Initiative requires countries to prepare PRSPs. The World Bank's handbook advising countries how to prepare such documents includes hundreds of pages that cover such varied topics as macroeconomics, gender, environment, water management, mining and information technology.

Expanding the HIPC Initiative to full debt forgiveness is practically inevitable. Towards that end it is necessary to ensure that the conditional debt forgiveness would indeed lead to effective reforms that improve the living conditions of the poor. The only way to actually meet *all* these conditions is to abrogate the government sovereignty and to effectively manage the country as a surrogate government. This will not be acceptable, however, to the legitimate government of that country. A possible alternative could be to use this money to directly finance local expenditures of schools, clinics, roads, etc. Even this alternative has many deficiencies: First, it will reduce the government's own obligations to finance and carry out these programmes. Second, it will replace the local authorities in determining the country's priorities by outside bureaucrats that will be in charge of implementing local programmes.

This is the background for the new approach that has evolved and is gradually implemented under the PRSP. This was clearly an evolutionary process in which all the parties – the governments of the countries and the international development organizations – were and still are involved. Unfortunately, this was a process of learning-by-doing, in which the learning process exacted a heavy toll from the countries and the local people who were part of this process (Hewitt and Gilson, 2003; IMF/World Bank, 2005).

In September 1999, the IMF and the World Bank endorsed the new approach to strengthen the focus on poverty reduction by replacing the concessional lending facility – formerly the ESAF – with the Poverty Reduction and Growth Facility. Under the new approach, each eligible country will prepare a Poverty Reduction Strategy Paper (PRSP) that will outline the country's anti-poverty strategy over the medium and long term. The PRSP will be drawn up by the government after broad-based consultations with representatives of civil society and development partners with assistance from World Bank and IMF staff. Long-term poverty reduction goals will be specified in terms of annual targets that include primary school enrollment, immunization rates, etc. The PRSP will provide the basis for debt relief under the HIPC Initiative and all World Bank/IMF concessional lending operations. The PRSP will diagnose poverty in the country and quantify the resources needed for various poverty reduction programmes.

Although the new policies that are expected to emerge from the PRSP process differ from those of the past adjustment programmes by establishing close cooperation with the governments of the affected countries and giving more attention to social outcomes, it is not at all clear how the traditional macroeconomic policies will change as a result of implementing more pro-poor public expenditures and how the structural reforms will be adjusted to the vastly different conditions.

The following five principles guided the PRSP:

- Country-driven: The initiative to formulate a PRSP should originate from the countries themselves with active participation of Civil Society Organizations (CSOs) and the private sector.
- Results-oriented: The PRSP should focus on the effects of the policies rather than on the policies as such.
- Comprehensive poverty: Poverty is multidimensional and poverty reduction cannot focus on increasing economic growth alone.
- Partnership-oriented: Partnership between bilateral donors and multilateral institutions, the national governments, domestic and international civil societies in the formulation and implementation of the programme and policies.
- Long-term consistency: The PRSP must be consistent over time to achieve a sustainable reduction in poverty.

Critics of the PRSP processes argue that the CSO participation is mostly formal, while the real influence over the process rests in the hands of just a few actors and is still guided by the IMF and the World Bank. In practice, the CSOs will have a very limited say on the agenda and the details of the PRSPs, not least because the documents discussed at the meetings are made available only at the last moment before the votes are cast. These documents are frequently written in English and not translated to the national language, and there is no time to prepare positions or to consult with wider constituencies.

CSOs often feel like hostages, token representatives in this process and they have almost no influence on the PRSPs that, in practice, contain some of the customary traits of the SAPs, starting with the emphasis on growth at the expense of redistribution, the requirement that public goods and services like the supply of electricity and drinking water are privatized, and a negative opinion on regulating trade flows.

The conclusion of the critics is therefore that since the PRSPs replicate core components of the traditional SAPs, no real local ownership can exist. Since the conditionalities of the IMF are more or less identical to those of the SAPs, the conclusion so far has been that key concepts like ownership, participation and transparency do not sit well with the PRSPs. An evaluation of the trade component of 17 PRSPs reached essentially the same conclusion: Only in a few cases there were significant deviations from the customary SAPs.

5 Why Have the Structural Adjustments Failed in SSA?

Nearly all SSA countries were affected by the debt crisis of the early 1980s that led to the SAPs which influenced their economic development since then. Their efforts during the 1960s and the early 1970s to realize very ambitious economic programmes and build an industrial base failed partly due to a series of military coups that toppled the central governments in many countries, and leadership crises that aggravated the economic decline and affected many other countries through what has become known as the 'neighbourhood impact' in which an economic or political crisis in one country spills over to its neighbouring countries. The result was decline and economic stagnation in the entire African subcontinent, with the exception of South Africa, that lasted throughout the 1980s and the 1990s. The rise in poverty and the fall in per capita income (Table 4.3) led to the fall of most SSA countries to the bottom of the world's income distribution and they became

Table 4.3. Sub-Saharan Africa: selected macroeconomic indicators, 1965–1997 (ratios, averages, %). (From African Development Bank, various years.)

Indicator	1965–1974	1975–1984	1985–1989	1990–1997
GDP growth	5.8	2.2	2.6	1.9
Per capita growth	3.1	−0.4	−0.1	−0.8
Agricultural growth	3.9	0.6	3.4	4.0
Industry growth	–	2.5	1.8	1.1
Investment/GDP	20.2	23.0	17.5	18.1
Inflation	5.8	13.8	15.8	16.3

the world's LDCs. Foreign investments dried out and private local investments were very minute and concentrated in small-scale manufacturing.

The prolonged stagnation of these African countries can only partly be attributed to the debt crisis, however, even though their large debts saddled their economies with correspondingly large payments for nearly three decades. The SAPs that were aimed to revive their economies and reduce the debt burden with credit at concessional interest rates also had very limited success and the majority of the population remained poor.

Per capita income continued to decline during the 1990s, even though the world economy was booming and the EA countries were flourishing. Even most LA countries managed to overcome their debt crises and the continent's largest economies, Brazil, Argentina and Chile, entered a period of rapid growth. The exports of the SSA countries, in contrast, continued to depend mostly on natural resources of oil and minerals, and traditional agricultural exports, primarily coffee, cocoa and cotton.

The main reason for the continued physical and financial ruin during these years was the seemingly incessant series of conflicts and civil wars that stifled the entire system of government with internal rivalries between tribes and ethnic groups. Box 4.6 summarizes some news reports of the past year on the civil war in Somalia, which is similar to conflicts in other African countries and illustrates the huge damage that these conflicts cause to the entire economy.

Box 4.6. The fragmentation of central controls: the Somali experience. (From: *IHT*; *NYT*; *The Economist* (2006)

– Somalia is an archetype of many Africa's ills. A newly elected government was established not so long ago, after the UN invested millions of dollars into propping it up and American officials provided all political, financial and even military support they could for the sake of regional stability. American forces landed on Somali soil to hunt down the opposition leaders who still held a firm grip on large areas. But whether Somalia manages to recover or explodes into bloodshed again depends not on American troops, foreign peacekeepers, investment or aid. It depends on clans.

– 'Clanism, said Ali, once a contender for president, is our national cancer.' It was clan animosities that tore down Somalia's last government in 1991, humiliated American troops in 1993, bringing a troubled aid mission to a hasty end, and resulting in clan warfare that has consumed thousands of lives in the capital city Mogadishu.

– The new government, which took the capital for the first time last month, is trying to address the clan problem head on. It is trying a mathematical formula based on rough estimates of the population to allocate parliamentary seats and ministerial posts on a clan basis. But this approach is hardly original and it is the 14th attempt in the last two decades to form a clan-based government; all the previous attempts failed and ended in more violence.

– The Green Line in Mogadishu is a monument to fratricide. It is a dividing line between two clans, which are actually parts of the same family.

'We had a stable government in this country in the past and this was our only hope. This government is a government of clans and it is only going to bring more war by creating clan competition.'

The infrastructure of African countries was in shambles and practically no public investments were made in road infrastructure and water systems. Most villages still have no road that connects them to the urban centres and they are unable to develop any trade or diversify production. Foreign investors were deterred by the low profitability and high risks of investments in the local industries despite the low wages of local workers, and the investments they made concentrated in oil production and exploration.

Nearly all countries were involved at one time or another in these wars, ranging from Algeria and Angola to Zaire and Zimbabwe. Some military conflicts were about natural resources, others were ethnic, sectarian or tribal rivalries. The most tragic conflicts were between the Tutsis and the Hutus in Rwanda and Burundi, the massacre in Biafra, and more recently the massacre in Darfur. Many of these wars reflected the heritage of unsettled borders that had been established by the colonial regimes, arbitrarily dividing ethnic groups, religions and languages. These animosities have not abated over time, and new wars continue to erupt in one region or another, bringing new miseries and economic chaos. For more than three decades, these conflicts have exacted a terrible human toll, deterred private investment, destroyed the local infrastructure, and hindered development.

The collapse of the system of governance, the lack of law and order, property protection and contract enforcement, and the corrupt court system are the Achilles heel of many African countries, and these problems were also a reflection of the collapse of the central governments. In recent years, donor countries and the international development organizations concentrated a larger share of their aid to Africa on the improvement of local institutions and their systems of governance. The SAPs also concentrated more on efforts to improve governance.

However, high transaction costs, poor infrastructure, unreliable energy supplies, and undeveloped financial systems still prevent the development of local industries despite the abundance of natural resources and cheap labour. Thus, for example, the cost of registering a business in Africa is the highest in the world, and surveys among businesses in various African countries reveal that bribes take away most of their profits (World Bank, 2004).

Against this background, the focus of the reforms on economic measures alone in the early stages of the SAP was not very productive; indeed, this was the main reason why so many programmes failed. Trade liberalization aimed at promoting exports by opening up the economy to trade cannot contribute to the development of export industries since these industries must grapple with many administrative and bureaucratic hurdles that make their exports unprofitable. As a result, by opening up their economies, the African countries only worsened their trade balance and ruined many local industries that relied on the local markets and could not compete with cheap imports. Even the textile industries that were developed during the 1990s thanks to the preferential access given to African exports by the EU and the USA are now at risk when the restrictions on entry to the EU market will be removed and the Indian and Chinese textile industries will be able to increase their exports.

Since the early 21st century and for several years now, the performance of quite a few African countries has improved and the political landscape has become more stable. The rates of growth in quite a few African countries have increased with the rise in the world price of their main export products, primarily oil and minerals. Duly elected presidents came to power in Senegal, Ghana, Kenya, and even in the Democratic Republic of Congo, raising hopes that democracy will be strengthened in the entire region. Ethiopia and Eritrea signed a peace accord, and the long and devastating conflicts in Angola and Sierra Leone seem to come to an end. The calming influence of South Africa on the entire region and its involvement in the region restrained many armed conflicts between countries. This experience with relative peace and stability is far too short, however, and the long experience of the past suggests caution in drawing conclusions about the future.

Indeed, in other parts of Africa, the political situation remains unsettled and violent conflicts continue. While in Côte d'Ivoire and in Liberia fighting has stopped and a fragile cease fire is maintained by an international peace-keeping force, civil wars or border conflicts continue in other parts. So far, less than a handful of African countries experienced peace, stability and a sustained economic growth in the past two decades. Proposals for emergency economic measures abound, ranging from a monetary union and deeper regional cooperation projects to a new round of 'conditionalities' that put greater emphasis on governance, controls over the endemic corruption and greater emphasis on strengthening non-governmental organizations and local authorities. Their implementation would require not only a new round of SAPs, but also a wider cooperation and more generous support by the international development organizations and the donor countries. Indeed, the donor countries should have strong interests in supporting these programmes, since they may otherwise be flooded with a wave of migrants who will spare no effort and avoid no risk in their desperate attempts to leave their countries and seek a better future.

5.1 Could the SAPs in Africa be better structured?

Many World Bank and IMF reports and studies (see, e.g. the World Bank report of 2002, pp. 110–111) argued that, in the final analysis, many of the SAPs were quite successful and did make a positive contribution to growth in many of the countries where they were implemented. An equally large number of studies claimed that their actual performance was mixed, at best. The study of Barro and Lee (2002) is just one example of a study that found a significantly negative effect of IMF lending on growth. Easterly found that 'none of the top 20 recipients of repeated adjustment lending over 1980–99 were able to achieve reasonable growth and contain all policy distortions. About half of the adjustment loan recipients show severe macroeconomic distortions regardless of cumulative adjustment loans' (Easterly, 2005, p. 1).

The statistical analysis of the performance of the SAPs is bound to be affected by a selection bias, however, since the countries that received the adjustment loans and needed the SAP were already in dire condition and their economies were on the verge of collapse after a long period of stagnation. These countries were therefore less likely from the outset to make effective use of the loans or exhibit a positive response in a short period of time. Judging the performance of these programmes by comparing the *ex post facto* results to the *ex ante* objective is therefore prima facie wrong.

In addition, the donors themselves have not been sufficiently selective, prudent and demanding in their lending, and this was actually acknowledged by the World Bank (1998, p. 48). The 26 structural adjustment loans that were given to Cote d'Ivoire to support the government's programme of structural adjustment are perhaps an extreme, but not entirely an atypical example. The goal of the programmes in Cote d'Ivoire was to improve public savings and efficiency in the use of public resources, to restructure the agricultural sector through more careful planning, and to design an extensive investment programme that would raise the sector's production, employment and productivity. In practice, the adjustment loans had very limited impact on any of these government policies.

A World Bank study of 10 African countries shows that they all received large amounts of aid and loans that had quite similar conditions, but ended up with vastly different policies (Devarajan *et al.*, 2001). In the debate over these findings, the effects of the SAPs that were evaluated in this and other studies were rightly brought as another reason to contest the 'one-size-fits-all' policies that arguably guided the reforms. Moreover, in later programmes, considerable adjustments were made in order to take into account the countries' specific socio-economic and political conditions and constraints.

The criticism of the structural adjustment loans must also take into account the fact that these loans had to be given to countries that needed them to avert their total collapse. It was recognized both by the donor countries and by the international development organizations that a collapse of these countries' economies would be highly damaging first and foremost to their poor, and could bring many of them to the brink of starvation. In other words, there was no way for the international development organizations not to give these loans. The conditions that were attached to the loans were perceived as the only sensible way of using them as leverage in order to pressure the recipient governments to change their manifestly poor and self-damaging policies and implement new programmes. With these programmes, the international organizations hoped to bring about a material improvement in the living conditions of the populations. A more blunt way of stating this aim is to say that the conditions attached to the SAPs were the only way to save these countries from their own governments.

The question whether the conditions were appropriate and whether the SAPs were successful must therefore be seen in a different light. An evaluation of the programmes by estimating their impact on the growth and poverty reduction in the recipient countries is bound to be affected not only by the programmes, but also by all the pre-existing conditions. Instead, an evaluation of the extent to which the programmes actually changed these countries' performance must be made by comparing the actual outcomes to *what would have happened* without these interventions. This is the method that Easterly (2005) used in his very methodical analysis.

> His main conclusion was that 'structural adjustment did not succeed in adjusting macroeconomic policy and growth outcomes very much . . . Structural adjustment loans were repeated many times to the same country, which itself is suggestive of limited effect of the earlier adjustment loans. There were some successes, but also some big disasters. The prevalence of extreme macroeconomic distortions did not diminish as adjustment lending accumulated, and there is no evidence that per capita growth improved [. . .] there is not much evidence that structural adjustment lending generated either adjustment or growth' (p. 20).

Table 4.4 provides a succinct summary of his conclusions.

However, this counterfactual analysis is also problematic, even after using various statistical techniques to overcome the selection bias. It is obvious that, without this intervention, many of these countries, particularly the most extreme cases that failed time and again, would have faced a catastrophe, and this had to be prevented. Easterly's conclusion that 'putting external conditions on governments' behavior through structural adjustment loans has not proven to be very effective in achieving widespread policy improvements or in raising growth potential' (p. 20), therefore ignores the fact that these loans *had* to be given to these countries and their immediate reason was *not* to improve government policies, but to prevent a catastrophe.

The counterfactual analysis of 'what would have happened *without* these interventions' was not an option that the international development organizations and the donor countries could have accepted. Perhaps the more meaningful counterfactual analysis in this case is to evaluate whether the SAP as actually implemented was the most effective way of making the necessary reforms or whether there could have been another way of giving the loans but implementing other programmes that would have achieved better

Table 4.4. Per capita growth rates in African countries following intensive structural adjustment programmes (ranked from worst to best performers). (From Easterly, 2005.)

Niger	−2.3
Zambia	−2.1
Madagaskar	−1.8
Togo	−1.6
Cote d' Ivoire	−1.4
Malawi	−0.2
Mali	−0.1
Mauritania	0.1
Senegal	0.1
Kenya	0.1
Ghana	1.2
Uganda	2.3

results. In other words, the relevant counterfactual analysis is the comparison of 'what would have happened *without* these interventions' with 'what could have happened *if better* interventions had been implemented'.

In a multi-country African research project on the SAPs, two dozen economists carried out some 30 separate studies.[3] Most of these studies concluded that the SAPs were 'deeply flawed'. Their main criticism was directed against the World Bank's 'deep-rooted anti-industrial-policy position'. Economic liberalization and the removal of state protection in Africa came too fast to allow the local industry to adapt. Unable to cope with the stiff global competition, many industries were forced to scale back or shut down and the exposure to world competition led to a steady deceleration of industrial growth.

At the same time, the authors of this research assert, the SAPs encouraged the export of raw mineral and agricultural products, thus basing Africa's trade even more solidly on primary commodities. In their view, the African countries should have adopted export-oriented policies that promote the exports of manufactures in exports, but for this to happen, African countries should have adopted explicit industrialization policies, along the lines of the successful policies of the EA countries. These programmes must therefore be managed with appropriate state-interventionist policies to resolve pervasive market failures and it should include protection for infant industries to gradually allow them to mature and enhance their international competitiveness.

5.2 Later adjustments in the SAPs: better governance and fighting corruption

The focus of the SAPs changed in the late 1990s from strict economic reforms to measures aimed at improving governance. The main motivation for this change was the realization that a successful implementation of economic reforms depends first and foremost on the efficiency of a country's institutions and the quality of governance. One of the lessons from the experience of the SAPs was that weak institutions in the majority of the African countries were a major obstacle for a successful implementation of the adjustment programmes. In the past, the reforms, guided by the principles of the Washington Consensus, were too narrowly focused on macroeconomic and trade policies and the new agenda was designed to stress effective governance, the fight against corruption, greater transparency and better social safety nets.

The World Bank and the IMF worked with the governments in order to implement codes of 'best practices' in a variety of technical and administrative areas such as banking regulations and supervision, transparency in institutions activities and accounting, etc. In all the African countries, the malfunctioning institutions and endemic corruption were obstacles that heavily taxed the entire economy by raising the costs of public services and private businesses. The efficiency costs of corruption have been far higher

[3] Their findings have been summarized in the book edited by Thandika Mkandawire and Charles C. Soludo (1999) *Our Continent, Our Future: African Perspectives on Structural Adjustment*. Africa World Press, Trenton, New Jersey.

than their direct costs, and in many countries they were a major cause of sluggish growth.

Endemic corruption penetrated all levels of the local administrations in both public and private institutions: At low levels, employees are paid far too little and their mere survival therefore depends on small bribes; at higher levels, ministers often sign false contracts with suppliers, politicians take cuts from the deals they approve, including the disbursement of aid, and local and foreign business executives pay bribes to improve their deals to promote their businesses. Foreign investors, especially in Africa's oil industry, have long paid massive bribes to politicians to win contracts. In too many countries, developed and developing, too many actors have strong interests and incentives to keep the corruption practices going, and the politicians are under enormous pressure to ensure that financial rewards are passed on to their supporters.

There is no consensus on how best to reduce corruption. Becker and Stigler (1974) suggested a combination of monitoring and punishments. In practice, however, the individuals in charge of monitoring are themselves often corrupt and monitoring thus results in more elaborate transfers between officials rather than in effective control over corruption. An alternative approach is to increase the community participation in local-level monitoring. This approach is now regarded as the most promising not only to reduce corruption, but also to improve public service deliveries. The entire WDR of 2004 was devoted to the goal of 'putting poor people at the center of service provision: enabling them to monitor and discipline service providers, amplifying their voice in policymaking, and strengthening the incentives for service providers to serve the poor'. This new emphasis is clearly necessary in view of the high correlation between the level of countries' corruption and the depth of their poverty that the WDR has demonstrated.

The raison d'être of the grass-roots approach is that since community members are those who benefit from successful programmes and suffer the most from corruption, they would have stronger incentives and better information to monitor their officials and prevent corruption (Stiglitz, 2002). As Friedman (2002) noted, Stiglitz expressed great hopes that this approach would be effective, but we must also keep in mind that even grass-roots monitoring is controlled by local officials that depend, in turn, on officials above them in the administration, and the channels through which the aid money is transferred from the donors to the local communities must go from the top, the central authorities that receive the aid, down to the local levels through all these channels, and all these transactions, at any step of the way, are likely to be greased with bribes.

6 Should There Be Another Round of Structural Adjustments?

The debate on the SAPs has always been very contentious and the controversy whether or not they were successful has not abated even after the IMF–World Bank took a new and very different approach with the Poverty Reduction Strategy Programme. The discussion in this chapter on the performance of past SAPs and the changes that were introduced to the programmes over the years was in large part motivated by concerns that the current trends in the global economy and the increasing marginalization of the LDCs in general and of the African countries in particular would require these countries to make very comprehensive structural changes in their economies and perhaps undergo another round of structural adjustments.

The difficulties of the African countries to integrate into the global economy and the continued decline in their share in world trade inhibit their capacity to benefit from the rapid growth in the world's economy as an effect of globalization. In recent years, quite a few African countries gained from the rise in the world prices of oil and minerals and managed to increase their exports. But the local value-added of these exports is very minute and only a small share of their higher export earnings is likely to trickle down to the poor. The increase in local employment has been very small and poverty continues to rise.

That rise in poverty and unemployment raises also worries that the majority of the population in these countries will continue to live in dire conditions in the coming years and will not be able to find employment in their own countries that will enable them to improve their standard of living. The gap between their standards of living and the standards in most other countries will therefore continue to rise and their only way to break this vicious circle is with massive external assistance that will enable them to implement major structural adjustments in their economies.

Many of the causes for the stagnation and marginalization of the LDCs are external, unrelated to their own policies and beyond their control or capacity to change. Some of the causes are related to the policies of the developed countries, which, by refusing to reach an agricultural trade agreement at the Doha Round, heightened the wall that blocks exports of the LDCs' agricultural products. Some of the causes for this are related to the high quality and food safety standards that the WTO and the European countries determined for imported food products. The small African farmers will not be able to meet these standards without additional assistance of the local agricultural research and extension centres, and the standards imposed by the developed countries are therefore likely to further reduce the capacity of the LDCs to export their products.

For the African countries that do not have large resources of oil or minerals, and for the vast majority of the African population, the future holds very little promise to get out of their cycle of poverty and enter the new world of the 21st century. Only comprehensive and very thorough structural adjustments will enable them to create better employment opportunities, raise their incomes and reduce poverty.

Countries differed in their capacity to make the adjustments required by the SAPs because of wide differences in their social, economic and political conditions that made it difficult for them to implement programmes according to rather uniform and rigid guidelines. These programmes were designed according to a set of 'neoliberal' principles that did not take into account the capacity of countries to adopt these principles.

This is, perhaps, the most controversial and most widely criticized aspect of the reforms, and many accusations were levelled against the 'planners'[4] who came with their pre-planned set of policies and insisted on implementing them regardless of the countries' specific conditions, character, institutions and constraints. Easterly, one of the leading critics of the SAPs, was particularly opposed to the grand and complex programmes, which did not take into account the capacity of local institutions to implement them and the need to design programmes that will be accepted by the local population. Easterly was particularly critical of the 'shock therapy' to jump-start a market economy by completely dismantling all the existing institutions and imposing on these countries a system that was imported from the West and from classical textbooks.

Indeed, one of the most important lessons from the failure of many SAPs is the high likelihood that establishing difficult conditions without mobilizing a domestic constituency will again risk failure. The result was a huge waste of resources by the countries themselves and by the international development organizations, because the programmes were not adjusted to the country's existing institutions and social organizations.

Although I agree with the criticism of the SAPs, especially at their early stages, it must also be acknowledged that the reform programmes themselves went through many adjustments over time. The main dilemma and the main conflicts in the design of these programmes were about the roles of and the division of responsibility between the outside 'planners', the central government and the local institutions. In the majority of African countries, not only the outside 'planners' and their plans were not supported by the local population, but also the central government and its institutions were not trusted and accepted.

Instead of following the outside plan, the government pursued its own agenda according to its own priorities and used its own highly corrupt institutions to implement the programmes. The outside 'planner' could not realize the sweeping reforms of the SAPs unless the central government and the institutions that are in charge of carrying them out have the trust and support of the population.

The outside 'searcher' (to use the somewhat romantic but a bit Orwellian term of Easterly) seeks to gain this trust by adjusting the programme according to the needs and constraints of the local population. However, even the programme of the outside 'searcher' must be implemented by the central government and through its institutions, and the local population did not trust and accept them. The outside 'searcher' must therefore include the local institutions and must make a decision in his/her plan on the division of the responsibility between these three bodies. I did not find in Easterly's book a satisfactory answer to this problem.

[4] This terminology is taken from the very controversial and confrontational book of William Easterly entitled: *The White Man's Burden: Why the West's Efforts to Aid the Rest Have Done So Much Ill and So Little Good* (2006).

In his review of Easterly's book, Amartya Sen (2006) found many reasons to agree that the World Bank and the IMF 'often imposed terrible policies on the developing countries and that Easterly is right to criticize their sometimes insensitive agendas, overly grand designs, and lack of engagement'. However, he found that Easterly's sometimes extreme criticism 'includes a strong element of caricature', and much too 'simple dismissal'.

There is another reason to reject this 'simple dismissal': The World Bank and the IMF have not been as narrow-minded 'planners' as Easterly repeatedly claims. Throughout the more than two decades in which the programmes were implemented, they went through major changes, and the World Bank and the IMF were in constant search for ways to improve and adjust these programmes so that they better suit the social and economic conditions and constraints of the countries. The introduction of the PRGF in November 1999 was not merely window dressing. In my view, it was a serious effort to find a participatory, country-led mechanism that would focus on the countries' poverty reduction efforts. Obviously, the process of making the changes in the structure of the reforms is slow and often internally arduous, and the changes therefore came with long delays. But there was a search and changes have been made.

References

African Development Report, various years, Published for the African Development Bank by Oxford University Press.

Amsden, A.M. (1991) Diffusion of development: The late-industrializing model and greater east Asia. *The American Economic Review*. 81(2), 282–286. Papers and proceedings, May.

Barro, R.J. and Lee, J.W. (2002) *IMF Programs: Who is Chosen and What are the Effects?*, NBER Working Papers 8951, National Bureau of Economic Research.

Becker, G.S. and Stigler, G.J. (1974) Law enforcement, malfeasance, and compensation of enforces. *Journal of Legal Studies* 3(1).

Bhagwati, J.N. (1982) Shifting comparative advantage, protectionist demands and political response. *Import Competition and Response*. University of Chicago Press, Chicago, 153–184.

Bhagwati, J.N. (1987) *International Trade Selected Reading*, MIT Press.

Chenery, H. (1986) *Industrialization and Growth: a Comparitive Study*. Oxford University Press, World Bank, New York.

Chen, S. and Ravallion, M. (2001) How did the world's poorest fare in the 1990s? *Review of Income and Wealth* 47, 283–300.

Deininger, K. and Squire, L. (1996) A new data set measuring income inequality. *World Bank Economic Review*, 565–591.

Devarajan, Dollar, S.D. and Holmgren, T. (eds.) (2001) *Aid and Reform in Africa*. World Bank.

Dollar, D. (1992) Outward – oriented developing countries really do grow more rapidly: evidence from 95 LCDs, 1976–85. *Economic Development and Cultural Change* April, 40(3), 523–44.

Dollar, D. and Kraay, A. (2002a) Spreading the wealth. *Foreign Affairs*, February.

Dollar, D. and Kraay, A. (2002b) Growth is good for the poor. *Journal of Economic Growth* 7(3), 195–225.

Easterly, W., Levine, R. and Roodman, D. (1994) Aid, policies and growth: comment. *The American Economic Review* 94(3), 774–780.

Easterly, W. and Kraay, A. (1999) Small States, Small Problems? Policy Research Working Paper No. 2139. World Bank, Washington, DC.

Easterly, W. (2005) *Reliving the 50s: the Big Push, Poverty Traps, and Take offs in Economic Development* DRI Working Paper No.15, June 2005.

Easterly, W. (2006) *The White Man's Burden! Why the West's Efforts to Aid the Rest have done so much Ill and so little good*. Penguin Press.

Edwards, S. (1998) Openness, productivity and growth: What do we really know? *The Economic Journal*, March.

Friedman, B. (2002) Globalization: Stiglitz's Case. *New York Review of Books* 49(August), 48–53.

Hewitt, A. and Gillson, I. (2003) *A Review of the Trade and Poverty Content in PRSPs and Loan-Related Documents*. ODI, London.

IMF (International Monetary Fund) (2001) *World Economic Outlook* May. IMF, Washington, DC.

IMF/World Bank (2005) *Concept Note. Joint World Bank and IMF Report on PRSPs – 2005 PRS Review*. Washington, DC.

International Herald Tribune (IHT), *World News Report*, various issues.

Kaufmann, D., Kraay, A. and Mastruzzi, M. (2004) Governance matters III: governance indicators for 1996, 1998, 2000, and 2002. *World Bank Economic Review* 18, 253–287.

Klasen, S. (2003) In search of the holy grail: How to achieve pro-poor growth. In: Tungodden, B., Stern, N. and Kolstad, I. (eds), *Toward Pro-poor Policies: Aid, Institutions, and Globalization*, Proceedings from the Annual World Bank conference on developed Economics Europe conference 2003. Washington, DC.

Lee, You-il (1997) *Strategic Market Entry Perspectives in South Korea*. British academy of management, London, UK.

Little, I., Scitorsky, T. and Scott, M. (1970) *Industry and Trade in Some Developing Countries*. Oxford University Press for the O.E.C.D. Development center.

Luxembourg Income Study (2001) LIS Key Figures – Income Inequality Measures. Available at: http://lisweb.ceps.lu/keyfigures/ineqtable.htm

New York Times (NYT) *World News Reports*, various issues.

Radelet, S. (1999) Orderly workouts for cross-border private debt. *In: International Monetary and Financial Issues for the 1990s Volume XI*. UNCTAO publication, UNCTAD/GDS/MDPB/6, New York.

Rodriguez, F. and Rodrik, D. (2000) Trade policy and economic growth: a skeptic's guide. In: Bernanke, B. and Rogoff, K. (eds), *NBER Macroeconomics Annual 2000*. MIT Press, Cambridge, Massachusetts.

Rodrik, D. (1992) The limits of trade policy reform in developing countries. *The Journal of Economic Perspectives* 6(1), 87–105.

Rodrik, D. (1995) *Handbook of International Economics*. Elsevier.

Rodrik, D. (2002) *Feasible Globalizations*. Harvard University, Cambridge, Massachusetts.

Rodrik, D. and Rodriguez, F. (1999) *Trade Policy and Growth: A Skeptics Guide to Cross National Evidence*. NBER working paper, April.

Sachs, J.D. and Warner, A. (1995) Economic reform and the process of global integration. *Brooking Papers on Economic Activity* 1, 1–118.

Sachs, J.D., McArthur, J.W., Schmidt-Traub, G., Kruk, M., Bahadur, C., Faye, M. and McCord, G. (2004) *Ending Africa's Poverty Trap,* Brooking Papers of Economic Activity.

Sala-i-Martin, X. (2006) The world distribution of income: falling poverty and covergence period. *Quarterly Journal of Economics*, April.

SAPRIN (2002) *The Policy Roots of Economic Crisis and Poverty: A Multi-Country Participatory Assessment of Structural Adjustment*. Structural Assessment Participatory Review International Network, Washington, DC.

Sen, A. (2006) The Man without a plan. *Foreign Affairs,* March/April, 2006.

Stiglitz, J. (2002) *Globalization and its Discontens*. Penguin, London.

The Economist, Country reports, various issues.

UNCTAD (1997) *Trade and Development Report*. UNCTAD, Geneva, Switzerland.

UNCTAD (2002) *Economic Development in Africa. Debt Sustainability: Oasis or Mirage?* UNCTAD, Geneva.

UNDP (2002) *Human Development Report*. Oxford University Press.

Wade, R. (1990) *Governing the Market: Economic Theory and the Role of Government in East Asian Industrialization*. Princeton University Press, Princeton, NJ.

World Bank (1993) *The East Asian Miracle: Economic Growth and Public Policy*. Oxford University Press, New York.

World Bank (1998) *World Development Report: Knowledge and Development*, World Bank.

World Bank (2000) *World Development Report*. World Bank.

World Bank (2001a) *Globalization, Growth and Poverty: Facts, Fears and an Agenda for Action*. World Bank, Washington, DC.

World Bank (2001b) Global Poverty Monitoring Database. Available at: http://www.worldbank.org/research/povmonitor/

World Bank (2002) *Global Economic Prospects and the Developing Countries*, Washington, DC.

World Bank (2004) *World Development Report*. World Bank.

5 Trade and Growth Policies for Poverty Reduction

The Lessons of the 'East Asian Miracle' for the LDCs

1 Introduction

In the wake of the Communist Revolution, China went through a long period of extreme poverty and starvation during the 'Great Leap Forward' and the 'Cultural Revolution'; likewise, the economies of most other EA developing countries (with the exception of South Korea, Taiwan, Singapore and Hong Kong – the 'Asian tigers') remained poor, stagnant and secluded from the outside world. In these years, more than two-thirds of the world's poor concentrated in the EA countries, while only 15% were in the SSA and the SA countries.

Today, the majority of the LDCs are in SSA, but only three decades ago, the resource rich African countries seemed to be well positioned to develop flourishing economies by taking advantage of their relative proximity to Europe, their ties from colonial times, their fertile lands and unique climatic conditions, their huge resources of raw materials and their low labour costs. The extreme poverty and little hope to escape it, which are nowadays so characteristic of SSA, was then found in EA, but within the space of just one generation, most of the EA countries developed into 'tigers'. Thus, in 1973, average income per capita in China was 11% lower than in Africa (including North Africa); in 2000, income per capita in China was 380% higher than in Africa. For the East Asian LDCs of the 1960s, globalization clearly yielded huge benefits that even the most optimistic development economists could not have predicted.

In order to give the SSA countries another chance to build an industrial base, the MFA of 1995 envisaged them as the main beneficiaries. The agreement was aimed at transferring a larger share of the production of labour-intensive textile products from the developed to the developing countries; however, when it was implemented 10 years later, it was the Asian countries that attracted most of the textile and apparel enterprises, and the African countries had to rely on preferential access of their products to the EU and the US markets.

Although the majority of the world's poor still concentrate in the Asian countries, primarily in India, the centre of gravity of the world's poverty problem has now shifted to SSA. The World Bank's *World Development Report* of 2000/2001 (World Bank, 2000) highlighted the contrast between the steep reduction of the incidence of poverty in the EA countries during the 1980s and 1990s, and the stagnation in SSA, where the poverty rate remained nearly unchanged. These trends are summarized in Table 5.1.

The World Bank report underlined the growing concentration of poverty in rural areas in all countries, the rising urban–rural gap and the impoverishment of many resource-poor and remote areas. These developments accelerated during the 1990s and the early 21st century with the rapid industrialization in China and India, and they brought with them a steep rise in income inequality. In China, incomes in the poor and more remote provinces have increased at a rather low rate of 2–3% per annum, whereas in the more affluent provinces along the coast, where the country's industries concentrate, annual per capita income has increased at a rate of nearly 10%. The rapid integration of the EA developing countries into the world economy proved to be a powerful engine for promoting their economic growth and reducing their poverty.

The speedy global growth during these years at an average annual rate of 3% was driven in part by the unparalleled pace of technological innovations that raised labour productivity and in part by the fast growth in international trade at an average annual rate of 6%. That growth was accelerated by concerted efforts to reduce trade barriers with the GATT that, in a series

Table 5.1. Trends in poverty in the main regions, 1990–2001.[a] (From World Bank, 2001a,b.)

	Millions of people		Share of total population (%)		Share of poor people living in rural areas[b] (%)	Rural population as share of total (%)
Region	1990	2001	1990	2001	2001[a]	2001
East Asia	472	271	30	15	80	63
Eastern Europe and Central Asia	2	17	1	4	53	37
Latin America and Caribbean	49	50	11	10	42	24
Middle East and North Africa	6	7	2	2	63	42
South Asia	462	431	41	31	77	72
Sub-Saharan Africa	227	313	45	46	73	67

[a]Poverty lines of US$1.08 per person per day set in 1993 US dollar.
[b]The poverty line is adjusted for purchasing power parity.

of negotiations and progressively more advanced and wider agreements since 1948, managed to bring about a steep reduction in tariffs and other barriers for a large share of world trade.

Although quite a few developing countries took early on full advantage of the opportunities to boost their economic development by expanding their trade and integrating into the global economy, many others opted in the 1960s–1980s, until the Uruguay Round (1986–1994), to restrict their trade in order to first build their own local industries. The evidence that accumulated since then makes clear that the countries that adopted an 'outward-looking' trade policy – most notably the EA countries – were able to accelerate their growth, raise their living standards and reduce their poverty by successfully integrating into the global economy and benefiting from the global growth and from the advanced technologies that they were able to import. In contrast, most countries that adopted an 'inward-looking' trade policy remained stagnant and their industries remained inefficient due to the small scale of their production, the absence of competition, and their inability to access new technologies. As a result, large segments of their population remained poor.

The economic success of the EA countries is largely due to the fact that these countries opted for 'outward-looking' trade policies and their achievements have provided the most notable example for a successful application of this approach. An admiring World Bank study entitled *The East Asian Miracle: Economic Growth and Public Policies* (1993) praised their pro-trade outward-looking policies that enabled them to have these stunning achievements. This and several other studies on the EA experience estimated the gains that the developing countries could have by eliminating the barriers to merchandise trade; according to these estimates, the gains would have been more than twice the amount of aid that these countries received. The EA countries are the leading examples of the gains from globalization, and the climax of their success was the dramatic decline in extreme poverty in China from well over 50% in 1980 to well under 10% just two decades later.

This chapter examines the lessons that can be drawn from the spectacular achievements of the EA countries, the stages they had to go through in their transition, and the policies they implemented in order to develop from being the world's poorest countries in the early 1970s to become the world's emerging and most rapidly growing economies in the 1990s and a dominating economic power in the first decade of the 21st century.

For today's LDCs, however, these lessons must be considered with some caution, because they are not necessarily applicable to today's structure of world production and trade. During the last decade, there were massive changes in the world economy, in the structure of global production and trade and in the institutions that govern world trade. 'Open trade' or 'outward-looking' trade policies and growth strategies that were effective two decades ago may not be equally effective today, and some adjustments may have to be made in order to take advantage of the new conditions. The changes that occurred include a wider distribution of world production and supply across more countries as an effect of outsourcing and offshoring, the intensive use of highly advanced information and communication technologies, the spread of worldwide supply chains that know no national boundaries, and, at the same time, the growing power and influence of RTAs.

Obviously, these changes require considerable adjustments in the development strategies that the LDCs can and should implement. The chapter first provides an overview of the changes in the structure of world production and trade and then evaluates the implications for the design of development strategies today.

The chapter also addresses the difficulties that an LDC is likely to encounter in making the necessary structural changes in its economy in order to meet the high standards that are demanded by today's world market. Another important aspect of the structural adjustment is the much greater political sensitivity in all countries and all segments of society to the unavoidable pain – and its uneven distribution – that the adjustments involve. The much greater openness of both society and the political system today, the spread and exchange of information, the greater awareness of all population groups of their rights, and their capacity to organize and join forces to oppose policies that may inflict an undue burden require policy makers to be much more sensitive to the needs and demands of different social groups, which may constrain decision-making powers. One manifestation of such constraints is the 'reform fatigue' in many lower- and middle-income countries that leads to a reduced support for market-oriented reforms, and sometimes even to doubts about the legitimacy of the market reforms. The rise of 'revolutionary' regimes in LA that demand more rights and more consideration for the needs of the poor epitomizes this spirit, but similar trends prevail in all countries and, for better or worse, they are part of the 'zeitgeist' in the 21st century.

2 Trade and Growth in the LDCs: the Theory and Its Practice Today

The classical and neoclassical economic arguments in favour of free trade are widely known. They emphasize the gains from specialization which requires countries to allocate their scarce resources to the production of goods and services in which they have a comparative advantage, and the benefits of specialization are combined with the economies of scale in production that increase the country's production possibilities. Countries that have different allocations of resources or different productivities of their factor inputs can benefit from trade: for countries that have abundant labour and low wages, free trade provides the opportunity to specialize in labour-intensive products and thus raise the incomes of low-wage workers and reduce their poverty.

In the decades since the mid-1950s, these very optimistic predictions of neoclassical trade theory materialized, and it has been a period of extraordinary growth in world trade, and in the openness of the majority of the world's economies. Since the early 1960s, world trade has grown more rapidly than world income during nearly the entire half century. This rapid growth in world trade reflects not only the impact of the remarkable economic expansion of all the large countries and most regions, but also a much greater openness to trade that was due to far-reaching reductions in trade barriers, transport costs, costs of communications and a joint effort of most trading countries to conclude trade agreements designed to open all the world economies to trade and to establish institutions to supervise the evolving trade rules. Other effects included the increase in the importance and the share of manufactures in world trade and the large changes in the structure of world production and supply that combined most countries in the cycle of production and trade.

The rise in world trade was expected to bring about a steep decline in world poverty. Two explanations were given for this expectation. According to the static explanation, the central effect of trade on poverty would come from the effects on real wages of the unskilled workers in countries that have abundance of labour but no human or financial capital. The presumption following the Stolper–Samuelson argumentation is that these countries would use their comparative advantage to export labour-intensive goods and thus raise the income of their unskilled workers and help in the reduction of poverty. According to the dynamic explanation, the more rapid growth in these countries as an effect of trade would reduce their poverty.

Dollar and Kraay (2001) conducted a very thorough study that related growth and trade. They used decade-average panel data for the 1980s and 1990s for a sample of about 100 countries to estimate growth in real per capita terms (from the 1980s to the 1990s) as function of the corresponding growth in the 1970s and 1980s, the change in a set of control variables from the 1980s to the 1990s and the common change in all countries between these two decades. They used in their analysis the *change* in trade rather than the level of trade in order to circumvent the problems created by the high dependence of trade by factors such as geography that do not change over time (although transport costs changed very significantly during these years and were one of the factors that gave a push to international trade). Their goal was to estimate the rather high coefficient of the change in growth on the change in trade. One of the key control variables was the ratio of the total trade (exports + imports) in the gross domestic product (GDP).

Dollar and Kraay estimated the relatively high coefficient of growth on the change in trade, which indicates that a 10% increase in the ratio of trade to the GDP would raise the GDP per capita by 2.5% over the decade. With the use of instrumental variables for the initial income and trade ratio, that coefficient nearly doubles. Their study thus confirms the impact of trade on growth, thereby strengthening the empirical evidence that supports export-oriented, outward-looking trade policies. Their main emphasis was on the significant impact of the variable that shows the trade/GDP ratio, which they interpret as the impact of opening the economy to trade. Easterly (2005) provides additional evidence for the impact of trade on growth. The World Bank (2002) developed a computed general equilibrium (CGE) model that incorporates the impact of trade on productivity and growth. The model more than doubles the estimate of free trade on the world's income.

The advantages of an 'outward-looking' trade policy have been widely recognized and such a policy is therefore regarded as a key component of a growth-enhancing development strategy. Openness to the global economy offers countries the potential to exploit their comparative advantage, enhance the utilization of their production capacity and available resources, and to reduce costs with economies of scale in production. Exports also provide a source of foreign exchange and access to advanced technologies. FDI can provide employment and encourage technological upgrading. Openness enables global technical change to increase domestic productivity with imports, and competitive pressures stimulate enterprises to increase their efficiency.

2.1 'Inward-looking' versus 'outward-looking' trade policy

In the developing countries, trade liberalization proceeded at a much slower pace. There was strong support for import-substitution policies amongst policy makers, particularly in the newly independent countries of SSA. One goal was to protect infant industries, particularly given the extensive plans for industrialization that were perceived to be necessary for development. These countries were concerned that free trade would increase their dependence on exports of primary commodities, particularly given the cyclical decline in their prices that brought a long-term decline in these countries' terms-of-trade. Import-substitution policies also had strong support from vested interests that gained either from the protection provided by, or from access to, rents created by NTBs.

Commodity exporting countries often granted a monopoly to selected firms. To protect their industries, they used a mixture of tariff and NTBs, such as quotas and licenses. Foreign exchange restrictions frequently imposed large additional taxes on trade. The use of NTBs was widespread even though they were frequently inefficient and subject to corruption. Frequently the protective effects of tariffs, quotas and licenses in developing countries were reinforced by distortions in foreign exchange markets. Foreign exchange restrictions on current accounts involve an overvalued official exchange rate.

Prior to the Uruguay Round (1986–1994), the political attractions of import-substituting policies for policy makers in developing countries were not effectively countered by arguments about the multilateral trading system. First, in these years, many policy makers in developing countries generally subscribed to the theory of import substitution, and focused their efforts in the GATT on obtaining unreciprocated improvements in their access to markets in industrial countries under the rubric of 'special and differential treatment', which gave them preferential market access by improving their terms of trade. However, this policy made it difficult to bargain for improvements in their market access for the products of greatest interest to them. Second, such access was frequently constrained by quantitative restrictions and the risk of preference erosion or removal. Third, this policy approach reduced the incentives of domestic exporters who had no need to lobby for reductions in domestic protection in order to improve their access to partner markets.

However, the fundamentals of trade theory raised great hopes that the exchange of tradable goods and services would narrow down over time the gap in their prices, even though that convergence would be only partial, because not all products are tradable. A country would also gain from productivity gains that typically concentrate in the tradable sector, even though the gap between the domestic relative price of tradable and non-tradable goods is likely to increase. In economic trade theory, this is defined as the Balassa–Samuelson effect which states: 'The traded goods sector has a higher productivity growth than the non-traded goods sector, leading to higher relative non-traded goods' prices.'

The explanation for the Balassa–Samuelson hypothesis of the varying price differentials is that productivity gains are faster in tradable than in the non-tradable sectors because the production of tradables can be automated more completely than non-tradables, their manufacture is more capital intensive and uses more advanced technologies. As a result, the productivity in local-services and non-traded goods is more uniform than is the productivity (in foreign earnings per year) of people working in the export sector.

There are a number of additional points that need to be taken into account:

- Certain labour-intensive jobs are less responsive to productivity innovations than others; these jobs are typically in services and they are performed locally.
- The low-productivity sectors are also the ones that produce for the local market and their products are non-tradable.
- Workers in the non-tradable sectors in the high-income country receive much higher wages than workers in the same sectors in the low-income country.
- The large and growing wage difference was the main driving force for offshoring many jobs in the service sectors that in the past were non-tradable to the developing countries.

The GATT and later the WTO were created in order to open up markets and promote international trade based on the 'free trade' paradigm. The WTO was established to, among other tasks, monitor the trade agreements and ensure that trade is conducted *on a level playing field* in which no country and no corporation will take advantage of their monopolistic position to distort world trade in their favour.

Despite the great promises, in the last two decades there has been vehement and growing criticism of free trade in general and of the WTO in particular. The main argument has been that world trade is not really free and the world trade playing field has not been levelled. Developing countries that have rich natural resources often feel exploited, but this is in large measure because their own autocratic regimes do not distribute the income from exports, and they spend huge resources on their army in order to stabilize their regime and fight frequent wars, which are often over the control of these natural resources. The populations in many developing countries harbour intense resentment because they feel exploited on account of the controls exercised by multinational corporations.

In many developing countries, critics of globalization oppose the 'imperialistic powers' that take control over their countries under the pretext of free trade. These 'imperial powers' arguably control the developing countries with their capital like colonies by using this dependency to their advantage (this is sometimes referred to as 'dependency theory'). In many developing countries, particularly in SSA, it is argued that free trade narrows down the production base of the economy, making them dependent on few commodities and fostering a narrow specialization in which they have a comparative advantage. As a result, their export earnings vary widely with the variations in world demand for these products, but their specialization makes it more difficult for them to diversify their production, particularly in industry. In these countries, as well as in developed countries, the dependence on natural resources, particularly oil, increases their vulnerability. This vulnerability is also one of the reasons for the domestic production of key agricultural products, even if this means higher production costs.

In the developed countries, there is also considerable opposition to free trade and to trade agreements on the grounds that they led to a reduction in wages, particularly of the low-skilled workers, and to many job losses with the growing trend of offshoring. In today's global economy, where capital movements are essentially unrestricted and the returns to investments are rising, the incomes of the 'capitalists' are rising much faster than the incomes of the workers, which leads to growing income inequality. Trade also requires difficult adjustments, forcing thousands if not millions of workers to change professions. Indeed, the most outspoken opposition to free trade comes from labour unions. During the adjustment period, local workers are forced to change careers and often their place of residence. Coping with these transitions can be difficult, especially for the middle-aged and the elderly.

Free trade gives companies the possibility to outsource the production of goods, components or segments of the production process like assembly to countries that have lower wages and thus a comparative advantage in their production. In addition, when environmental and labour standards are lower in foreign production, it increases the profitability of transferring the production of components that cause damage to the environment or problems with labour standards to locations where such regulations are weaker or do not exist; companies can then circumvent the more demanding regulations in their home countries. Weak or un-enforced labour and environmental protection laws will thus create a race to the bottom that gives advantage to countries where these laws are barely enforced. It is argued that, as free trade increases, the balance of power shifts in favour of the transnational corporations and away from governments, even though all countries still have the right to restrict free trade.

In the past, the displacement and change caused by free trade fell on the poor and less-educated with the transfer of labour-intensive industries that employ low-skilled workers. But in recent years, with the plentiful supply of low-cost but well-educated labour in Asia and the former Soviet Union countries, more professional workers are also displaced or remain unemployed. In the USA, for example, offshore outsourcing and an increase of temporary visa workers resulted in a drop in the demand for computer programmers. The Internet also made it easier for dispersed groups to coordinate their work.

Other problems are caused when governments encounter difficulties because elements of the Western legal system, such as the imposition of property rights, are not taken well in tribal areas where property is held communally. Property rights in developing countries and their implications for free trade are important issues that have been raised by the Peruvian economist Hernando de Soto.

Another issue is the diversity of legal systems across countries that reduces their capacity to coordinate their regulatory and tax-collecting efforts; the absence of that coordination can create loopholes that may be exploited by corporations and private individuals, who can seek out havens from regulation and tax collection without violating the law. The freedom of capital to move outside the purview of a single authority can enhance tax avoidance, money laundering and obfuscation of corporate accounts. Multinational and regional agreements made considerable efforts, therefore, to develop methods of harmonizing international regulatory institutions, in particular accounting practices, to improve transparency in the financial markets and reduce the risk experienced by investors. In theory, globally or regionally harmonized regulations regarding wages, the environment, safety, human rights and other areas of economic control, would also prevent a 'race to the bottom'. Several regional agreements consider these regulations. However, with the increase in capital mobility there is also a competition between countries on tax rates and other regulations in order to attract capital.

2.2 The changes in the organization of world trade

The web of rules and regulations that a country joining the WTO or an RTA now must abide by has been formed over many years, and these rules became far more comprehensive and far more restrictive over the years. Although the multinational trade agreements within the framework of the WTO take utmost care to facilitate the accommodation of the LDCs, give them a much greater say than their share in world trade and cater to their special needs, these rules also require the LDCs to make considerable internal adjustments and implement far-reaching reforms that at times seem to curtail their national sovereignty.

This is an experience that practically all the developed countries had to undergo when they had to implement considerable reforms as members of the WTO; but it is doubly more difficult for poor developing countries. The dramatic political changes that are now taking place in many LA countries and the mounting opposition to the WTO are a reflection of these difficulties. The 'Development Round' that the WTO Ministerial Conference in Doha initiated in November 2001, in order to assist the developing countries with their integration into the global trade system, was the culmination of the efforts of the international community to aid the LDCs. The stated goal of the Doha Round was to 'fundamentally reform encompassing strengthened rules and specific commitments in order to correct and prevent distortions in world agricultural markets' (WTO, 1999). The purpose of the reforms was to make changes in the global trade system that would help the low-income countries to accelerate poverty reduction and enhance food security.

Since agriculture is the main economic sector and the primary source of income in these countries, the most effective way of increasing their income is to promote their agricultural production by enhancing their exports. There was a basic agreement among all WTO members on the goal of liberalizing agricultural trade, but they did not succeed in the various rounds to agree on a formula of trade rules that would be accepted by all and still accomplish their goal. The failure to reach an agreement is therefore a major setback for the LDCs.

The institutional changes in world trade are discussed in the third part of the book that provides a more elaborate analysis of the reforms that were planned at the Doha Round, their impact on the LDCs and on the WTO itself, and the reasons for the failure to reach an agreement. To evaluate the potential effects, positive and negative, of the reforms that were planned in this Round on the LDCs, the subsequent sections examine in more detail several key agreements that have been reached in recent years on trade reforms in sugar, cotton, textile and clothing products, and evaluate their effect on the LDCs.

With the stalemate at the Doha Round, the RTAs are now becoming a far more important factor in the institutional structure of world trade. These regional agreements have a major impact on the LDCs, since most of them are effectively excluded from active participation in the agreements, even though they are given 'special and differential treatment' by some of these agreements; however, this 'special treatment' is often more nominal than real.

Moreover, unlike the multinational trade agreements, the regional agreements are strongly influenced by the larger members that practically dictate the rules, leaving very little say to the smaller developing countries. These agreements further marginalize the role of the LDCs in world trade and reduce their capacity to be integrated into the global trade system, while the EA countries are now increasing their influence with their own RTAs and their growing share in world trade.

2.3 Is there a level playing field in today's free trade for the LDCs?

Although it is widely agreed that, over time, the expansion of a country's trade through trade liberalization and other measures will significantly contribute to its economic growth and the reduction of poverty, it has been recognized that, in the short run, when the country is at the initial stages of opening up its economy to trade, there is a difficult adjustment period that is bound to be particularly damaging to the sectors in which the country does not have a comparative advantage. Over time, many of the affected industries will have to close down and many of their workers will lose their jobs and suffer large income losses. Many middle-age workers may not be able to find new employment, and many others may fail to acquire the new skills that are necessary in the 'new economy'. Many enterprises may not be able to make the necessary changes due to factors beyond their control, like poor infrastructure, lack of credit, etc.

The structural changes that take place when a country is opened up to trade may therefore involve considerable pain and deepen the social divide between population groups that succeed in acquiring new skills and make the necessary adjustments and other groups that fail to acquire the needed skills and find new income sources. The structural changes may also deepen the divide between regions where sectors in which the country has a comparative advantage concentrate and regions where many industries are forced out of business, causing a large portion of the local population to sink into poverty.

These difficulties during the transition period may put pressure on the government to restrict trade and forego further liberalization of sectors that were weakened by trade liberalization. If there is strong opposition of workers and entrepreneurs, there may be demands that the government prevents any further damage from additional measures of liberalization, and the promise of future gains from free trade is not likely to moderate this opposition, which often evolves into a fierce political struggle against free trade and against trade agreements that remove the protection on local industries.

This was one of the main reasons why developing countries adopted different strategies of trade and development, with one group of countries, particularly the countries of LAC and SSA, implementing import-substituting policies to develop their local industries, and their policy makers did not take active part in the GATT prior to the Uruguay Round (1986–1994).

The subsequent analysis in this chapter is conducted against the backdrop of two heated debates among researchers and policy makers that have dominated the economic literature for some time: One is the debate on the pros and cons of globalization given its direction and impact in the past decade on different countries, sectors and income groups. The other debate concerns the desirability of open trade for the LDCs given the organizational and institutional changes in the world trade system that have all too often left them out of that system. The central questions in this debate today are:

- How relevant and desirable are the policies known as the 'Washington Consensus' for the developing countries today? This question does not refer to the entire list of policies in this category, but to two specific policies: open trade and free market.
- How beneficial is open trade for a developing country today when world trade is not really free but strongly dominated by monopolistic transnational corporations?

- How open is world trade when world market prices are influenced by various tariffs, administrative restrictions and subsidies that the big industrial countries with the largest share in world trade and the greatest influence on world prices are maintaining?
- How free is world trade given the many trade agreements that divert the flows of world trade, and how desirable is open trade for the LDCs that are not part of these agreements?
- Given all these constraints, what are the alternative strategies for a small (in terms of its share in world trade) economy in view of the unavoidable distortions that any other policy is bound to create?
- What alternative strategies are available for LDCs, given common constraints due to the lack of proper institutions and an efficient system of governance?

Let us consider these questions in a specific example that illustrates both the bias of the world trade system, even under the rules of the WTO, against the small developing countries, and the dilemma that these countries must face. The next section discusses in great detail the special incentives that the EA countries have given and continue to give to their export industries by, among other measures, reducing their import tariffs, creating export processing zones, and providing rebates or exemption of duties on imported materials for export production. Some EA countries have attracted significant FDI with various tax incentives, and gained the investors' help in upgrading their domestic technology and fill in gaps in management and marketing skills.

Selective subsidies to export industries are still common in the EA countries, even after they became members of the WTO, which forbids these incentives. Thus, for example, as recently as in early 2007, the USA started a legal action at the WTO against a wide range of Chinese subsidies. According to the complaint, the Chinese government is using support and tax policies to bolster Chinese firms in competing against US and other foreign companies in a wide range of industries, from steel to paper to computers. In 2005, the USA took a similar legal measure when China scrapped a rebate on a value-added tax on semiconductors. In 2006, the USA, the EU and Canada filed a joint complaint against China's policy on auto parts tariffs. According to the rules of the WTO, if the organization issues a ruling against China, this clears the way for the USA to impose economic sanctions against China. Obviously, if the complaint is filed by the USA, these economic sanctions are potentially damaging and China has a reason to retract, as it did in the past. If, however, the complaint is submitted by a small country or even by, say, a group of all the African countries, these sanctions are practically meaningless for China due to the small share of its trade with these countries, and it would not motivate China to change its policy. In other words, even under the WTO rules, these countries do not have a level playing field, and by opening their own economy, they cannot actually expect to benefit from free trade.

Rodriguez and Rodrik (2000) noted that the priority given to trade reforms in the development strategies designed by the Washington Consensus institutions may create the impression that these reforms by themselves are the panacea for growth and poverty reduction. However, comprehensive policy reforms must include reforms in the system of governance and in a country's institutions, and these reforms should come first in the design and implementation of the overall policy reforms. On these grounds, the role of institutions and the system of governance have been given a much more central part in many reports and studies in recent years, and these issues have been the focus of the recent annual reports of the World Bank and the IMF.

3 Open Trade and Public Policies: Lessons from the East Asia Experience

The success of the EA developing countries during the 1980s and 1990s to accelerate their growth and sharply reduce their poverty is the most widely quoted example in support of the merits of outward-looking, export-oriented policies and the benefits of globalization. The World Bank study *The East Asian Miracle* noted, however, that an outward-looking policy based on trade liberalization was only part of the reason for the rapid growth of the region:

> But these fundamental [outward-looking] policies do not tell the entire story. In most of these economies, in one form or another, the government intervened systematically and through multiple channels to foster development, and in some cases the development of specific industries. Policy interventions took many forms: targeting and subsidizing credit to selected industries . . . protecting domestic import substitutes, subsidizing declining industries . . . establishing firm- and industry-specific export markets, developing export marketing institutions . . . Some industries were promoted, while others were not.
>
> (World Bank, 1993, pp. 5–6)

That study and many others emphasized that, in practice, the important factors that contributed to growth in EA were not just open trade and free markets, nor even the support through government interventions to guide the economies and 'get the fundamentals right'. Instead, three groups of policies were identified as the pillars of the region's success: (i) industrial policies to promote particular sectors; (ii) strict government controls over the financial markets combined with active measures to stabilize the exchange rate, lower the cost of capital and subsidize direct credit to strategic sectors; and (iii) a variety of other active market intervention measures to promote exports and, at the same time, protect domestic industries from cheap imports. Crucial to these three groups of policies were strict controls of the central government over the entire system of private and public institutions.

The World Bank study also underlines the fact that, at their early stages of recovery, these countries' economic policies were not based on the free market but were accompanied by active interventions of the central government in trade, in the financial sector and in the introduction of advanced technologies (Wade, 1990; Chang, 1994). Government interventions played a crucial role in their development in a number of ways: First, with the adoption of 'outward-oriented' trade policies, they actively promoted both industrialization and exports through targeted and subsidized credit, tax incentives and other incentives to promote the structural changes (Dieter *et al.*, 1998). Second, by maintaining stable macroeconomic and exchange rate policies, they secured low inflation and a pro-investment environment, and by supporting a pro-business environment, they supported private sector-led growth. Third, by intervening to support export industries that could take advantage of the sharp reduction in transport costs and the very low wages of their own labour force, these industries still needed government support during their infancy stage. They received that support with a slow and gradual reduction of the trade barriers that protected the local industries and with direct incentives to promote production and exports that included, among others, export zones in which industries were exempted from government taxes and most controls.

3.1 A brief background: the diverging roads of the developing countries

In the 1950s, following the destruction and misery of World War II, the war in Korea and the Great Leap Forward in China, the EA region deteriorated to extreme poverty and millions died of starvation. Yet, the remarkable achievements of the EA countries had already begun in those years, with the rapid growth in Japan that later extended to the 'four Asian tigers' – South Korea, Taiwan, Singapore and Hong Kong. The war in Vietnam slowed down that progress, but since the late 1970s it continued to spread to an increasing number of other countries in the region. These countries gradually changed their policies, though not their regimes, and they implemented very methodically a series of measures that transformed the structure of their economies by accelerating industrialization and growth, while paying little attention to the human toll, especially among the rural population and in the agricultural sector that in most countries trailed behind.

Most developing countries in other regions, particularly in LA and SSA, went through a cataclysmic cycle during those years that took them from relatively rapid development and great optimism in the 1960s, following their liberation from colonial rule, to a decline, economic stagnation and an endless series of political crises and violent conflicts that halted their growth and brought them to the deep debt crisis of the late 1970s and early 1980s. Most SSA countries, with the exception of South Africa (still under Apartheid), as well as many LAC countries suffered considerable income losses. The SSA countries drifted to the bottom of the global income distribution and became the world's LDCs.

There was another group of developing countries, particularly in North Africa and LA that continued to grow at a slow pace and made only limited structural changes in their economies. Only since the mid-1990s, an increasing number of these countries started to implement more thorough changes in their economies, but their growth was still slow due to the low and declining terms of trade of their exports of natural resources and agricultural goods. These countries became known as the 'frontier economies'. In recent years, many of these countries benefited from the sharp rise in the world prices of oil, gas and many minerals, and an inflow of FDIs that started when their regimes became more stable and their natural resources became more valuable and attractive.

3.2 The sources of the economic growth of the East Asian countries

A notable attribute of the EA countries was their success in stimulating capital accumulation that assisted their industries to increase productivity, adopt new technologies and accelerate growth. Asian economies generally relied heavily on high rates of physical and human capital accumulation that have been and still are important contributors to their growth: large investments were initially made possible only by the high rates of domestic savings of up to 35–40% of income, even before any foreign investments started to arrive. Their high local savings rates, encouraged by the stability of local prices and the exchange rate, remained the main source of their large investments, and they were vital during the initial phase of their development.

An IMF (2006) study shows that during 1970–2005, growth differences – both across the world's main regions and within Asia – were driven mainly by labour productivity. In China as well as in most other EA countries and later in India, total factor productivity (TFP) – often used as an indicator of technological progress – became the principal engine of growth since the 1980s. However, the sources of economic growth of the EA economies were initially hotly debated. Young argued in the mid-1990s that EA's growth miracle was mostly due to increased inputs of labour and capital rather than to the growth in TFP. Krugman concurred that '[the] miracle turns out to have been based on perspiration rather than inspiration'. In fact, these studies underestimated the growth in TFP in the Asian emerging economies, which had been considerably faster than in the developed countries (see below).

In addition to perspiration, however, their sustained high rates of investment enabled their industries to import advanced machinery that embedded new technologies from the developed countries. These investments were initially financed by their own large savings and increasingly also by capital and direct investments of foreign companies. The growth of TFP in China has been even faster than in the rest of EA, and in the EA countries it has been faster than in all other regions (Fig. 5.1). According to World Bank estimates, TFP in these countries increased during the past quarter century by an average of 3% a year and accounted for roughly the same rate of growth of their GDP as their capital investments. In the USA, TFP grew during these years by an annual average rate of only 1%.

The productivity take-off in EA benefited heftily from the opportunities provided by international trade and globalization and from the support that governments there gave to private enterprises. FDIs were pivotal to the adoption of advanced technologies from the developed countries and the profitability of their exports attracted many new investors and industries that accelerated their technological progress still further.

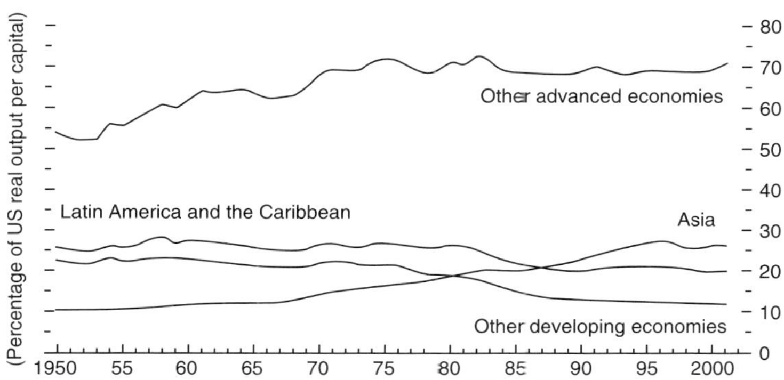

Fig. 5.1. Real output per capita in main regions. (From World Bank/World Development Report, 2006.)

The expansion in the operations of transnational corporations in the EA economies was motivated by the high profits from segmenting their production into subcomponents along the value chain. The resulting reduction in production costs and greater efficiency brought new industries and technologies to these countries. The corporations were thus able to restructure their production by utilizing the comparative advantage of different countries in the production of different components of the final product, and they took advantage of the low labour costs in the EA countries for outsourcing to them the production of individual parts in which production is labour-intensive.

As these corporations widened the geographical distribution of production and supply chains with outsourcing and offshoring, they changed very fundamentally the structure and organization of the global production and trade systems. They created an internationally integrated production system that generated large trade flows in intermediate products across national boundaries, and increased both intra-industry trade and trade between enterprises in similar products. With the expansion of their trade, the transnational corporations increased their investments in the developing countries, but most of these investments concentrated in a small number of the emerging economies.

It is by no means evident that the lessons from the success of the EA countries in the 1980s and 1990s are valid for other developing countries, particularly the LDCs, in the 21st century. It is also not certain that the policies that proved to be successful in the EA countries will be equally effective in the low-income countries of LA, SA or SSA.

This section examined the lessons that have been drawn with respect to these issues in several multi-country studies. Obviously, these studies tend to generalize the lessons by grouping all the developing countries, or all the LDCs, together, and by considering the impact of a single indicator or a small number of indicators that represent the main characteristics of the policies implemented by the EA countries. This generalization also has obvious limitations when the lessons are applied to individual countries. The subsequent sections therefore consider in more detail the specific lessons that are relevant for the main subgroups of developing countries.

The positive impact of free trade on growth has been widely recognized, but the contribution of globalization to accelerate growth and reduce poverty in all countries is still disputed. The reason is that there were large differences in the effects of globalization on different countries as they opened up their economies to trade. These differences were often due not only to globalization but also, and sometimes primarily, to the impact of other factors, including government policies, specific socio-economic conditions, the state of private and public institutions, legal and personal security systems, etc. The impact of these factors on the way and extent in which the economies of different countries were affected by globalization makes it difficult to establish a clear and significant *causal* effect of globalization on growth.

Although economic theory and numerous studies demonstrated that, in the long run, a country's integration into the global economy through open trade policies would contribute significantly to its economic development (provided that certain conditions hold), in the short and medium run, trade liberalization may well be harmful to certain economic sectors, geographical regions and segments of the population, making many people worse off. The rural areas and farmers are particularly vulnerable due to their lower flexibility to make the necessary adjustments in their production.

3.3 World Bank studies on the impact of globalization

Openness to free trade has always been a fundamental building block in trade theory, and the ability of countries to gain from their comparative advantage has been regarded as a positive-sum policy that enables all countries engaged in free trade to reach higher growth rates. Lucas (1993) emphasized that openness to trade accelerates growth because it allows a more rapid accumulation of human capital through learning by doing; Grossman and Helpman (1991) emphasized the accumulation of knowledge in generating endogenous growth by promoting production in sectors that are more knowledge-intensive. These transitions, although beneficial to the country in the longer run, are bound to sharpen in the shorter run the gap between those who can acquire the new skills and rise on the income ladder together with the economy and those who lack the basic education and may find themselves on the sidelines.

However, empirical studies that focused on the impact of liberalization on the developing countries did not yield conclusive results, though Dollar (1992), Sachs and Warner (1995) and Edwards (1998) highlighted the positive impact of liberalization. Rodrik and Rodriguez (1999), in contrast, argued that methodological problems with the empirical analysis leave the results open to diverse interpretations, largely because the openness indicators that were used were poor measures of trade barriers (see below). In developing countries where the economy is weak, there are also strong pressures on the government to avoid policies that create unemployment, even though this may be unavoidable during the transition period, and liberalization is bound to harm employment in certain sectors. The promise of future gains, after the end of the transition period when the gains from trade materialize, has generally not helped governments to calm protests against the exposure to foreign competition.

A comprehensive World Bank study entitled *Globalization, Growth and Poverty: Building an Inclusive World Economy* (2001b) evaluated the impact of globalization on the poor and the extent to which they actually gained from globalization. The evaluation was made on the basis of the experience in the last two decades, based on a survey in 73 developing countries.

These countries were ranked according to the extent to which they increased their trade relative to income over the period. The top one-third of the countries in this ranking was labelled 'more globalized' without in any sense implying that they adopted pro-trade policies, and the study recognized that the rise in their trade may have been due to other policies or even to pure chance.[1] The remaining two-thirds were labelled 'less globalized'; the one-third/two-thirds distinction was of course arbitrary. As many as 24 developing countries with some 3 billion people were included in the group of the 'more globalized' countries, and they doubled their ratio of trade to income over the last two decades. The group of the remaining 49 developing countries actually trades less today than 20 years ago.

The criterion used in the World Bank study is itself quite arbitrary since it is based on the rate of increase in the countries' openness to trade, but not on their actual degree of openness; the group of 'more globalized' countries can thus include countries with many 'anomalies'. Figure 5.2 shows the

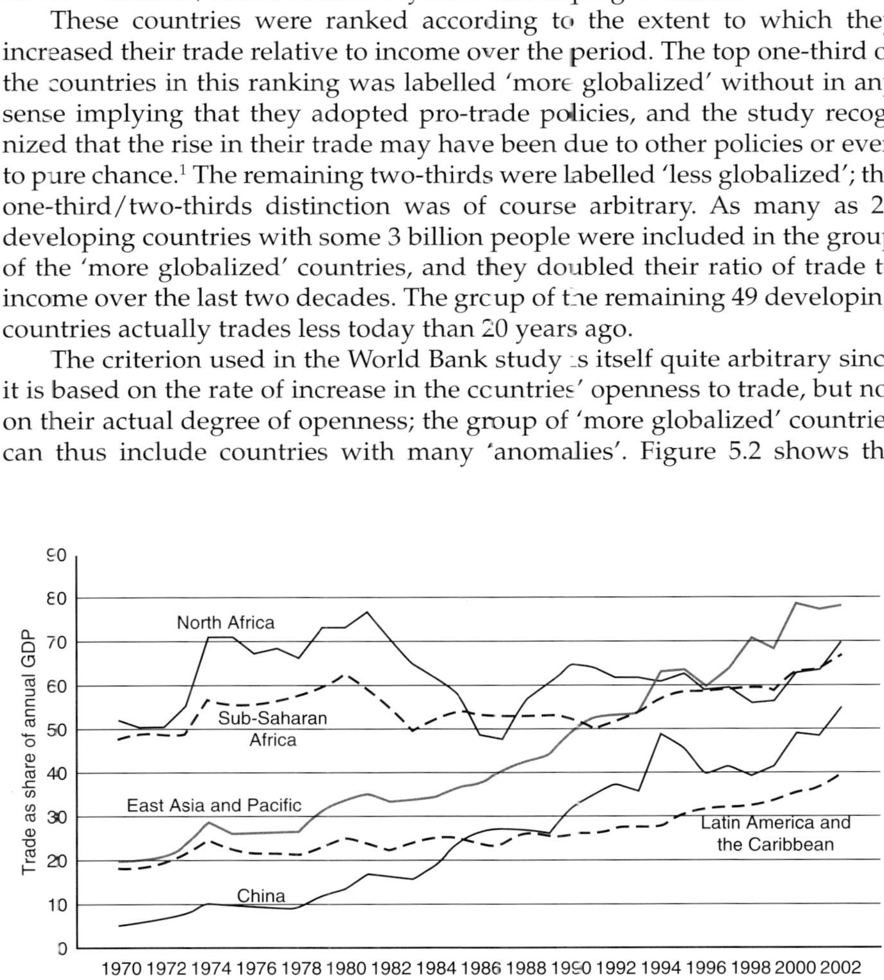

Fig. 5.2. Trade as a share of GDP in selected regions, 1970–2002 (%).

[1] The 'more globalized' countries include several 'anomalies': Côte d'Ivoire, the Dominican Republic, Haiti, Jordan, Mali, Nepal, Rwanda and Zimbabwe.

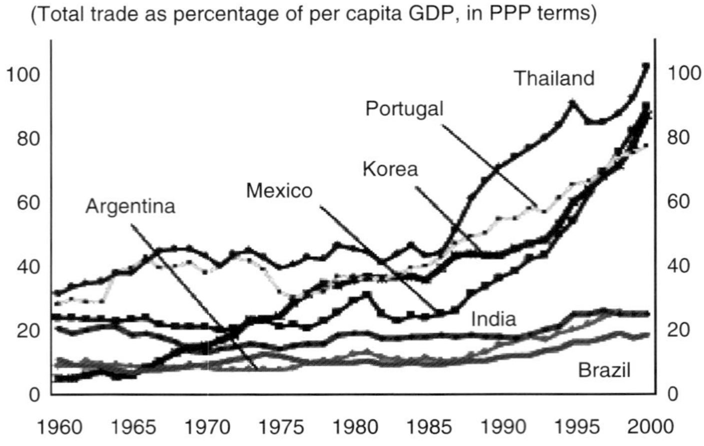

Fig. 5.3. Trade as a share of GDP in selected countries, 1960–2000.

differences in the level of openness between the different regions according to their *trade/income* ratios: In EA, the trade/GDP ratio rose from around 20% in 1970 (a rise that was then due to the rapid increase in the trade of Japan and the 'Asian tigers') to nearly 80% in 2002. In LAC, that ratio increased during these years from less than 20% in 1970 to around 40% in 2002. In SSA, the ratio fluctuated between 50% and 60%, but did not rise much during these years.

The 'more globalized' developing countries increased their per capita growth rate from 1% in the 1960s to 3% in the 1970s, 4% in the 1980s and 5% in the 1990s. The per capita income of the 49 'less globalized' countries increased during these years at an annual rate of 0.5%.[2] The 'more globalized' developing countries started to grow more rapidly only in the 1980s and accelerated their growth steadily from 2.9% in the 1970s to 5% in the 1990s.

Figure 5.3 compares the degree of openness between selected countries. Most of these countries were included in the group of 'more globalized' countries in the World Bank study. The figure shows the large differences between countries in the degree of their openness, and these differences increased very significantly since the mid-1980s.

The World Bank study suggests that, as an effect of their openness, the 'more globalized' countries found themselves in a virtuous circle of rising growth and penetration of world markets, possibly through 'policies of educational expansion, reduced trade barriers and strategic sectoral reforms' (World Bank, 2000, p. 36). While trade clearly reinforces growth, their educational expansion, even though it already started in the 1980s, most likely had an impact on these countries' growth only a decade or so later, perhaps only during the 1990s, with the rise in labour productivity. Moreover, as noted above,

[2] These growth rates are population-weighted and they do not include the rich economies (the original members of the OECD plus Chile, Korea, Singapore, Taiwan and Hong Kong).

most of the 'more globalized' EA countries reduced their trade barriers only marginally during the 1980s and most of the 1990s, and therefore this measure could not have an impact on their growth during most of these years.

On the basis of these and other findings, the Bank study emphasized the following conclusions:

> Whether there is a causal connection from opening up trade to faster growth is not the issue. In those low-income countries that have broken into global markets, more restricted access to those markets would be damaging to growth, regardless of whether industrialization was triggered by opening up. However, opening up integrates an economy into a larger market, and from Adam Smith on economists have suggested that the size of the market matters for growth. A larger market gives access to more ideas, allows for investment in large fixed-cost investments and enables a finer division of labor. A larger market also widens choice. Wider choice for high-income consumers is irrelevant for poverty reduction, but wider choice may have mattered more for firms than for consumers. For example, as India liberalized trade, companies were able to purchase better-quality machine tools. Similar effects have been found for the Chinese import liberalization. Finally, a larger market intensifies competition and this can spur innovation.
>
> (World Bank, 2000, p. 36)

These conclusions are, however, open to debate and some of the points are clearly less convincing than others. My main comments are on the following points:

1. While restricted access to the world markets of the 'more globalized' countries, particularly the EA countries, would indeed have been damaging to their growth and would have slowed down their industrialization, in practice their own markets did remain tightly restricted to the global markets. It can therefore not be concluded from the experience of the EA countries that opening up a country's own market to both exports and imports is essential to growth, as suggested by the Washington Consensus.

2. The advice in the report: 'Opening up integrates an economy into a larger market [. . . and as suggested by economists since Adam Smith . . .] a larger market gives access to more ideas, allows for investment in large fixed-cost investments and enables a finer division of labor' is contrary, however, to the policy that most 'more globalized' countries actually implemented. The policy of the EA countries until just few years ago was not a two-way integration with the world economy that opened both their exports and their imports to the larger world market. Instead, it was 'one-way' integration – if such a thing exists – that enabled them to increase their exports to the world market, but also allowed them to protect their own industries from competing imports during a prolonged adjustment period. Only very gradually and very partially did these countries open up their own markets to imports. Figure 5.4 shows that even during the 1990s, their import tariffs were higher than those of the 'non-globalizers'.

3. The refusal of some of the 'more globalized' countries to lower their tariffs more rapidly and remove all the restrictions on their imports was evident

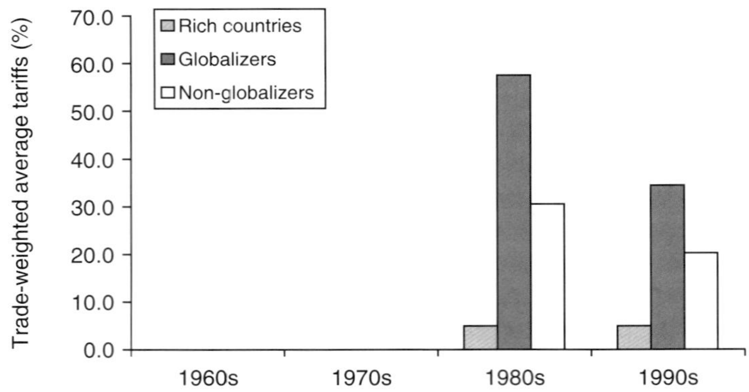

Fig. 5.4. Average tariff in globalized and non-globalized countries.

during the trade negotiations at the Doha Round. Their trade policy throughout these two decades suggests that they clearly did not accept the recommendations of the neoclassical, neoliberal economists since Adam Smith to resolutely and rapidly open their economies to trade. Their concerns about the difficulties that a 'finer division of labour' during the course of these adjustments would cause to their working population and the damage that would be caused to many state enterprises and some of the infant industries prevented them from adopting this advice. In China, this was, in part, a reaction of the leadership to the hardships that the population had to endure during the 'Great Leap Forward'. However, this cautious approach did not reduce their access to foreign investments that flowed to their export industries, and, with them, to new ideas.

4. The wider choice of products that trade makes possible indeed matters for firms, and the EA countries enabled their local industries to import new machinery and advanced technologies as long as they did not compete with their own industries, but they imposed strict restrictions on imports of competing consumer goods.

5. Figure 5.3 and Table 5.2 show that trade liberalization in both India and China was very gradual and tightly controlled and their tariffs remained high. With the unsuccessful conclusion of the Doha Round, these two countries are still far from being examples of open economies and free trade.

Table 5.2. Average weighted tariff rates in selected countries.

	1985s	1990s	1995s
China	38.8	39.9	20.9
India	99.4	61.9	38.4
Argentina	27.5	13.9	11.0
Brazil	45.8	21.0	11.5

The World Bank study attributes the stagnation of the 'non-globalizers' to their failure to change their production base, open their economies to trade and diversify their agricultural production and exports. As a result, their share in world trade declined sharply, and it still consists mostly of traditional crops, oil and raw materials. The latter two groups of export products have very small value-added and contribute very little to these countries' employment and growth.

The EA miracle was driven not only by trade liberalization, however, but also by pro-trade macroeconomic and industrial policies. The macroeconomic policies were aimed at maintaining local stability and a stable exchange rate that secured price stability and gave incentives to local savings and investments. The industrial policies included incentives to selected sectors. These policies were combined with disciplinary measures applying to government ministries, the financial sector and the industries themselves through strict monitoring, close supervision and the use of export performance as the main productivity yardstick. In the World Bank study, the criterion of the *increase in the trade/income ratio* defines all these policies according to their outcome and not according to the inputs, and it is by no means clear that a combination of similar policies today would lead to similar outcomes in other countries.

Since the 'more-globalized' countries included both China and India, they included the majority of the population in the developing countries. By dividing all the developing countries into two subgroups according to the very general criterion of the *increase in the trade/income ratio*, they included among the 'globalizers' several countries that may have not been classified as 'globalizer' according to any other criterion. In other words, with this rather mechanic indicator and arbitrary classification, important information that is contained in many other indicators that define the policies and can explain the differences between 'globalizers' and 'non-globalizers' may have been 'classified out'.

Even this general classification, however, highlights the miraculous growth of the 'globalizing' countries that were among the world's poorest countries in the 1960s and 1970s, but accelerated their growth rate per capita from an average annual rate of less than 2% during the 1970s to over 6% during the 1990s. These countries' rapid growth was mainly driven by the fast expansion of their exports that enabled them to build a strong industrial base and achieve a major reduction in poverty. The 'globalizers' concentrated in the EA countries and their share in the world's moderately poor population declined from 82% in 1970 to around 39% in 1999.

3.4 The Impact on income inequality

The trends in income inequality in these two groups of countries are more complex: when the developing countries are combined together, inequality rose between 1960 and 1975, declined to 1985 and then started to rise again since then. One important reason is that China and India were the poorest developing countries in 1960s and the 1970s (Table 5.3). When China started

Table 5.3. Poverty among the 'globalizers' and the 'non-globalizers', 1970–1999.[a] (From World Bank, 2001a,b.)

	'Globalizers'			'Non-globalizers'		
		Poverty rates (%)			Poverty rates (%)	
	Population	US$1/day	US$2/day	Population	US$1/day	US$2/day
1970	1516	25	61	454	18	42
1980	1986	19	51	589	11	32
1990	2373	9	33	759	12	30
1999	2657	4	19	906	18	30
Share in world's poor (%)						
1970		82	83		18	17
1999		39	65		61	35

[a]Income per capita from the national accounts.

to accelerate its growth, the income gap between the poorest and the most affluent countries within the group of developing countries was reduced and income inequality declined.

Later on, when a larger share of the Chinese population moved to higher income groups and out of poverty, a larger share of the population in SSA dropped to the bottom of the global income distribution. However, the weight of China in the world population brought about a decline in the total number of the poor and in the income gap between the poorest and the least poor developing countries. The decline in the world's income inequality between *individuals* was due to the faster growth rate of per capita income of the poorest one-fifth of the world population at an annual rate of nearly 4% from 1980 to the early 1990s (up from 1.8% in 1960–1980), whereas the per capita income of the top one-fifth of the world population grew during these years at an annual rate of 1.7% (down from 3.3% in 1960–1980). Since the early 1990s, the growth rate of per capita income at the lowest quintile was dominated by the stagnation in the SSA countries, and it was then slower than that of the developed countries.

In the study of Dollar and Kraay (2002) that is also discussed in Chapter 2, they tested directly the impact of greater openness, or 'globalization', on growth, poverty and income inequality. Their criteria of globalization included, however, not only the rise in the share of trade in GDP and the countries' cuts in tariffs. In order to eliminate the geographically determined component of trade, they estimated the effect of trade on growth using decade-over-decade changes in the countries' trade as shares of their GDP, thus removing the spurious effect of geography on trade and growth. To control the effects of other contemporaneous changes in policies and institutions that may affect growth, they included measures of the stability of monetary policy, financial development and political instability.

They showed that both the increase in trade and the increase in FDI are positively related to the accelerated growth in income, and found a statistically significant and economically meaningful effect of trade on income growth: an increase in trade as a share of GDP of 20% points increased growth by between 0.5% and 1% point a year. The 'globalizers' increased the share of trade in their GDP from 16% to 33% and cut tariffs by 22% points from 57% to 35%. By contrast, the share of trade to GDP of the 'non-globalizing' countries declined and the cut in their tariffs was by only 11% points. Growth in the 'globalizing' countries increased from 2.9% per annum in the 1970s to 3.1% in the 1980s and 5.0% in the 1990s, while in the 'non-globalizing' countries growth fell from 3.3% per annum in the 1970s to 0.8% in the 1980s and 1.4% in the 1990s (Fig. 5.5).

Comparing the growth performance of these two groups of countries and their rates of poverty reduction, Dollar and Kraay also found that greater openness that accelerated growth did not bring about an increase in income inequality within countries or disproportionately hurt the poor. Their finding that there was a one-to-one relationship between the growth rate of income of the poor and the growth rate of per capita income, so that the percentage changes in incomes of the poor were on average equal to percentage changes in average incomes and there were no systematic changes in income inequality, is, perhaps, the least intuitive. As shown in Chapter 2, the reason is not that the poor had an equal share in the growth rate enjoyed by the general population. Instead, it was due to the large differences in the growth rates of the countries at the bottom of the world's income distribution and the resulting changes in the identity of the poor over time.

What I found most surprising in both the World Bank study and the study of Dollar and Kraay is that, by lumping together all the 83 countries in their sample and dividing them into two groups on the basis of a performance criterion, which is sensible in itself, they ignored the fact that many of these countries, including some of the countries that were in their list of 'globalizers', were in fact countries that went through a very comprehensive SAP since the early 1980s. These countries were not entirely free to determine whether they would be 'globalizers', since the condition on open trade was one of the conditionalities they were required to meet. The results for the globalizers were primarily determined by the developments in China, and

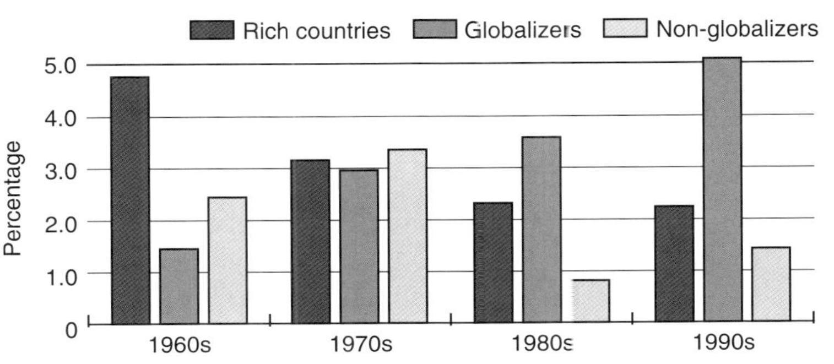

Fig. 5.5. Real per capita GDP growth in main subgroups.

its success to reduce poverty and accelerate growth determined the 'average' for the entire group. In the 1990s, India joined the group of rapidly growing economies, and, together with China, it biased the results even further. Figure 5.4 above shows the tariff rates in different regions. On the basis of this criterion, one cannot include these countries among the globalizers, and although they reduced their tariffs very sharply – at least relatively – in the last two decades, by the end of the 1990s their tariffs were still higher than in most other countries. Nevertheless, the EA globalizers were not included among the countries that had to meet the Washington Consensus conditionalities, whereas a large group of the non-globalizers were not free to determine their growth, trade and anti-poverty and most other policies.

One of the most prominent features of the world economy in the last 20 years has been the liberalization of international trade and payments under the auspices of the GATT (and now the WTO), the IMF and the World Bank (as part of conditionality and SAPs) that contributed to a rise in world trade nearly five times faster than world output.

There has been a good deal of work on the relation between trade liberalization and economic growth, but much less work has been done on the impact of trade liberalization on income inequality in the developing countries and on the necessary institutional conditions for a successful liberalization. Alan Winters (2003) argued that trade liberalization by itself is unlikely to boost economic growth very much, unless it is accompanied by improved macroeconomic policy making, better functioning institutions, predominant law enforcement and reduced corruption. Econometric studies that disregard these necessary conditions cannot therefore provide a complete explanation of the strong causal link between trade liberalization and growth. Rodriguez and Rodrik (2000) expressed concern that the priority given to the implementation of trade reforms may create expectations that these reforms, by themselves, are the panacea for growth and poverty reduction and can therefore substitute other policy and institutional reforms. Greater openness, trade and foreign investments are highly complementing and strongly interrelated, but effective governance, well-functioning institutions and the rule of law are necessary conditions for each of these components.

Santos-Paulino and Thirlwall (2004) use a panel of 22 developing countries that have undergone extensive trade liberalization since the mid-1970s to examine the impact on their export growth, import growth, the balance of trade and the balance of payments. They found that export growth rose by about 2% points, but the growth of the imports was much larger, at an average rate of 6% points, leading to a deterioration of the trade balance of at least 2% of GDP. However, the impact on the balance of payments was lower, which suggests that while liberalization may have, on balance, improved growth performance, countries were forced to adjust in order to reduce the size of their payments deficits to sustainable levels, thus reducing their growth below what it might otherwise have been. This conclusion suggests that the sequencing of liberalization is as important as the sequencing of internal and external financial liberalization for reducing the constraints on the growth process.

Dickerson *et al.* (2001) consider the impact of trade liberalization on the level and structure of wages in developing countries, with particular reference to Brazil, and the difference in performance between the traded and non-traded goods sectors. Contrary to the predictions of the Heckscher–Ohlin theorem, several studies show that the wages of unskilled workers have actually fallen relative to skilled workers through a process that the authors call 'skill-enhancing trade' associated with greater inflows of FDI and modern technology. This pattern has also been observed in Brazil, where significant liberalization was introduced in 1990. A number of interesting and significant results were observed. First, while the overall real hourly wage hardly changed between 1981 and 1999, the ratio of wages in the traded goods sector relative to wages in the non-traded goods sector fell in the 1990s, which suggests increased competition and reduced rents in the traded sector. Second, if improvements in education and work experience are controlled for, real wages fell by 15.9% in the traded sector and 8.1% in the non-traded sector. Third, this fall was more marked the more exposed was the traded sector to foreign competition. Fourth, workers in the four lowest education categories were affected the most. For workers in the two top educational levels, the marginal returns to further education increased after the trade liberalization. The authors believe that this pattern is quite common in a fairly large number of developing countries that have liberalized their trade. In short, standard international trade theory that predicts a convergence of the wage distribution does not seem to be borne out in practice when developing countries liberalize their trade because it ignores the flow of international capital and the technology transfer which increase the demand for skilled labour.

3.5 Impact on poverty estimates: based on income data

The World Bank (2000, 2001a,b) study is based on household consumption data collected in the income and expenditure surveys; Sala-i-Martin (2004) showed with a larger data set of the national income the differences between the changes in poverty in the main regions where the developing countries concentrate. According to his findings, the incidence of poverty in China was reduced from 46% in 1970 to 24% in 1998, and in India from 24% in 1970 to 15% in 1998; in Africa, in contrast, the incidence of poverty increased during these years from 11% to 38%.

The study thus concludes that globalization has been a key factor in the reduction in global poverty, and economies that were more outward-looking and export-oriented in their policies and managed to integrate into the global economy grew faster, and their growth was generally widely distributed across income groups. On these grounds, the study proposed an agenda for action based on pro-trade policies in order to integrate the developing countries into the global economy and accelerate their growth. Accordingly, since growth is the key to the reduction in poverty and open trade is the most effective policy to accelerate growth, open trade should be at the centre of the LDCs' development strategy and offers the best way of reducing their poverty.

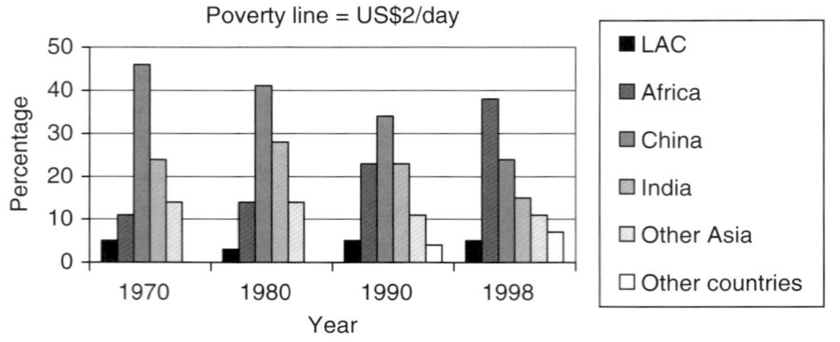

Fig. 5.6. Share of main regions and countries in the world's poor population. (From Sala-i-Martin, 2004.)

The study by Sala-i-Martin (Fig. 5.6), like that of the World Bank, draws its conclusions on the basis of the outcomes, namely the success in increasing the share of trade in the national income. However, in order to propose an agenda for action, it is necessary to know what were the *input variables*, namely the policies that brought about the increase in that ratio. It is also not clear whether an increase in that ratio is desirable in all countries. In some countries, that increase can represent an increase in the share of the manufacturing sector in their GDP; in others it may represent an increase in the exports of natural resources that may have been due to an increase in their price. An agenda for action should specify the sectors that would increase output, the means to achieve that increase and the action that should be taken in the other sectors and with their employees. It should also clarify how relevant this agenda is for all countries.

3.6 Potential pitfalls in the lessons from multi-country studies

The World Bank study *Globalization, Growth and Poverty* (2001b) recognized the limitations of statistical analysis that associates trade and growth in an attempt to establish links of causality. It is possible to have in this analysis identification problems, since trade is not the only factor that brought the increase in growth; the assumption of ceteris paribus that is necessary to identify causality does not hold in these comparisons. Moreover, trade not only affects, but it is also affected by these countries' growth.[3]

The main criticism of that study was that an explanation of the rate of economic growth for a group of countries on the basis of a *single* indicator, which thus defines their 'degree of globalization' by means of a single and rather ad hoc criterion, is bound to be problematic. Rodriguez and Rodrik (1999) argued that the methodological problems with the empirical analysis

[3] Other studies conducted a more complete econometric analysis in order to net out the effects of other factors, but came to essentially the same conclusions. See Frankel and Romer (1999), Dollar and Kraay (2001).

in the multi-country studies leave the results open to diverse interpretations, largely because the openness indicator that was used is a poor measure of the reduction in these countries' trade barriers and a poor measure of their trade policy.

Rodrik also commented:

> This is why studies such as David Dollar and Aaart Kraay's 'Trade, Growth, and Poverty' (Development Research Group, The World Bank, March 2001), which purport to show that 'globalizers' grow faster than 'non-globalizers,' are so misleading. The countries used as exemplars of 'globalizers' in these studies (China, India, Vietnam) have all employed heterodox strategies, and the last conclusion that can be derived from their experience is that trade liberalization, adherence to WTO strictures, and adoption of the 'Washington Consensus' are the best way to generate economic growth. China (until recently) and Vietnam were not even members of the WTO, and together with India, these countries remain among the most protectionist in the world.
>
> (Rodrik, 2001, p. 8)

However large the explanatory power of the single indicator that determines the degree of globalization (that in the statistical analysis is determined by the standard error of that ratio), one cannot make any conjectures or recommendations on this basis about the policies that led to this outcome. Moreover, the results that were derived from a large sample of countries cannot be effectively used to make any predictions about future policies for an individual country. Indeed, as noted earlier, the EA 'globalizers' kick-started their industrialization by targeting specific sectors for learning-by-doing, but given the speedy changes in technology and in the global economy today, there are high risks in using the same strategy in other regions, particularly in SSA, in the future.

Another important factor that is averaged out in the multi-country study is the large differences between the trade policies of the different countries that were included among the 'globalizers'. Moreover, the trade policies of the two largest countries in this group, China and India, do not even fit the simple specification of a 'globalizing' country as defined in this study. Both countries combined, and are still combining, many of the orthodox guidelines of outward-oriented policies and measures to promote their exports with inward-oriented and very restrictive policies that included extensive government interventions in nearly all sectors, very active industrial policies that were aimed at promoting selected industries, high rates of trade protection on many products, limited privatization and so forth. As noted earlier, this was one of the subjects on the agenda of the Doha Round, but it remained unresolved when the negotiations reached a stalemate.

Another cautionary note on these studies concerns the fact that they are based on the experience of the EA countries in the last two decades and on data from the 1980s and the 1990s – in most studies until 1997/98. An econometric analysis of these data or an analysis that is using CGE models can identify the factors that contributed to the successful performance of these countries in the past, but they have obvious limitations when they are used for predicting their contribution in future years, because there have been comprehensive structural changes in the world's economy since then.

Studies that use CGE models usually try to make corrections that take these changes into account as much as possible, but the predictive power of the models is inevitably more limited. These changes are bound to affect the relations between the explanatory variables, particularly variables that are determined by government policies, and the dependent variables that predict the outcomes of these policies. These changes must also include the reorganization of the institutions that govern world trade, ranging from the WTO that started to operate only in 1995, to the many RTAs that proliferated mostly in recent years and are still in the process of determining the rules and the participating countries. Moreover, these changes have very different effects on different countries and economic sectors, and the lessons from the EA countries with respect to, say, 'outward-looking' trade policy, may not be equally suitable for other countries where the social, economic and political structures are very different.

Estimates that are derived from these models without taking all these changes into account are particularly vulnerable to what has been termed the 'Lucas critique'. This critique emphasizes the limitations and potential bias of predictions on the response of the economy or a given sector to a new set of policies when the predictions are made on the basis of estimates that were derived before these structural or policy changes took place and new policies were introduced.

Another critique or potential pitfall is the failure of these predictions to take into account the impact of changes in policies or in the external conditions that are still taking place due to the dynamics of past policy changes. In other words, some institutional and organizational changes are likely to be made in reaction to past changes, and it is difficult to predict what their impact will be 2, 3 or 5 years from now. One example is the failure to reach an agreement in the Doha Round, which raises the possibility of different ways to organize world trade through RTAs, including the much talked about possibility of a 'Trans-Atlantic Free Trade Area' (TAFTA). These future changes that are a reaction to past developments are likely to have very different effects on our predictions.

In today's world, the professional duty to take into account structural changes that have already taken place must therefore be combined with the art of estimating the impact of changes that are about to occur or are very likely to happen during the prediction period. In a simulation analysis, these changes can still be taken into account by simulating different future scenarios, but even this analysis is constrained to a very small set of simulations about future scenarios.

All these critiques are, in fact, well known, but they are not always taken adequately into account in the interpretation of these predictions. Despite the difficulties of making predictions, especially, as the saying goes, about the future, they seldom stop us, or other fortune tellers, from proceeding to crunch the numbers and make the predictions. All these reservations notwithstanding, there are some important lessons from the experience of the EA countries that are relevant for the design of development strategies in the other developing countries today.

3.7 The impact of China and India

In the World Bank study on *Globalization, Growth and Poverty* (2001b), the statistical results for the 'globalizers' are dominated by the data of China and India. The steep reduction in poverty in the 'globalizing' countries, as well as in global poverty, is due to the steep reduction in poverty in China. Thus, while the reduction in poverty among the 'globalizers' was one of the key conclusions of the study, this was in fact a result of the reduction in poverty in China, even though China implemented a very heterodox trade policy that included many restrictions on trade. The rapid growth of the countries in that group is also mostly due to the rapid growth in China and India that overshadowed the impact of the financial crisis and the steep drop in output in the late 1990s and early 21st century, which affected quite a few other 'globalizers' – their globalization notwithstanding – that forced them to restrict their imports and thus also their total trade.[4] In a commentary on the World Bank (2001a,b) study, Rodrik argued: 'Rapid integration into global markets is a consequence not of trade liberalization or adherence to WTO strictures per se, but of successful growth strategies with often highly idiosyncratic characteristics' (Rodrik, 2001, p. 1).[5]

Even the high rates of tariff reductions among the 'globalizers', at twice the rate of reduction among the 'non-globalizers', were primarily due to the very high tariffs that prevailed in China and India at the initial stages of the process. During the 1990s, and even after China became a member of the WTO in 2001, imports into these countries were still restricted by a maze of high tariffs, complex regulations and various other restrictions. Many of these restrictions still exist today.

In India, economic reforms started only in 1991, and the country still maintains high protection on many local industries. In both China and India, poverty already started to decline during the 1980s, well before they started to open up their economies to trade (Fig. 5.7). The main driving force for the reduction in poverty in India during the 1980s was the Green Revolution and during those years, incomes rose more rapidly in the rural sector and poverty declined.[6]

[4] Interestingly, though, the share of trade in the GDP in Argentina was not affected by the crisis since the sharp reduction in trade was accompanied by a sharp reduction in the GDP.
[5] In the case of China, however, the more extensive trade liberalization was largely (though not entirely willingly) due to adherence to WTO strictures when it sought acceptance to the organization. Until the late 1990s, both China and India maintained considerable restrictions on imports.
[6] The estimates of poverty in India that are based on consumption surveys vary widely between studies: The World Bank's *WDR* 2000/2001: 'Attacking Poverty' shows that the number of people living on less than US$1 a day grew from 1.18 billion in 1987 to 1.20 billion in 1998 – an increase of 20 million. The World Bank publication *Globalization, Growth, and Poverty* showed that the number of people living in poverty fell by 200 million from 1980 to 1998. Agnus Deaton estimates that the headcount ratio shows a fairly steady decline from 1987/88 through 1999/2000 and poverty fell from 36% in 1993/94 to 28%. Chen and Ravallion estimated that poverty declined from 54.4% of the population in 1981 to 42.1% in 1990 and 34.7% in 2001.

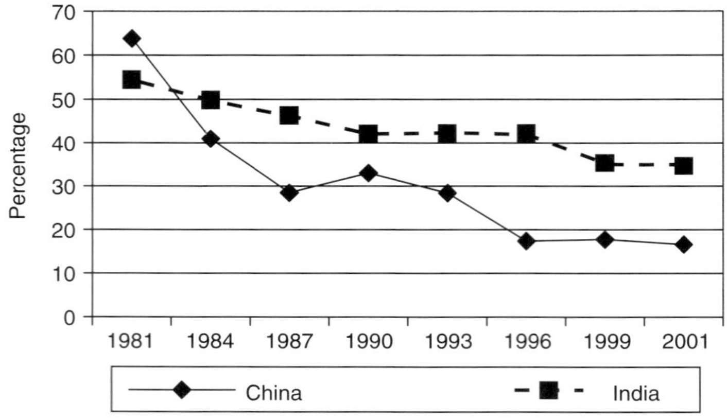

Fig. 5.7. Poverty measures in China and India. (From Ravallion and Chen, 2004.)

Only after the economic reforms in 1991, urban incomes started to rise more rapidly, whereas the rural sector was barely affected. That growth concentrated in the service sector and among skilled workers and it raised the country's income inequality and the rural–urban gap. The reduction in rural poverty was at a much slower pace and it was partly due to the rural–urban migration and partly due to remittances of the urban workers to their families in the rural areas.

In China, after a steep decline in poverty in the early 1980s (Fig. 5.8), poverty reduction stalled and recovered only in the mid-1990s, but stalled again in the late 1990s, with large variations between provinces. Inequality was rising, and absolute inequality increased appreciably over time between and within both urban and rural areas, with higher inequality in urban areas. From 1985 to 2001, the headcount measure of poverty declined from 17.5% to 8%. The growth process concentrated on labour-intensive industries that required low-skilled workers, which led to an increase in the number of migrants from the rural areas.

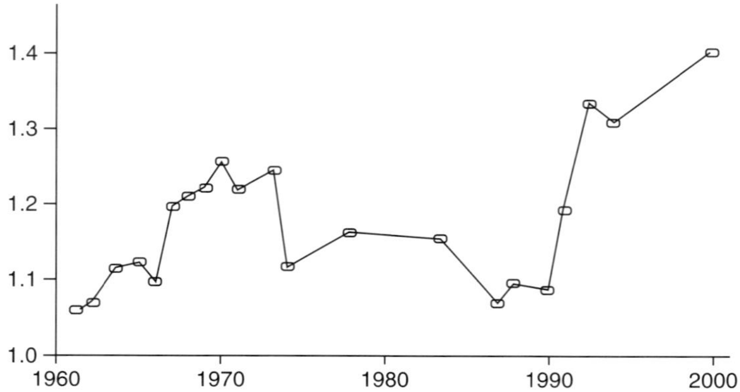

Fig. 5.8. The ratio of rural to urban poverty in India. (From Ravallion and Chen, 2004.)

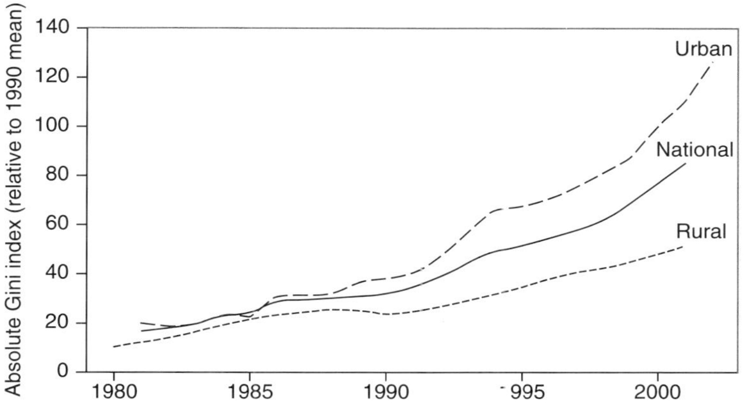

Fig. 5.9. Income inequality in India. (From Rava lion and Chen, 2004.)

Income inequality between the rural areas in China, measured by the Gini coefficient, increased from 25.7 in 1985 to 36.5 in 2001, and in urban areas inequality increased during these years from 17 to 32.3 (Fig. 5.10). The high-income inequality within the rural areas in the 1980s reflects the large income differences between regions. For rural areas, proximity to the urban centres was a key factor for reducing their poverty and raising their incomes, because a nearby urban centre made it easier for the people in villages to find employment given the strict controls that the government applied on migration. In addition, these villages were also able to send their products to the urban centres and thus to diversify their agricultural production.

For more remote villages, particularly villages in the western provinces, the distance from the eastern part of the country prevented them to attract any industry and they were also unable to have much trade with the urban centers, not even in their own provinces, due to the poor infrastructure; consequently, most of their production was for self-consumption. These provinces were much less affected by the country's industrialization and rapid growth. The large income gap between them and the rural

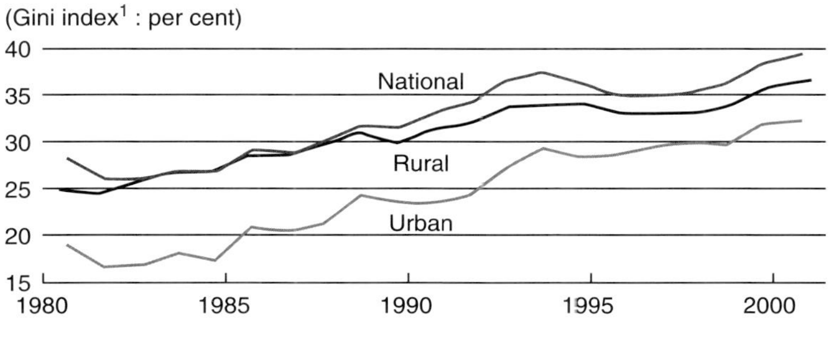

Fig. 5.10. Trends in income inequality in China, 1980–2001. (From Ravallion and Chen, 2004.)

population in the coastal provinces is the main reason for the high-income inequality in rural areas. In urban areas, inequality has been smaller because the vast majority of the urban population concentrates in the coastal provinces and income differences between regions have a smaller impact on income inequality in the urban areas as a whole. Nevertheless, the *rise* in income inequality in urban areas was faster than in rural areas and the income inequality in the country as a whole rose at a higher rate since the mid-1990s.

3.8 Impact on poverty in the LDCs

The steep reduction of poverty in China and, to a lesser degree, in India, was therefore the driving force for the reduction of poverty in all developing countries; however, in the LDCs, poverty actually increased. On this point, Sala-i-Martin (2004) argued that, although this was true, it was irrelevant: 'Of course when we exclude those countries where poverty declines, poverty in the remaining countries increases.' This, however, does not answer the criticism. The questions today are not about the changes that took place in the past, but about the changes that are likely to take place in the future, and whether the experience of China, successful as it has been, is relevant for the analysis of future trends in the LDCs in SSA, Central America and Central Asia.

Can these countries expect a similarly rapid growth and a steep reduction in poverty if they follow the footsteps of the 'globalizers' and implement similar policies? As noted earlier, regression analysis that demonstrates the successful experience of the 'globalizers' in the past may fail to provide an adequate and complete answer if it does not take into account the massive changes in the basic economic conditions in world production and trade. The experience in India also shows that, even though pro-trade reforms can generate growth that concentrates in the urban centres, this may have a limited impact on poverty in the rural areas. Although the incidence of moderate poverty in the LDCs was high, the size of the poor population in China and India in 2000 was considerably higher.

The structural changes in China and in the other EA countries since the late 1970s and the economic reforms in India since 1991 enabled them to make the adjustments in their own economies that prepared them for the new structure of world production and trade. The sheer size of these two countries enabled them, together with the other emerging economies, to gain considerable influence in shaping the emerging institutional organization of world trade through the WTO and the RTAs. These countries also have the conditions to attract foreign investments of transnational corporations. In the LDCs, in contrast, the structural changes in their economies and the changes in their policies or in their regimes were minimal and were largely enforced by the 'conditionalities' of the international development organizations (Table 5.4).

Table 5.4. Poor population and the incidence of moderate poverty in the LDCs in 2000 (poverty line US$2 per day per person). (From World Bank, 2001a.)

Year 2000	No. of countries	Population (millions)	Poor population (millions)	Poverty incidence (%)
UN list of LDCs	48	644	476	74
HIPC countries	41	615	418	68
SSA countries	46	642	469	73
All LDCs[a]		1006	714	71

[a]Countries included in either one of the above lists.

3.9 The impact on the trade of the developing countries

The structure of world production and trade changed also the composition of exports of the developing countries: For nearly two decades and until the early 21st century, demand for minerals and natural resource-based manufactures in the world market was sluggish and prices declined, which negatively affected the exports from the developing countries. Since the early 21st century these trends were reversed; prices started to rise quite sharply, most notably the price of oil and natural gas, due to the massive increase in the demand of the emerging economies, particularly China. The price rise brought a sharp rise in the value and volume of the exports of the developing countries in SSA, LA and Central Asia, that were the prime exporters of these products. That rise in world demand and price attracted large foreign investments to these countries.

The stagnation or decline in the exports of primary products and resource-based manufactures until the early 21st century stood in sharp contrast to the rapid rise in the exports of manufactured products based on advanced technology such as electrical appliances and equipment for the information and communications industry. The share of the entire group of developing countries in world exports of low-tech manufactures rose by 50%, whereas their share in high-tech manufactures rose by 120%. These changes mirror, in part, the large structural changes in the production of the EA countries and their rapid growth (Table 5.5). The exports of manufactured products were primarily from the 'globalizing' developing countries that accelerated their industrialization.

The efforts of the LDCs today to build and expand their industrial base and their exports of manufactured products are bound to meet stiff competition from countries like India, China and other emerging economies that already have similar industries that make use of their own abundant and cheap labour force and have an advantage in production which is partly due to the experience and expertise they accumulated, but also because of their higher productivity, more developed infrastructure and their connections to the global supply. At the initial stage, the LDCs and other low-income countries may therefore have to rely mostly on regional trade, on a wider diversification of

Table 5.5. The share of developing countries in world exports. (From calculations based on World Bank indicators, various years.)

	1985	2000
Primary products	59.6	62.0
Resource-based manufactures	31.3	31.8
Low-tech manufactures	33.6	50.3
Medium-tech manufacture	10.8	21.4
High-tech manufactures	16.8	36.6

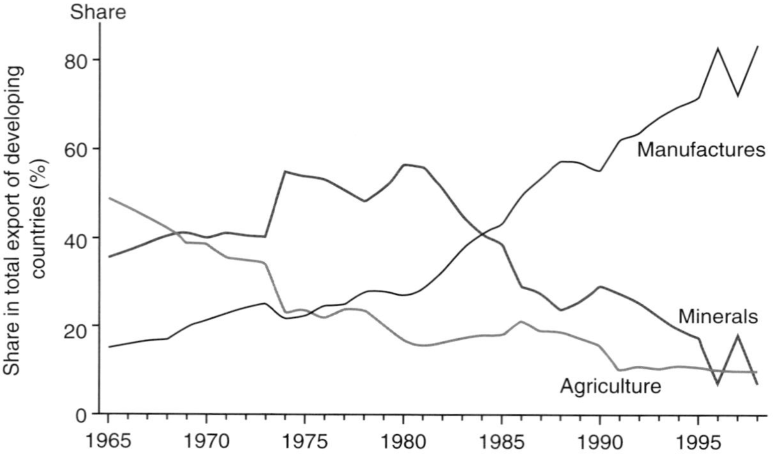

Fig. 5.11. Share of main products in total exports of developing countries. (From World Bank, 2001b.)

their agricultural production and on processing industries that will make use and increase the value-added of their resources.

Some of the recent changes in the global market conditions may assist the LDCs in adjusting to the new structure of world production and trade. The most important development is the steep rise in the prices of oil, many minerals and several key agricultural products (Fig. 5.11). Despite the failure to reach an agreement on agricultural trade at the Doha Round, the global market for these products is going through fundamental changes that may benefit the countries that have these resources. The conditions that are necessary to make effective use of these developments are discussed in Chapter 6.

4 The Role of the Government in Planning Trade, Growth and 'Pro-poor' Policies

The economic difficulties and the political pressures that many countries, developed and developing, encountered during the negotiations on the trade agreement during the Doha Round, and the ultimate failure to reach an agreement despite the benefits that it offers, are testimony to the power of the political opposition and the need to take into account in that process both the long-term gain and the short-term pain that trade liberalization is likely to bring.

The spectacular rise of the EA countries was driven by their rapid industrialization that followed the footsteps of the Asian tigers. Their openness to trade did not play a decisive role in these stages and throughout the 1980s and most of the 1990s, both China and India maintained very high import tariffs. Only in the mid-1990s, China started to cut its tariffs down to an average rate of 20% by the end of that decade. India cut its import tariffs gradually since the onset of economic reforms in the early 1990s by a total of two-thirds during the 1990s to an average rate of 30%.

In practice, therefore, the increase in openness was much less dramatic and many of the local industries remained protected in later years either by tariffs or by various regulations and restrictions. The reduction in the *average* tariff rate reflects primarily the reduction in tariffs on primary products and intermediate inputs that were aimed at assisting the rapidly growing export industries by lowering their costs.

Only in the late 1990s did China start a more substantial process of tariff reductions, largely aimed at fulfilling its obligations for entry to the WTO. Another important policy instrument in China was its exchange rate policy. Since the early 21st century, China maintained the yuan effectively fixed to the US dollars despite its growing trade surplus and mountains of foreign exchange reserves; with this exchange rate policy, the Chinese authorities gave strong support to their export industries, protected their import-substitution industries and prevented exchange rate risks on FDIs.

The expansion of exports and local industries in most other EA countries was driven by their relative political and economic stability despite the currency crisis in 1997/98; other important factors were the improving infrastructure made possible by large government investments, large local savings and the rising FDI, improving labour productivity due to great progress in the education system and the imports of advanced technologies with the FDI, as well as the low wages that remained low despite rapid growth due to the continued flow of migrants from rural areas.

4.1 Restrictions on trade liberalization and the role of government

Today, the difficulties during the transition period have become a major constraint on trade liberalization in both developed and developing countries. Labour unions and enterprises in sectors that are likely to suffer losses from the reduction in import tariffs and the removal of other restrictions on imports oppose the plans to liberalize trade and put pressure on governments to forego further liberalization or to put other restrictions on that trade in order to protect the local industries. Consumers' gains from the liberalization and the promise of future gains from free trade for the economy are not sufficiently strong incentives for the general public to apply counter-pressure in favour of trade liberalization. The economic difficulties and the political pressures that many countries encountered in the negotiations during the Doha Round are testimony to the power of this political opposition and the need to take it into account in the design of a country's trade policy. In other words, the short-term pain of trade liberalization may prevent the long-term gain altogether.

As a result of these constraints, progress in trade liberalization during the last two decades has been highly uneven between countries. In quite a few developing countries, the expansion of trade was precipitated by bilateral agreements or RTAs or by outsourcing, offshoring and large flows of FDIs. But progress has been quite slow in 75 developing countries, particularly in Africa, South and Central Asia, as well as in the Middle Eastern countries (except for the oil exporters), including virtually all the LDCs.

When the government is fully aware of the short-term pain it can ease the transition and reduce the opposition by taking steps in advance to mitigate the difficulties of vulnerable population groups and protect those who may lose their source of livelihood. Relevant measures include social safety nets, adequate training programmes, public investments in the weaker regions and investments to support the sectors in which the country has an advantage in order to accelerate the transition and create new employment. When these measures were not taken, there was a strong opposition to globalization and it was particularly belligerent in countries where the gains from trade during the transition period concentrated in the hands of relatively small segments of the population and few regions, while the losses were widespread.

Under the current structure of world trade, where over three quarters of the trade, other than oil, is between countries that are either in the category of developed countries or in the category of emerging economies, the LDCs have very little advantages to offer for easing their integration into the global trade system: Their wages are not much lower than in the emerging economies, but their poor infrastructure, poor financial and other private institutions and cumbersome public administration make their production much more expensive than in most emerging economies, even in labour-intensive industries (Tables 5.6 and 5.7).

The obstacles and difficulties that the LDCs would encounter in the implementation of outward-looking economic policies are therefore less likely to secure for them the same advantages in today's global economy as the EA countries achieved with such policies a decade or two ago. The policy recipes and development strategies that brought about the EA miracle may

Table 5.6. The shares of developing regions in world exports (%). (From UNCTAD, Handbook of Statistics, various years.)

Region	1980	1990	2000	2002
Developing Africa	5.9	3.0	2.2	2.0
Sub-Saharan Africa	3.7	1.9	1.5	1.5
Developing Asia	17.9	16.9	24.3	23.3
Developing America	5.5	4.2	5.5	5.9

Table 5.7. Imports of developed countries from the developing countries in 2000[a] (in US$ millions). (From World Bank, 2001a.)

	World	USA	EU
All developing countries	1717	434	416
China	249	52	39
India	44	10	11
Categories of less developed countries			
LDCs (UN classification)	35	8	10
HIPC	57	10	16
SSA	83	20	18

[a]Including oil.

not be suitable for the LDCs of today given their own conditions and constraints and given the enormous changes in the global economy. The constraints of the LDCs also require a re-evaluation of the measures that the international development organizations and the donor countries would have to take in order to facilitate the integration of these countries into the global economy and help them accelerate their growth and achieve a meaningful reduction in poverty.

Moreover, in today's global economy, abundance of labour and low wages are much less decisive factors in determining a country's comparative advantage, particularly in the more advanced industries and stages of production. Far more important are the accumulation of human capital and rising labour productivity, which primarily depend on improvements in education and training of the workforce, the adoption of advanced technologies, the improvements of local infrastructure, primarily with the introduction of advanced information communication and technologies, and the connections to the global supply chains.

At the initial stages of development, transnational corporations outsourced to developing countries mostly the production of labour-intensive products, components or stages of production; gradually, however, they outsourced or offshored to these countries the production of technologically more advanced products. High and rising profits of foreign investors in these economies provided incentives to further increase investments in the emerging economies and to restructure the entire production by outsourcing more segments of the production process or by offshoring the entire production to these countries. As a result, the share of the emerging economies, particularly in EA, in world exports more than doubled, while the share of the African countries fell from over 3.5% in the early 1980s to 1.5% in the second half of the 1990s.

4.2 Impediments to trade in the LDCs and the role of government

The country-specific structural, institutional and legal organization of the system of governance in most LDCs, particularly in SSA, that dominate all aspects of their economic, social and political life, prevent a straightforward adoption of the policy recipes of growth and poverty reduction that were successful in EA. Even arguably universal principles of economics and international trade as well as social organization, such as property rights and the rule of law, often fail to function in many of these countries.

The unique characteristics of traditional social institutions in many African countries that were effective in the organization of social life in earlier times can now pose considerable constraints and impediments on their adjustment to the structures of a modern nation state. After independence from colonial rule, the organization of the social and economic life within the framework of a nation state and in borders that were drawn during colonial times, often mixing together many ethnic groups and different religions, required major transformations in the traditional social institutions and a complete reorganization of the economy that was still primarily rural. When these countries embarked on the process of changing the structure of their economies, building institutions of governance and implementing policies such as free market policies, they therefore faced an immense challenge.

In many countries, the evolving institutions were either a mixture of traditional and modern institutions or, more often, imported or superimposed institutions that frequently failed to function in their environment (Fig. 5.12). The organization of a modern legal system and a system of governance were slow and painful processes. As these countries now face another round of adjustment to the new structure of the global economy, which requires them to adopt new rules and regulations in order to abide by the rules of the international organizations and to be integrated into the global economy, the challenges are almost insurmountable. Their difficulties in making necessary adjustments created a vacuum during the transition period that often led to the collapse of the country's system of law and order, rampant corruption and, in some countries, an open abuse of the political institutions and the legal administration.

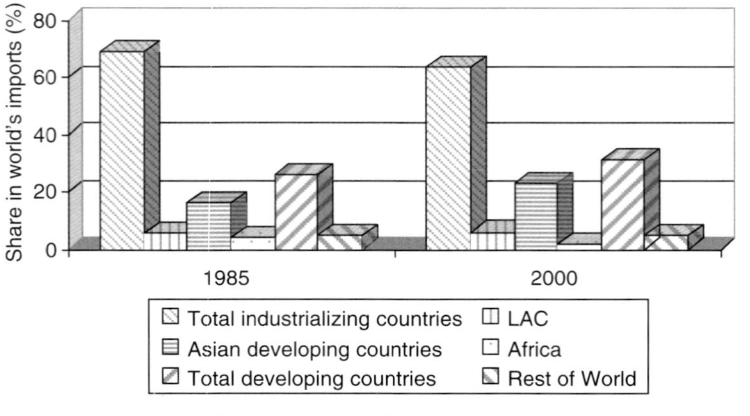

Fig. 5.12. Share of main regions in the world's imports. (From OECD, 2001.)

Given these obstacles and constraints, both the governments of these countries and the international development organizations must calculate very carefully and methodically their steps in the adjustment process and the application of general principles of economic decision making, like those recommended by the Washington Consensus or the specific lessons from the experience of the EA countries, which have to be translated into precise and definite guidelines for policies in the SSA countries. These policy guidelines must be designed to suit these countries' local conditions and unique characteristics, since it makes no sense to impose demands that these countries cannot meet.

However, these considerations have not always been taken into account. Although there have been disclaimers about the Washington Consensus, its spirit still prevails in many formal documents of the IMF and the World Bank. Indeed, the challenge of drawing those lessons from the experience of the EA countries that are relevant for the LDCs may sometimes seem to be a 'mission impossible', as the lakes of ink that have been spilled on claims and counterclaims amply demonstrate. Keeping in mind all these reservations and qualifications, several observations on some of the implications of key factors that contributed to the successful experience of the EA and that may be relevant for the LDCs should be noted:

4.3 Pro-trade macroeconomic policies

- Following the example of Japan and the 'four tigers', the EA developing countries adopted a wide range of policies that gradually transformed them into market economies. Sound macroeconomic policies were put in place and provided a stable and solid base for the business sector to grow. Later on in the reform process, greater emphasis was given to the liberalization of the trade and financial sectors and to orienting the entire economy outward by means of strong and well-targeted incentives to selected export industries.
- Political stability was essential for the stability of the economy at the early stages of development; later on, the strong economic growth itself contributed to further enhance the political stability, making many of the EA dictators nearly invincible for many years. The political instability in some EA countries after the financial crisis of 1997/98 severely shattered the confidence of foreign investors and local savers alike, and some of these countries are still struggling to restore stability.
- However, prudent macroeconomic policies were only partly supported, by efficient institutions and an adequate regulatory framework to supervise business practices and impose discipline on the market. The lack of reliable checks and balances and proper supervision led to 'crony capitalism' as more wealth was accumulated in which politicians, bankers and businessmen working together made unsound, and often corrupt, economic decisions.
- The banking system in nearly all EA countries, including China, also remained weak, risk management capabilities were underdeveloped,

supervision systems were meagre and vulnerable to political interference, and banks seemed to be confident that the good times would continue indefinitely and the bubble would never burst.

- The high investment rates in the EA countries had been possible owing to high savings rates that consistently exceeded 30% of the GDP during the 1965–1990 period. These investment rates were significantly higher than in nearly all other developing countries, and due to these investments the share of private investment in these countries' total investment was more than 40%.
- Exchange rates in the EA countries, and later also in some LA countries, were maintained at stable levels against the US dollar for an extended period of time. This stability made an important contribution to stabilize the domestic economy and attract foreign investments, but it also proved to have clear limitations and ended rather abruptly and disastrously in many (but not all[7]) of these countries. Among the LA countries, Argentina was harmed the most from the collapse of its currency (see below). The lesson is clear and well known: The exchange rate cannot and should not be maintained at a fixed but unsustainable level.[8]
- As a result of weak banking systems and untenable exchange rates, the entire financial system in most EA countries was highly vulnerable and unable to confront the financial panic. Initially, the panic gripped investors when the Thai government removed the baht's peg to the dollar, later the loss of confidence of local and foreign investors spread to other EA countries and immobilized the region's entire banking systems.
- The spread of the crisis to LA, particularly to Argentina that for quite a few years pegged the peso to the US dollar, was also partly due to the sharp decline in the real exchange rate as an effect of the local inflation and partly due to the spread of panic.
- The policy of maintaining a stable exchange rate is still disputed, but it is clear that a stable exchange rate can only be a supporting instrument to stabilize the economy and that long-term stability first and foremost requires very disciplined and prudent macroeconomic and monetary policies. Without these necessary conditions, the system is highly vulnerable to financial and economic crises.
- The EA experience exposed the limitations of the formula 'Get your macro balances in order, take the state out of business, give markets free reign', which seems to be taken from the ten commandments of the Washington Consensus. In fact, this important lesson from the EA experience has been already recognized in the World Bank report mentioned

[7] Taiwan, Malaysia, Singapore, China and Vietnam were less affected by the crisis, in part because they maintained tight controls and severe restrictions on capital flows.

[8] Many countries, as well as the IMF, are now trying to convince China to adopt this lesson with respect to the Yuan, but there is a large difference between a country that maintains its currency at an undervalued exchange rate and a country that maintains its currency at an overvalued exchange rate. In principle, China may seem invincible, as long as it is willing to take the risk of huge losses if the dollar one day collapses.

earlier on, *The East Asia Miracle*, which noted that in most of the EA economies, the government intervened systematically to foster development in general and the development of specific industries in particular. In addition to direct support to local industries through subsidized credit, they also developed export marketing institutions and provided direct assistance to selected sectors. The Chinese government also made large investments in infrastructure and in professional training.

In 'Making Openness Work: The New Global Economy and the Developing Countries', Rodrik lists the complementary policies that must be in place in order for a developing country to benefit from trade:

- 'Domestic investment strategy' to 'kick-start growth', as was the case in EA. With these investments, domestic industries are greatly strengthened, forming a strong base to build further growth and to compete in international trade. South Korea, Taiwan and Singapore heavily subsidized private investment. With domestic investment, there is an accumulation of technology as well as human and physical capital.
- Macroeconomic stability is needed in order to secure the success of trade liberalization. Social and political institutions must be strong enough to absorb and adjust to shocks, unstable conditions, conflict and other types of change.
- Strong institutions are essential in dealing with instability and change.
- A functioning judicial system and enforcement of the rule of law.
- Social safety nets to smooth the transition and strengthen the macroeconomic instability.

Since this list of conditions is very demanding and very few developing countries are likely to be able to meet all of them, many economists question the wisdom of making a rapid transition to laissez-faire liberalization policies.

4.4 Selective Interventions

The World Bank report classified the entire set of policies that the EA countries implemented into two broad groups: (i) fundamentals; and (ii) selective interventions. Selective interventions are in the commodity and financial markets and include mild financial repression, directed credit, selective industrial promotion and targeted incentives to promote nontraditional exports. The fundamental interventions focus on a prudent macroeconomic policy. This policy can only provide the necessary conditions for growth, however, whereas the direction of that growth to specific sectors has to be given by the direct and selective interventions. In EA, these interventions included:

1. All countries began with a period of active restrictions on imports and a strong bias *against* exports in order to promote import-substitution industries, but they gradually moved to reduce the bias against exports and establish a pro-export regime and outward-oriented trade policies. The export promotion policies compensated the exporting enterprises for the import controls and moderated the protection of the domestic market by giving exporters automatic access to duty free imports. The reduction in import protection was therefore more gradual and was accompanied by direct support to exporters from the government ministries in charge. Additional incentives to the export industries were given in the form of export credit, broadly based market incentives and measures to attract FDI.

2. Intensive and well-targeted intervention to promote exports seemed to be preferred over the option of leaving the production decisions entirely in the hands of the private sector and letting the free market reign. These selective and targeted incentives included deliberate distortions in the costs of capital and various credit subsidies to entrepreneurs.

3. Growth of exports, particularly in manufacturing, was facilitated by capturing a large share of the preferences granted to the developing countries under the general system of preferences (GSP) and they received a significant share of the textile and garment trade subject to the MFA (Fig. 5.13).

4. Open access was given to transfers of private foreign capital to the local banking system in order to promote competitiveness.

5. FDI played a critical role by accelerating technology transfer, creating employment, training the labour force and generating foreign exchange.

6. Targeted acquisition of technology by governmental institutions through a variety of mechanisms that included licenses, imports of capital goods and foreign training was an important tool to promote certain industries and import advanced technologies (Fig. 5.14).

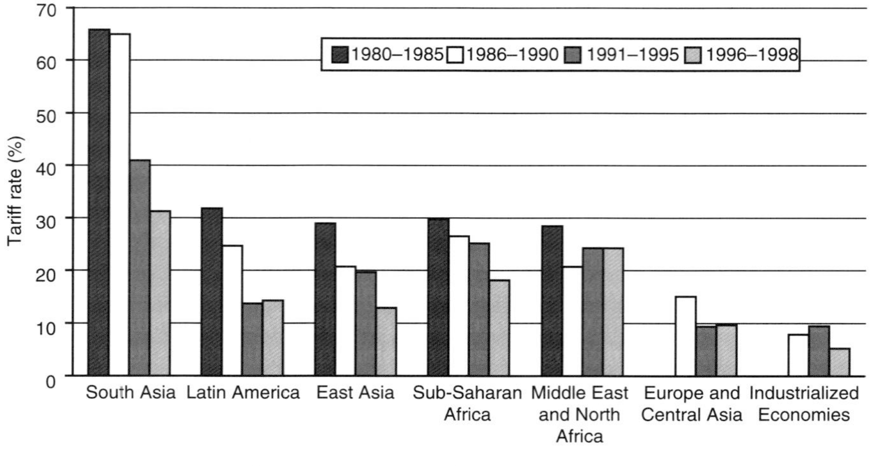

Fig. 5.13. Changes in tariff rates since the early 1980s.

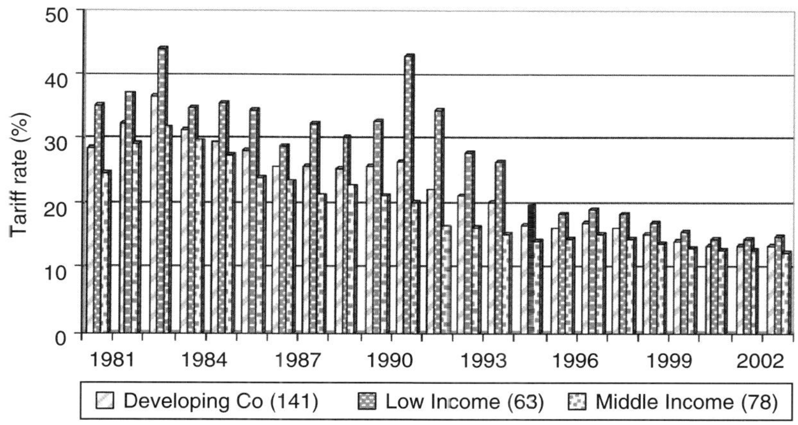

Fig. 5.14. Average tariff rates in developing countries.

Some economists summarized these forms of government intervention as a way of remedying the market failures and guiding investments by deliberately 'getting the prices *wrong*' and altering incentives in order to boost industries that would not have otherwise survived (Amsden, 1989).[9] The World Bank report (2000, p. 10) also acknowledged that governments intervened extensively in the local and the export markets in order to guide the resource allocation of the private sector.

The lesson from this experience is the contribution and the positive impact of, and sometimes the need for, well-designed selective interventions. However, the limitations of these interventions must also be recognized: the basic concept of 'well-designed interventions' transfers the decisions on the allocation of a country's resources from the free market forces to the hands of a bureaucracy that, well-intentioned as it may be, is bound to be arbitrary and more often than not highly inefficient. Many selective interventions of the government bureaucrats in the EA countries were indeed inefficient and often outright biased by corruption and political pressures.

Ironically, an advantage of the EA countries was that a transition to a free market economy was not really an option since there was no free market in these countries and the only experience they had in the past was with arbitrary decisions of the central authorities with respect to each and every aspect of economic development. In China, the climax of the government decisions was the collectivization of the entire economy, particularly all agricultural lands; all economic decisions, including even the details of everyday life, were made by the central authority – at the level of the village, the district, the province and the country. The transition to selective interventions that

[9] On these grounds, some economists and political scientists also argued that the EA miracle was due to the high quality and authoritarian nature of the region's institutions, describing the EA political regimes as 'development states' in which powerful technocratic bureaucracies, shielded from political pressure, devise and implement well-honed interventions (Dieter *et al.*, 1998; Hae, 1996; World Bank, 2001, p. 13).

maintained the power of the central authority, but gradually transferred economic decisions to individuals, was their only way to make the reforms.

Many other LDCs, including Vietnam, Cambodia, Nepal and Mongolia, are now going through a very similar process and face similar choices and dilemmas. Structural adjustments that transfer the decision-making process to the free market are not an option in countries that never had a free market and the only way for the 'market forces' to survive in this environment was by means of endemic corruption, that enabled mid-level bureaucrats to take some initiatives and make some decisions, that inevitably were self-serving.

5 How Relevant Is the East Asian Experience for the LDCs Today?

The lessons from the EA miracle are obviously clouded today by the events that followed the bursting of the bubble in the capital and real estate markets and the currency crisis at the end of that miracle in 1997/98. The relatively rapid recovery in most EA countries restored the reputation of the miracle only partly. However, the seemingly unstoppable progress in China, which achieved nearly equal status to that of the USA not only in the economic arena, but also in the political arena, as well as the relatively rapid recovery of many of the EA countries that were affected by the crisis of 1997/98, and the miraculous progress of Vietnam during the past decade, left many of the lessons from the EA miracle still relevant. In hindsight, the past decade suggests several additional observations:

1. By many accounts, the high rates of growth were not only due to the accumulation of physical and human capital, but also due to TFP growth and, according to some estimates, a full one-third of the overall growth was due to TFP increase (see Chapter 6, this volume). As noted earlier, Krugman (1994) argued that his research led him to conclude that the rapid increase in the quantities of physical capital and labour, rather than in TFP, was the main reason that explains the region's rapid growth. Later studies, based on a wider set of data, confirmed the significance of the TFP growth. The issue of TFP remains high on the agenda of economic development, not only in developing but also in developed countries, and it will be discussed in more detail below.

2. Despite strict restrictions of the Chinese government, a large flow of migrants from the rural areas was the main reason for the rapid expansion in low-skilled and labour-intensive industries. The influx of rural migrants maintained the urban wages low and stable, enabled a rapid expansion of these industries and attracted foreign investors. These workers did not contribute to the rise in TFP, and it took them quite sometime to acquire the necessary know-how and expertise to move to higher positions on the income ladder. In the past decade, the share of college graduates among the migrants increased substantially and their higher education increased their productivity.

3. Only since the late 1990s the pressure of workers in manufacturing to raise their wages became slightly more conspicuous, despite their countries' rapid growth, the rise in income inequality and the reduction in the share of labour in the national income. The main reason was the flow of migrants to the urban centres that continued unabated and the growing rural–urban gap that was driving that migration. The low wage increase was partly due to large foreign investments in more advanced and capital-intensive industries that led to a more rapid rise in output per worker but curbed the demand for trained workers.

4. With the transfer of more advanced industries or stages of production to the EA countries, that was accelerated by the transfer of services, particularly in the ICT sector, there has also been an increase in the demand for skilled and experienced workers. Newcomers from the rural areas cannot meet these demands and fill the positions, and these sectors are therefore more amenable to increased wages: In 2005, wages for a factory job of a low-skilled worker in India were around US$2 a day compared with up to US$4 in some regions in China. To moderate the demand for workers and the pressure to raise wages, industries are gradually changing their production technologies, making them more capital-intensive. However, these changes increase the demand for skilled workers and raise the wage and income inequality between workers.

5.1 Should government interventions be prohibited?

1. The EA miracle introduced to the lexicon of the orthodox economic school of the Washington Consensus the 'four-letter' words: 'government intervention'. There are obvious risks in taking this lesson from the EA miracle and adopting it in other countries. First, there are no rules about the extent and nature of these interventions and too much government intervention can do more harm than good. Second, there is also a danger that governments will take these interventions as the formula for good economic policy, and invest much smaller efforts in implementing all the other policies conforming to the principles of the Washington Consensus that are necessary for the economy to function properly.

2. Rodrik (2004) made a comparison between the policy measures that were implemented in the 'miracle economies' and the principles recommended by the consensus of orthodox economists. Box 5.1 lists these EA anomalies. However, the criticism of the orthodox economic principles that is highlighted by this comparison should be seen in perspective. These principles did not and could not work in the EA countries, not because they are wrong or 'anomalous' but, as noted earlier, because these countries did not have the basic conditions needed to make the free market work. There was no way, for example, to enforce property rights or the right of shareholders by the rule of law if this law did not exist; there was no way to establish prudential supervision through regulatory oversight over the financial system in countries that never had these regulations or the institutions to implement this supervision. For the same reason, the orthodox economic principles are not likely to work in most LDCs, and 'conditionalities' based on the orthodox principles in these countries' SAPs have often failed.

Box 5.1. The 'East Asia Anomalies'.

Institutional domain	Standard ideal	'East Asian' pattern
Property rights	Private, enforced by the rule of law	Private, but government authority occasionally overrides the law (especially in Korea)
Corporate governance	Shareholder ('outsider') control, protection of shareholder rights	Insider control
Business–government relations	Arms' length, rule based	Close interactions
Industrial organization	Decentralized, competitive markets, with tough anti-trust enforcement	Horizontal and vertical integration in production (*chaebol*); government-mandated 'cartels'
Financial system	Deregulated, securities based, with free entry. Prudential supervision through regulatory oversight	Bank based, restricted entry, heavily controlled by government, directed lending, weak formal regulation.
Labour markets	Decentralized, de-institutionalized, 'flexible' labour markets	Lifetime employment in core enterprises (Japan)
International capital flows	'Prudently' free	Restricted (until the 1990s)
Public ownership	None in productive sectors	Plenty in upstream industries

3. In the early stages of economic reforms, the governments in the EA countries were generally less vulnerable and less responsive than governments in many other developing countries, particularly in LA, to demands of organized labour. In part, this was due to the stream of rural migrants that stemmed the pressure to raise urban wages.[10] Another reason was the rapid expansion of employment in the new industries that reduced the pressure to form labour unions. Perhaps the main reason, however, was the iron hand of their central regimes that effectively forbade the formation of labour unions. As a result, the 'miracle' economies had fewer labour market regulations than most other developing countries, and that, in turn, gave their local employers much greater flexibility. At later stages, with the advent of democracy and political pluralism, many programmes and policies that were feasible under authoritarian regimes had to be adjusted or totally changed.

4. In SSA, LA and SA, where unemployment remained high or even increased during the 1980s and the slow growth of local industries did not contribute much

[10] During the years 1965–1990, the agricultural labour force in these countries grew at an annual rate of 1%, compared with 1.7% in all other developing regions, at nearly half the annual growth rate of the general population.

to boost the demand for workers, wages in the urban formal sector were frequently pushed up none the less by legislated minimum wages and other labour laws or even by administrative constraints on entry to the formal sector.

5. In India, manufacturing was hobbled for years by restrictive labour laws, frequent strikes and powerful labour unions that still deter foreign investors, even though local labour costs are considerably lower than in China. Only in the late 1990s, the government, in collaboration with foreign investors, took more active measures to open the labour market and reduce the stifling labour laws in order to increase employment and mitigate the rise in wages.[11] Nevertheless, the income gap between urban workers in the formal sector who are members of the labour unions and urban workers in the informal sectors is still high.

5.2 Limitations of the East Asian 'formula'

The EA experience and the formula for rapid growth adopted by these countries proved to be very successful during the 1980s and until the financial crisis in the late 1990s. But their unique characteristics and distinct geographic, ethnic and religious conditions became apparent when they were overwhelmed by and coped with the currency crisis and the political crisis that followed. In practically all these countries, the system of law and order had been controlled until then by the iron hand of highly centralized, though very opaque, secretive and corrupt regimes (Krugman, 1998).[12]

The failure of these regimes to adjust the legal system and the organization of governance, as their economies expanded and became more decentralized and as the role and say of the private sector and the more educated labour force increased, weakened the central government and destabilized the entire social order.

Among the least developed EA countries, Vietnam is perhaps the most obvious candidate to follow the experience of the EA emerging economies and possibly even the Chinese success One of the few remaining EA countries that is still in the process of making the transition to the new world of market economy and globalization, Vietnam is now swept by a fervor for capitalism and a race to achieve rapid economic growth. The success of the Chinese model of economic growth through market reforms and pro-trade policies gave strong incentives to Vietnam to follow that path and harness the benefits of trade and investment, making it now the fastest growing economy in Asia after China, and, relative to its size, attract more foreign investments than China.

[11] Labour unions now charge, however, that when workers want to join a union they are being sacked from their job and some are required to sign a pledge not to join a union when they are hired. Rapid increase in employment in India reduced the power of the unions in recent years.

[12] Paul Krugman described how politically connected individuals or institutions – Thai finance companies, members of the Suharto family, *chaebol*-controlled banks – were widely perceived to be backed by implicit government guarantees, yet were not subject to any effective supervision.

Despite careful efforts to put its economic house in order in preparation for its accession to the WTO and despite the measures to attract foreign investors and to restructure its legal infrastructure, banking system and other regulations, Vietnam is still overwhelmed by corruption at all levels of government, by demands for greater democracy and by a deepening social divide between the rural poor and the urban rich that threatens the control of the Communist Party and the country's stability. It would require great care and caution for Vietnam to successfully implement the policies and renovate the system of governance in order to secure its achievements in the coming decade.

6 Concluding Remarks

As a result of all these changes in the world economy, the LDCs will now find it much more difficult to realize the benefits from trade liberalization and to remove the regulations and constraints in order to free their markets and allocate their resources more efficiently. In today's world, even the basic principles of orthodox economics, namely free and competitive markets, open trade, a decentralized industrial organization and deregulated financial system, are not likely to work when world trade is not really free, because world production is dominated by oligopolistic transnational corporations and global financial centres often collude. World trade is also more regulated by a maze of rules established by the various trade agreements that increasingly exclude the LDCs and raise obstacles to their integration into the global trade system.

Against this background, the emphasis on assistance to and conditions on economic reforms in the LDCs has moved from economic conditionalities to institutional conditionalities aimed at improving the operation of institutions in these countries. These institutions include both central government institutions, such as the central bank and the legal system, that guide and regulate the economy, and private institutions, primarily in the financial system, that provide a wide range of centralized services to individuals and businesses. These institutional reforms that are combined under the title of 'better governance' have become the focus of attention of the World Bank and the IMF as well as the entire community of donor countries that provide professional expertise on how to implement these reforms and often condition their aid on their implementation. The transition reflects the realization that economic reforms by themselves cannot bring economic growth or improve the living conditions of the local population unless the country's system of governance is reformed.

Econometric studies that disregard these necessary conditions and concentrate directly on statistical measures that estimate the causal link between, say, trade liberalization and growth, are therefore not likely to provide the right signals for future policies and by itself, trade liberalization is not the panacea for growth in the developing countries. Openness to trade and investments, by itself, is not sufficient to attract private investors if their investments are not protected. A country's comparative advantage depends much less on the relative costs of labour and capital and far more on the relative risk of investments and the quality of the infrastructure and institutions. The limited experience or sheer incompetence of many private and public institutions in most LDCs is their main handicap in building up their economy and attracting foreign investments. This should therefore be the focus of reforms. Although the removal of distortions in trade and adequate macroeconomic policies are important ingredients of a country's overall economic reforms, measures to create more effective governance and well-functioning institutions are not merely important components of these reforms, but necessary preconditions.

In their extensive study on 'Institution Rule', Rodrik *et al.* (2002) demonstrated in a sample of 140 countries that the role of institutions, particularly the role of property rights and the rule of law, is far more important in determining a country's economic development than the role of other economic factors and policies that were regarded in the past as essential to drive the process of economic growth. These results are by no means surprising. Without the safety of property rights and the rule of law, production, investments and trade cannot develop.

These measures are necessary, but even when they are all duly implemented, the developing countries will encounter great difficulties to reap the benefits from free trade that the neoclassical model is promising. Trade restrictions are essentially measures that redistribute income between countries and from domestic consumers to domestic producers within countries. This redistribution may be hard to justify, but it is even harder to cancel. The agricultural policies that the EU and the USA are implementing to protect their local producers can serve as an example: due to the domestic subsidies and trade restrictions, the domestic price of the affected products in the EU, for example, is much higher than on the world market; tariffs and protection benefits local producers by billions of dollars, but the cost is borne by the European consumers. The benefits go to just 2% of the population. The agricultural subsidies are not just an economic, political and environmental inequity, but they prevent desperately poor producing countries from exporting to the EU and the USA. Efforts to reach an agricultural trade agreement have already been going on for 6 years, but the subsidy programmes are politically extremely difficult to change because the benefits accrue to a limited number of specific companies and states, while the costs are spread over a broad group of consumers.

The multilateral trade agreements that have been reached after long and hard rounds of negotiations in the framework of GATT and, since 1995, the WTO have meant major progress for the establishment of a collaborative framework of world trade and a level playing field. But with all that progress, and despite the attention devoted particularly in the recent 'Development Round' to the special needs and disadvantages of the LDCs, world trade is still not conducted on a truly level playing field. The agreements that have been reached on products which are of major concern for the LDCs – the MFA and the ATC – prove this point, since the main beneficiaries turned out to be the large emerging economies and not the LDCs.

By and large, trade proved to be a positive sum game, as Ricardo had predicted, but the distribution of the positive sum between the different players proved to be a difficult process in which the larger players demanded a larger share. What kept players in the game, Harsanyi noted, is not only the gains from trade but also the cost of breaking the collaborative structure of trade (or the trade agreement) and entering into a conflict with the other players.

The LDCs are obviously facing an uphill climb on this supposedly 'level playing field', despite all the special and differential considerations, and their share in the positive sum of the global gains from trade is clearly disproportionally low. However, it is also clear that for these countries the costs of breaking the rules and leaving the 'game' are prohibitively high. Nevertheless, on this playing field, open trade and the removal of all trade restrictions, sometimes as a 'shock therapy', as the Washington Consensus advocated, is not always the optimal strategy, and more careful considerations and meticulous planning may be necessary in order to ensure that a country can benefit from trade, but also protect its own economy and its own workers as much as possible.

Appendix: The Impact of Offshoring and the Transfer of Capital from Developed to Developing Countries

Diagrammatic illustration[13]

The objective of the diagrammatic illustration in the appendix is to explain the impact of capital mobility on commodity prices and input prices. In the neoclassical trade model, free trade is a perfect substitute for the movements of factor inputs: Countries that have abundance of labour will specialize in the production of labour-intensive products; countries that have abundance of capital will specialize in the production of capital-intensive products. In the process of trade, each country will export the products in which it has a comparative advantage and import the products in which it does not have an advantage. This trade will bring about factor–price equalization, in which the price of all factor inputs and all products will be equalized in all markets. Is there any advantage then in the neoclassical world to free movements of the factor inputs?

The appendix will examine this question both diagrammatically and mathematically, in the traditional structure of the neoclassical trade model. This model inevitably makes a number of simplifying assumptions that are no longer relevant in today's structure of the global economy. I will not change these assumptions, but I will comment on their potential impact on the results. Some of these assumptions, particularly those concerning the differences in productivity in developed and developed countries are relaxed and discussed in Chapter 6 in a more detailed analysis of factor productivity. One specific assumption that the model does make is that the removal of

[13] In this illustration I benefited from the article of Robert Mundel (1957).

restrictions on the free flow of factor inputs applies only to the free flow of capital across national borders, whereas labour mobility is still not allowed.

The starting point is the impact of free trade in the traditional neoclassical model on product and input prices illustrated in Fig. 5.15. Without trade, the developed country is producing both the labour-intensive product – textile – denoted in the figure as T, and the capital-intensive product – machines – denoted as M. In the absence of trade, the production possibility frontier of the developed country is denoted in the figure by YY, and the autarky (economic self-sufficiency) point will be at M.

Suppose now that both the developed and the developing countries open their economies for trade. The developed country has abundance of capital but a scarce supply of labour. The developing country has scarce supply of capital but a very large, practically infinite (in the sense defined below) supply of labour. If factor inputs are immobile, but there are no other restrictions on trade, the equilibrium point is determined at P. At that point, commodity and factor prices in the two countries are equalized. The large supply of labour in the developing country means that the wage rate in the developing country does not change with an increase in the demand for labour. In other words, the wage rate in the developing country is determined at a certain minimum level (below which workers would prefer to stay in the rural sector), and any increase in the demand for labour would be met by additional migrants so that the wage rate would remain at that minimum level.

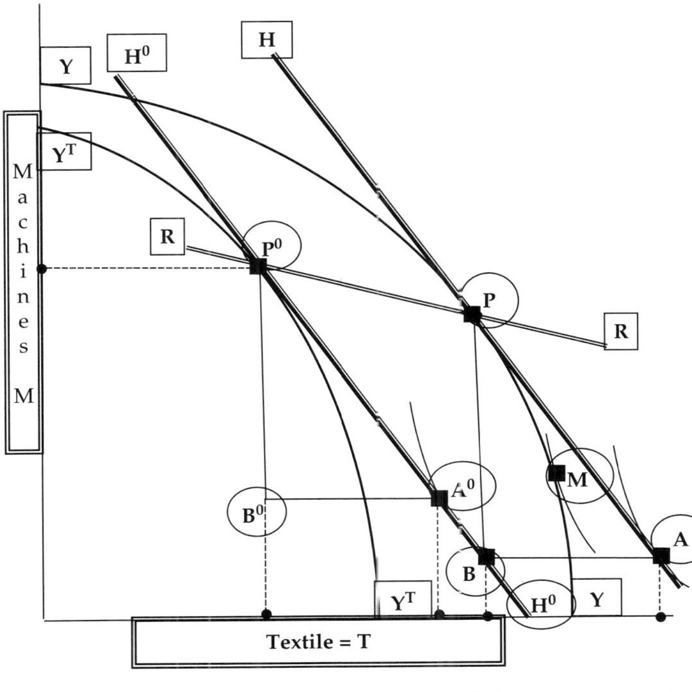

Fig. 5.15. The impact of trade between developing and developed countries.

If the capital is immobile, then at the free trade equilibrium point production in the developed country is at P and consumption is at A. The developed country's import of textiles is AB, and its export of machines is PB. With trade, the developed country thus imports the labour-intensive product – textiles, exports the capital-intensive product – machines and trade equates the prices of the two factor inputs in the two countries. The low-wage rate in the developing country implies that the rental rate capital is high. In addition, the equilibrium commodity prices are determined at the point at which the trade balance of the two countries is balanced.

Before we proceed to the mathematical analysis of this model, two comments on the two basic assumptions. First, the law of one price indicates that wages are equated in the two countries. In a more realistic world, the wage rate in the developed country is pressured down due to the cheap supply of textiles from the developing country, whereas the interest rate is pushed up by the large demand for machines. Wages in the developed country are likely to be restricted, however, in their movement downward by the minimum wage rate in the developed country, by the existing wage agreements, by the long-term wage contract and by the various social security payments to which workers are entitled. As a result, the wage rates will not be equalized and a large gap will remain. The wage rate will also be different due to the large difference in the productivity of workers in the developed and the developing countries. This issue will be dealt with separately in Chapter 6. Second, the rate of return to capital will also be equated due to the higher risks in the developing country, by the high transaction costs. In the mathematical formulation we disregard these differences and maintain the ordinary assumption that the two countries are identical in all respects except for the differences in the abundance of the two factor inputs.

In the two-countries, two-commodities, two-factor model, factor–price equalization is both necessary and sufficient to ensure commodity price equalization if the two countries have identical and homogeneous linear production functions. Along the *unit* production possibility curve, the input requirements of labour and capital per *unit* of output are denoted by a_{LT} and a_{KT} for textile and by a_{LM} and a_{KM} for machines. At the point of cost minimization, the iso-cost line, with the slope given by (minus) the ratio of factor prices, $-(w/r)$ is tangent to the isoquant – with slope given by: da_{KT}/da_{LT} for textile, and the same conditions holds for the isoquant of machines. In other words, cost minimization implies the familiar conditions:

$$wda_{LT} + rda_{KT} = 0 \tag{5.1}$$

$$wda_{LM} + rda_{KM} = 0 \tag{5.2}$$

In the developing country, the wage rate is determined exogenously by the minimum wage for labour and it remains unchanged. The competitive profit conditions are given by

$$wa_{LT} + ra_{KT} = p_T \tag{5.3}$$

$$wa_{LM} + ra_{KM} = p_M \quad (5.4)$$

In the normal case (i.e. when the wage rate is determined endogenously), these four conditions determine the four equilibrium prices. Another condition determines the trade equilibrium that guarantees that the balance of payment is indeed balanced:

$$p_T X_T = p_M X_M \quad (5.5)$$

This condition, the equilibrium prices and the economy's production possibilities determined the quantities traded between the two countries. In Fig. 5.15, with the opening of the two countries to trade, the iso-cost is determined by the exogenous wage rate and is given by the line denoted as H. When capital and labour remain immobile, the trade equilibrium is determined at point P and the trade balance is determined by the triangle ABP.

Suppose now that all restrictions on the mobility of capital are removed. Will that trigger a movement of capital – offshoring – from the developed to the developing country? Under normal conditions, the input prices are equal in the two countries and there is no incentive for movements of factor inputs. That incentive will exist only if input prices are not equal.

Suppose, however, that the wages in the developed country are higher than in the developing country for the reasons listed above. In this case, the rate of return to capital in the developed country will be lower than in the developing country. If free movement of capital is possible, there will be an incentive to transfer capital to the developing country and invest it in the labour-intensive textile industry where wages are very low and the rate of return on investments is therefore high.

From which of the two sectors in the developed country will that capital come? Here we can select one of two alternative assumptions:

1. If capital is, in the terminology of the capital theory of the 1950s, fully malleable (like a mechano set), and could be transferred from either one of the two sectors in one country and fully and costlessly adjusted to either one of the two sectors in the other country, then the investor's choice would be to transfer capital from both sectors so as to minimize the reduction in the value of output. Capital would then be reduced from both sectors and both would suffer a reduction in output. This is a direct interpretation of the assumption that is implicitly made in most analyses, and this is also the simple interpretation of the diagrammatic illustration in Fig. 5.15.
2. If 'capital' or the machine is designed for a specific industry, then the transfer of capital can take place with the transfer of the entire enterprise (or of the specifically designed machines). This is, in fact, the meaning of 'offshoring.' With free movements of capital, an entire textile enterprise will move from the developed to the developing country. The loss of output in the developed country would be only in the textile industry. If the wage difference is very large, some machine-producing enterprises may decide to move, but this is not the assumption made in the above analysis.

In the diagrammatic illustration, we take a simplifying assumption and when the economy of the developed country is opened to free flow of capital the production possibility frontier will shift inward from YY to YTY T. The decline from Y to Y T along the horizontal axis represents the loss in the capacity to produce machines in the developed country due to the movement of some capital to the developing country. The movement of capital offshore will continue until the rental rate in the two countries is equated. As noted earlier, the equalization of the rental rate is after taking into account the differences in risk and in all transaction costs between the two countries.

The wage rates will still not be equated despite the movement of capital, since the wages in both the developed and the developing countries are already at their minimum levels. Some textile industries will prefer not to move or maintain some segments of their production of textile in the developed country due to the higher productivity of labour in the developed country which is particularly important in these segments of production.

The iso-cost line has the same slope since both the wage rate and the interest rate are determined by Eqs (5.1) and (5.2) and the cost minimizing output shifts to P_0.

To see the impact of this transition, write the conditions in Eqs (5.1) and (5.2) in terms of percentage changes weighted by the relative shares as

$$\xi_{LT}\ \hat{a}_{LT} + \xi_{KT}\ \hat{a}_{KT} = 0 \tag{5.6}$$

$$\xi_{LM}\ \hat{a}_{LM} + \xi_{KM}\ \hat{a}_{KM} = 0 \tag{5.7}$$

where $\hat{a}_{LT} = (da_{LT}/a_{LT})$ and $\xi_{LT} = (wa_{LT}/p_T)$. ξ_{LT} and ξ_{KT} are the distributive shares of labour and capital.

These equations indicate that for each industry, the relative change in the unit costs involves a substitution of one input for another.

For a given quantity of capital in the developed economy, K, the amount of output of, say textile, that can be produced, is given by

$$x_T = K/a_{KT} \tag{5.8}$$

With a change in the amount of capital, assuming that the amount of labour remains unchanged, the change in the country's output of textile is given by

$$dx_T/x_T = dK/K - \hat{a}_{KT} \tag{5.9}$$

The quantity of labour that will be required to produce the amount of output x_T is given by

$$L = a_{LT}\ x_T = [a_{LT}/a_{KT}]\ K \tag{5.10}$$

Hence, with the change in capital, the percentage change in labour that would leave output unchanged is simply given by the percentage change along the unit isoquant:

$$(\hat{a}_{LT} - \hat{a}_{KT}) + dK/K = dL/L \tag{5.11}$$

The latter equation determines the additional amount of labour that would be needed in the developed country when the amount dK of capital is transferred from the developed to the developing country in order to keep output unchanged.

References

Alcala, F. and Ciccone, A. (2004) Trade and productivity, *The Quarterly Journal of Economics* 119(2), 612–645.

Amsden, A. (1989) *Asia's Next Giant – South Korea and Late Industrialization*. Oxford University Press, London and New York.

Bigman, D., Dercon, S., Guillaume, D. and Lambotte, M. (2000) Community Targeting for Poverty Reduction in Burkina Faso, *World Bank Economic Review* 14(1), 167–194.

Blinder, A. (2005) *Fear of Offshoring*. CEPS working paper No. 119, December, Princeton University, Princeton, New Jersey.

Boltho, A. and Toniolo, G. (1999) The assessment: the twentieth century – achievements, failures, lessons. *Oxford Review of Economic Policy* 15(4), 1–17.

Chang, R. (1994) Endogenous currency substitution, inflationary finance, and welfare. *Journal of Money, Credit and Banking* 26(4), 903–916.

Chen, S. and Ravallion, M. (2001) *How Did the World's Poorest Fare in the 1990s?* Global Poverty Monitoring Database, World Bank, Washington, DC.

Cornia, G.A. and Court, J. (2001) *Inequality, Growth, and Poverty in the Era of Liberalization and Globalization*. United Nations University World Institute for Development Economics Research (WIDER), Helsinki, Finland. Available at: http://www.wider.unu.edu/publications/policy-brief.htm

Deininger, K. and Squire, L. (1996) A new data set measuring income inequality. *World Bank Economic Review* 10(3), 565–579.

Dickerson, A., Green, F. and Saba Arbache, J. (2001) *Trade Liberalization and the Returns to Education: A Pseudo-panel Approach*. Studies in Economics, Department of Economics, University of Kent, UK.

Dollar, D. (1992) Outward-oriented developing economies really do grow more rapidly: evidence from 95 LDCs, 1976–1985. *Economic Development and Cultural Change* 40, 523–544.

Dollar, D. and Kraay, A. (2001) Trade, growth, and poverty. *Finance and Development* 38(3).

Dollar, D. and Kraay, A. (2002) Growth is good for the poor. *Journal of Economic Growth* 7(3), 195–225.

Easterly, W. (2005) What did structural adjustment adjust? The association of policies and growth with repeated IMF and World Bank adjustment loans. *Journal of Development Economics* 76, 1–22.

Firebaugh, G. (1999) Empirics of world income inequality. *American Journal of Sociology* 104, 1597–1630.

Frankel, J.A. and Romer, D. (1999) Does trade cause growth? *American Economic Review* 89(3), 379–399.

Grossman, G. and Helpman, E. (1991). Trade, innovation and growth. *American Economic Review* 80(2).

Krugman, P. (1994) The myth of Asia's miracle. *Foreign Affairs* 73(6), 62–78.

Krugman, P. (1998) *Will Asia Bounce Back?* Speech for Credit Suisse First Boston, Hong Kong, March.

Lee, Ha Yan, Ricci, L. and Rigobon, R. (2004) Once again, is openness good for growth? *Journal of Development Economics* (2004).

Lucas, R. (1993) Making a miracle. *Econometrica* 61(2), 251–272.
Melchior, M., Telle, K. and Wiig, M. (2000) *Globalization and Inequality: World Income Distribution and Living Standards 1960–1998*. Studies on Foreign Policy Issues, Report 6B, Royal Norwegian Ministry of Foreign Affairs, Oslo.
Milanovic, B. (1999) *True World Income Distribution: 1988 and 1993*. World Bank, Washington, DC.
Mundel, R. (1957) International trade and factor mobility. *American Economic Review* 47(June), 321–335.
OECD (2001) *Employment Outlook*. OECD Publishing, June.
Radetzki, M. and Jonsson, B. (2000) The 20th Century of Increasing Income Gaps. But How Reliable are the Numbers? *Ekonomisk* 1, 43–58.
Ravallion, M. and Chen, S. (2004) China's (Uneven) progress against poverty. *Policy Research Working Paper 3408*. World Bank.
Rodriguez, F. and Rodrik, D. (2000) Trade policy and economic growth: a skeptic's guide. In: Bernanke, B. and Rogoff, K. (eds) *NBER Macroeconomics Annual 2000*. MIT Press, Cambridge, Massachusetts.
Rodrik, D. and Rodriguez, F. (1999) *Trade Policy and Growth: A Skeptics Guide to Cross National Evidence*. NBER working paper 7081, April.
Rodrik, D. (2001), Globalization, Growth and Poverty: Is the World Bank Beginning to Get It? December.
Rodrik, D., Subramanian, A. and Trebbi, F. (2002) *Institute Rule: The Primacy of Institutions Over Geography and Integration in Economic Development*. NBER WP9305, November.
Rodrik, D. (2004) *Rethinking Growth Strategies*. WIDER Annual Lecture 8, UNU–WIDER, Helsinki, Finland.
Sachs, J. and Warner, A. (1995) Economic reform and the process of global integration. *Brookings Papers on Economic Activity* 1, 1–118.
Sala-i-Martin, X. (2002) *The World Distribution of Income*. NBER Working Paper 8933, May.
Sala-i-Martin, X. (2004) *Why are the Critics so Convinced that Globalization is Bad for the Poor: A Comment*. Columbia University and NBER, Washington, DC.
Santos-Paulino, A. and Thirlwall, A.P. (2004) The impact of trade liberalisation on exports, imports and the balance of payments of developing countries. *Economic Journal* 114(493), F50–F72.
Schultz, P.T. (1988), Inequality in the distribution of personal income in the world: how is it changing and why? *Journal of Population Economics* 11, 307–344.
Stiglitz, J. and A. Charlton (2006) *Fair Trade for All: How Trade Can Promote Development*. Oxford University Press, New York.
Wade, R. (1990) *Governing the Market: Economic Theory and the Role of Government in East Asian Industrialization*. Princeton University Press, Princeton, New Jersey.
Winters, A. (2003) Going alone: The case for relaxed reciprocity in freeing trade. *The Economic Journal* 113(491), F656–F658.
World Bank (1993) *The East Asian Miracle: Economic Growth and Public Policy*. Oxford University Press, New York.
World Bank (1987) *World Development Report*. Washington, DC.
World Bank (1994) *Adjustment in Africa: Reform, Results, and the Road Ahead*. Oxford University Press, New York.
World Bank (2000) *Attacking Poverty: World Development Report 2000/2001*. Oxford University Press, New York.
World Bank (2001a) *World Development Indicators*, Washington, DC.
World Bank (2001b) *Globalization, Growth and Poverty: Building an Inclusive World Economy*, Policy Research Report, Washington, DC.
World Bank (2002) *Global Economic Prospects and the Developing Countries*, Washington, DC.
World Bank (2006) *World Development Report 2006*, World Bank.
World Trade Organization (1999) *General Agreement on Tariff and Trade*, Geneva, Switzerland.

6 Will Africa Be Left Behind?

1 Introduction

Among the developing countries that have remained on the margins of the world economy and have not benefited from globalization are all the countries in SSA with the exception of South Africa. Their deepening economic stagnation is due to a wide variety of factors; some are externally determined, such as their geographical conditions or developments in the global economy, but others have their roots in the policies pursued by these countries.

For the countries in SSA, the continuous drift downhill after more than a decade of great optimism in the wake of their liberation from colonial rule was first and foremost due to violent internal conflicts and wars that ruined their economies and destroyed their communal foundations as a nation. Many of these countries went through a series of military coups and, governed by dictatorial leaders, found themselves embroiled in incessant wars and internal conflicts, and a huge portion of their resources were wasted on maintaining a strong army. On top of all that, many governments implemented irresponsible and wasteful economic policies that misused not only their own resources, but also most of the aid provided by donor countries and the international development organizations. As a result, SSA became the centre of gravity of today's poverty problem, and the region's poverty is marked not only by poor living conditions, but also by the prevalence of hunger, the spread of diseases, from HIV/AIDS to malaria, and also by poor education.

Unfortunately for these countries, integration into the world economy has become more difficult for latecomers like them because they face an increasingly more ordered and therefore less flexible global economic system that is dominated by a small number of leading trading countries, a close-knit web of RTAs, increasingly more powerful and more monopolistic transnational corporations and rather rigid rules of the WTO that govern the entire global trade system. Countries that are joining this system face the challenge of adjusting their own production and trade to this structure of world trade. Moreover, the abundant and seemingly endless supply of cheap labour in China, India, Bangladesh, Pakistan and other Asian countries leaves very few advantages for SSA countries to offer.

Since the late 1990s, however, the tide has changed dramatically for the better. In an increasing number of African countries, leaders are now being selected in democratic elections and the process of policy making is becoming more participatory by involving more segments of the society, thus contributing to reduce ethnic tensions and civil conflicts. The number of countries where conflicts still persist has declined from more than 20 in the early 1990s to less than 5 in the mid-2000s. Yet, the continent still has a long way to go, since many regimes remain rather unstable, violent conflicts are still erupting and the genocide in Darfur is allowed to continue. In Nigeria, the region's second largest economy, continued conflicts and attacks on the oil infrastructure led to a fall in oil production by 25% during 2006, slowing down the country's growth. In several other countries, the risks of political turmoil and conflicts remain high and continue to undermine growth; in 2005/06, the conflict countries included Chad, Côte d'Ivoire, the Democratic Republic of Congo, Eritrea, Lesotho, the Seychelles, Somalia, Swaziland and Zimbabwe.

Nevertheless, a growing number of countries realize the great significance and the great benefits of political and macroeconomic stability and their populations are more determined to support the government in maintaining stability. Countries that pursue economic reforms also benefit from debt relief, and the donor countries and the international development organizations increasingly concentrate their aid and assistance on the reduction of poverty rather than on structural adjustments. All these reforms are bearing fruit, and many SSA countries have grown in recent years at a more rapid and steady pace than at any time since their independence.

The wheel of fortune continued to turn with the sharp rise in the price of many primary resources, most notably oil, that are the main exports of many African countries. Although improvements in technology and new discoveries in the developed countries are anticipated to preclude any major disruptions to future growth, commodity prices are expected to continue to rise in the coming years. The sharp rise in oil prices has received most of the media attention, but in fact the rise in the price of metals and minerals during 2004–2006 has been equally strong or even stronger. The prices of agricultural products have trailed behind the prices of oil, minerals and metals and, despite an average rise in prices by 35% in real terms since the cyclical low levels in 2001, their terms of trade and purchasing power have actually declined; however, since 2002/03, also the prices of several key agricultural products have started to rise quite sharply.

Africa's growth performance in recent years is indeed impressive (Fig. 6.1). After the dismal growth during the 1990s, GDP has grown, in real terms, at an average rate of over 3% since the late 1990s, and in 2004–2006 annual growth has risen to over 5.0%. Africa's oil exporters, including Angola, Equatorial Guinea, the Republic of Congo and Nigeria, have gained from the

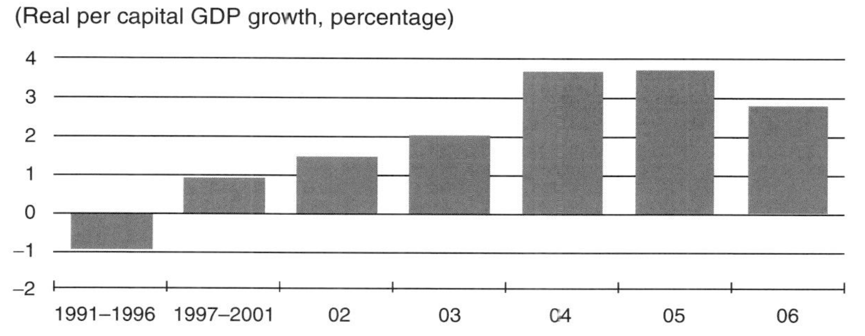

Fig. 6.1. Average growth rate per capita in SSA (population-weighted). (From IMF, World Economic Outlook, various years.)

higher oil prices and increased production, but growth was equally strong in the oil-importing countries. Growth has also been broad-based: half the oil-importing countries grew at an annual rate of over 5% in 2006 and are expected to sustain this growth also in 2007. As a result, the SSA economies expanded in 2004–2006 at an average rate of 4–5% and the oil-exporting economies grew during these years at an annual rate of 6–7%. South Africa, the region's largest economy, grew even faster, driven by large household expenditures.

The objective of this chapter is to review the factors that determined the course of the subcontinent's development in the last two decades and the prospects of accelerating its growth so that Africa will not be drifting to the margins of the world economy, but will instead be able to make the fundamental structural changes that will enable its people to embrace and be part of the global economy of the 21st century.

2 From the Optimism of the 1960s to the Pessimistic Outlook Today

> The Conference calls for the founding of African Freedom Day, a day to mark each year the onward progress of the liberation movement, and to symbolize the determination of the People of Africa to free themselves from foreign domination and exploitation.
>
> (The First Conference of Independent African States, Accra, Ghana, 15 April 1958)

During the 1950s and the 1960s, the developing countries on the two sides of the two oceans – the Pacific and the Atlantic – had very different perspectives and aspirations. In those years, globalization was not even on the agenda; instead, attention was focused on the challenges presented by the Cold War and the efforts of the European countries to recover from the wounds of World War II. However, in Africa, the mood from the mid-1950s to the mid-1960s was upbeat. After their liberation from colonial rule and the emergence of free and independent countries, charismatic and highly patriotic leaders came to power, from Kwame Nkrumah in Ghana, Jomo Kenyatta in Kenya,

Sékou Touré in Guinea, Julius Nyerere in Tanzania to Gamal Abdel Nasser in Egypt, and the future seemed bright.

During these years, the SSA countries experienced rapid growth as they increased agricultural production and exports of natural resources that were fuelled by booming commodity prices. The wave of liberations that spread to more countries, the promising growth and their charismatic leaders seemed to put these countries on a sure path to democracy and sustained rapid growth. In May 1963, the leaders of 32 independent African states met in Addis Ababa to form the Organization of African Unity (OAU). By then, more than two-thirds of the African states had gained independence. Their low labour costs, rich natural resources and close ties to Europe gave them the keys to develop competitive industries, expand their exports and increase their agricultural production to provide enough food to their populations and raise living standards.

But already in the mid-1960s, things took a turn for the worse in SSA, when a series of military coups toppled the leaders in charge and installed army generals, colonels or even corporals as heads of states. Most SSA countries taken over by military regimes were governed despotically, the heads of states spent huge amounts of money, much of it borrowed from foreign banks, on their military forces, primarily to secure their loyalty, and they looted huge amounts for their private coffers, mostly in foreign banks. As a result, the economies deteriorated, large debts accumulated and the oil crisis brought these countries to the brink of a catastrophic crisis. Perhaps most seriously, the nation state started to fragment according to loyalties for warlords, clans, tribes and ethnic groups, and the central regime lost effective control of large parts of the country.

Considering the contrast between the heady optimism in SSA during the 1950s and 1960s and today's grim reality and feelings of hopelessness, we cannot avoid asking why the region's development has taken such a diametrically different course than the one envisaged in the hopes of the early days. Originally, the concept of 'soft states' – states that lack the social discipline to carry out policies – was used to describe the Asian governments. But Asia has been largely successful at famine prevention and a number of formerly food-deficient countries are now food exporters. If Asia could do it, why can't Africa? Why have the Africans failed to achieve each year 'an onward progress', and instead were led to today's abyss of poverty, hunger and despondency? Why have practically all the African countries failed to achieve the progress that they had so confidently predicted half a century ago? Did they pursue so very wrong policies? Or were they the innocent victims of ruthless exploitation by the industrialized world and American capitalism – as many of them persistently claim? An unavoidable but highly charged question in this context is: Is there anything in the 'culture' of these countries that is simply not conducive to economic success?

At the time of independence, most of SSA was self-sufficient in food. In less than 40 years, the subcontinent has gone from being a net exporter of basic food staples to reliance on food imports and food aid. Since independence, agricultural output per capita has stagnated and, in many places, declined. SSA is the only major region where cereal production per capita was lower in 2001 than in 1961. The stagnating or falling per capita cereal production in SSA over the last 40 years is in great contrast to the development in East, South-east and South Asia. Comparing the first and last 5-year annual averages during the entire period, 1961–2001, per capita output in Asia grew 24% and decreased 13% in SSA. One question that presents itself is whether there are lessons to be learnt from the Asian experience that could benefit food security in SSA. It is a fact that 30–40 years ago the Asian food situation was depicted in much the same alarmist terms as is that of SSA today. The 'population bomb' was ticking in the 1960s, especially for Asia.

The second question is perhaps even more difficult, and its implications are so far reaching that they have often been avoided altogether. Simply stated, the main question is: What does the future hold? A World Bank report entitled *Global Economic Prospects 2007: Managing the Next Wave of Globalization* predicts that, despite the recent outburst of more rapid economic growth, Africa may fall behind in the coming three decades. An earlier World Bank study entitled *Can Africa Claim the 21st Century?* comes to even more pessimistic conclusions. However, there are also more optimistic projections, and these evaluations will be discussed later on in this chapter. Another question must concern the human consequences of these projections. What are the implications of the World Bank study which came to the conclusion that Africa may well fail to claim the 21st century? What will happen to the people who live in a continent that does not seem to have a future in the world of the 21st century? What would be the consequences for the people if Africa falls behind? Can the world let Africa drift out of the global economy?

So far, none of these questions has been answered satisfactorily in my opinion. The discussion below presents several views and opinions as well as an overview of studies that tried to support specific viewpoints with the relevant factual evidence. The connection between the two sets of questions mentioned above is, however, self-evident: We cannot design a proper strategy for the future if we do not understand the past. We cannot come, for example, with a proposal to 'push' more aid to Africa unless we understand why aid programmes have often failed and how both aid donor and recipient countries can prevent similar failures in the future.

It has sometimes been argued that the geographic disadvantages of many SSA countries, particularly the over 15 landlocked countries, present challenges that even with good policies, institutions and infrastructure will not allow these landlocked, malaria-infested countries to become competitive in manufacturing, in services such as tourism or agriculture; and even the exploitation of their natural resources is more difficult and expensive. Under these conditions and in the highly competitive global environment, the benefits of good policies and institutions, and the gains from investments in infrastructure are arguably extremely low, and it is argued that these countries therefore had little incentives to make any reforms. In addition, in many of these countries, transport costs alone make their products non-competitive.

2.1 The deteriorating conditions in SSA

For decades after their independence, practically all the SSA countries suffered from deteriorating economic conditions, resulting in deepening poverty and malnutrition. Violent conflicts, unstable governments and poor and erratic economic policies effectively prevented public and private investments and left the economy completely devastated. Local infrastructure rapidly deteriorated, leaving roads, railways, ports and the electricity systems in shambles. Despite their rich natural resources such as oil, gold, copper, diamonds and many others, the rising costs of excavating and transporting these resources and their falling prices in the world markets in the 1980s and 1990s reduced the profitability of exports and the potential of increasing their value-added through local processing in the African countries themselves.

In the 1990s, the deteriorating conditions in a group of very poor and heavily indebted countries required the World Bank and the IMF to launch the HIPC Initiative as a comprehensive debt relief effort. The first HIPC programme, already negotiated in 1996, focused on a large reduction of the debts in order to achieve debt sustainability and remove the disincentive effects of the debt overhang on private investment. For the first time, the debt relief included a reduction of debt owed to the World Bank, the IMF and the regional development banks. In 1999, the HIPC Initiative was enhanced and targeted more clearly to provide additional funds for social spending. By the beginning of 2004, 42 countries were HIPC eligible: 34 in Africa, 4 in LA, 3 in Asia and 1 in the Middle East.

The policies in these relief efforts differed from past adjustment programmes by establishing close cooperation with the governments of the countries concerned and by emphasizing the social outcomes of the public expenditures. These policies had the following key principles:

- Poverty reduction is a central objective of international development cooperation.
- The national governments must take greater responsibility for poverty reduction in their countries by formulating and implementing nationally owned poverty-reduction strategies in participatory processes which involve the local civil societies and the non-governmental organizations (NGOs).
- Donor countries will be more selective in focusing their aid and debt relief programmes on those countries that have good poverty-reduction policies and a good system of governance for implementing these policies.
- Policy conditionalities should be derived from the national strategies and be tailored to the countries' conditions and constraints.
- Donors should increase the coordination of their financial support.
- Rich countries should provide greater market access for products from poor countries.

The new approach was agreed in Brussels in May 2001. For most LDCs, it was put into practice through the preparation and implementation of PRSPs, which are meant to be country-specific and vary between countries according to their conditions and constraints. They were perceived as the vehicle through which governments are expected to elaborate their nationally owned poverty-reduction policies, through which the 'Bretton-Woods' institutions, namely the IMF and the World Bank, identify satisfactory policy environments, and through which donors align their assistance for poverty reduction. In 2002, 34 LDCs were engaged in the process. Of the 15 LDCs that were not engaged in the process, six are small island states and six of the others were externally sanctioned or strongly affected by conflict in the recent past (Afghanistan, Haiti, Liberia, Myanmar, Somalia and Sudan).

The PRSP/HIPC Initiative taken by the Bretton-Woods institutions constituted an important and much needed effort to bring the poorest and most highly indebted countries into the world economy and to try to prevent their exclusion from the global growth process. Actual debt levels have been reduced very substantially. In the first stage of the HIPC programme, the country works in coordination with the Bretton-Woods institutions to establish a record of implementing economic reforms and poverty-reduction policies. The HIPC countries need to submit PRSPs that are prepared after consultation with civil society representatives and describe their proposed economic and social policies and programmes to reduce poverty.

Although there has been progress in some countries, the overall situation has not changed. The majority of the poorest countries, most of them in Africa, are still essentially excluded from global growth. The assessment in the spring of 2004 was that the majority of HIPC/PRSP countries will not attain the MDGs by 2015, or even a decade later, unless they have access to much more foreign resources to complement their domestic savings. However, if these foreign resources are provided as debt, even on concessional terms, most of these countries will not be able to service that debt again (Hewitt and Gillson, 2003).

The central problem is that most of these countries are not even able to provide security or offer protection from internal and external threats, prevent internal conflict or provide to their populations the most basic public services, such as basic health services and education. These countries have therefore been categorized as *failed states*. Most of these countries are as poor today as they were 40 years ago and their rapidly growing population makes it practically impossible for them to organize their economy in order to accelerate growth.[1]

2.2 The prospects of recovery

Since the early 21st century, the African countries have started to see a lever out of that predicament. Many African countries began to benefit both directly and indirectly from the strong international demand for oil, metals and minerals. The global demand was driven primarily by China and India that continued their relentless growth at unparalleled speed and raised, in turn, the demand

[1] Martin Wolf 'A Fair Globalization: Creating Opportunities for All,' Financial Times 2/22/06.

for oil and minerals. Countries that do not export these commodities have also benefited from the favourable regional and global growth and from the inflow of aid – including debt relief. But while the average growth rate per capita of about 3% was well above the region's average growth of 0.8% during 1997–2001, and came after years of decline since the early 1980s, this growth still falls short of the growth rate of 5% that SSA needs in order to achieve the central MDG of halving by 2015 the proportion of the population that in 1990 lived on less than US$1 a day (Baldwin, 2003). In addition, nearly half of the SSA countries are not expected to meet the health and education goals in the MDGs.

With the sharp rise in oil prices and the hectic search for suitable and more environment-friendly substitutes, the prices of several key agricultural products are also rising very rapidly. The price of sugar, which is being diverted to ethanol for automotive fuel, more than doubled since late 2004, the price of natural rubber (a substitute for synthetics produced from petroleum products) rose 60% in the first half of 2006, and the price of maize, which is used as the feedstock for ethanol production in the USA, rose by about 8% in 2006. As the production of biofuel becomes economically more viable and the prices of sugar and maize are rising, many farmers may try to substitute their current production of grains and other produce for the production of sugar and maize. Already 20% of the US maize crop and 50% of the Brazilian sugarcane crop are used for the production of ethanol. Should this trend continue, the price of other commodities, especially grains, may also rise due to the substitution in production and the increased use of crops for biofuels.

However, the high oil prices also contribute to an unwelcome increase in agricultural production costs by raising the price of fertilizers and energy and thus leading to a reduction in yields because farmers prefer to use less fertilizer. The rising production costs of energy and fertilizer-intensive crops such as grains are expected to lead to a reduction in their production and raise their prices in 2007 by up to 30%. On average, however, agricultural prices are expected to decline by about 1% in 2007 and by nearly 3% in 2008 due to the decline in demand and a weaker dollar.

The regional expansion has been driven primarily by only one-third of the SSA countries that had growth in excess of 5%. Among the oil exporting countries, growth was particularly strong in Angola (16.9%), Sudan (11.8%) and Mauritania (17.9%). A number of countries that recently emerged from prolonged conflicts also entered a cycle of rapid growth after prolonged stagnation and decline (Burundi, the Democratic Republic of Congo, Liberia and Sierra Leone). Countries that have resources of metals and minerals had large revenues from the rise in world prices and gained from the investment plans of large international corporations that were lured by the high prices of these resources and by the economic stability in many African countries. More investments are reaching even SSA's poorer countries, including oil-importing countries that are benefiting from the HIPC Initiative and the Multilateral Debt Relief Initiative (MDRI).

Countries that had macroeconomic stability and greater openness to trade, as measured by the share of their trade (exports + imports) in their GDP tended to grow more rapidly, in part due to a more rapid growth in TFP. In the second half of the 1990s, the rapidly growing countries also had TFP growth of 2.3%, compared with 0.7% in the medium growers and negative TFP growth in slow-growing countries. The financial sectors of the fast-growing countries also became better established, and, with lower and stable inflation, banks were used more widely. Many other SSA countries remained embroiled in conflicts and civil wars, and their governments were very unstable. In these countries, growth was very meager and their economies were gradually ruined.

A longer-term perspective that examines the impact of globalization on the African countries must evaluate the prospects for their continued recovery, their ability to accelerate their growth and reduce their poverty and food insecurity. The evaluation in this chapter is based on several extensive studies that were conducted in the last 2 years in order to assess the capacity of the SSA countries to embrace the 21st century and integrate into the global economy. However, the studies came to very different conclusions, and the chapter examines the reasons for their widely divergent outlooks about the future of Africa and its prospects to emerge out of its current low point.

According to the pessimistic outlook, SSA will be increasingly marginalized in the coming decade and many countries will effectively drift out of the world economy as they return to the stagnation of the 1980s and 1990s, while the rest of the world economy will flourish at an unprecedented rate. The optimistic outlook sees the route of Africa's future development charted by the MDGs until 2015, and that development will be followed by additional steps and more rapid progress later on that will gradually lift the entire African population out of poverty.

At the midterm point of the MDGs, between 2000 and 2015, it is hard to see a realistic option that the MDGs will indeed be realized according to that schedule. Cold realism suggests that some variant of a more pessimistic outlook is practically inevitable. At the present point in time, when the goal of halving the number of the poor by 2015 does not seem realistically within reach and all too many countries continue to deteriorate, the question is how far and how long will the world community allow Africa to drift into deeper poverty and hunger.

Cynically, it must be acknowledged that ethical concerns alone may not yield much more than another UN declaration and new MDGs that will merely postpone a more serious commitment. On the other hand, pragmatic, if not outright selfish considerations may force the developed countries to take a more drastic course of action as early as possible: The people in the African countries themselves will simply refuse to accept the verdict and acquiesce in the pessimistic outlook that dooms them to a life of intolerable misery; if they cannot change their conditions in Africa, they will do whatever they can in order to leave that seemingly doomed continent and search for a better place to live.

An escalating wave of migrants that may well become a human tsunami is likely to engulf Europe and the USA well beyond their capacity to absorb. The writing is already on the wall: at the shores of Senegal, at the Mediterranean sea and along the 2000 miles of fence that was erected to 'defend' the USA against the flood of migrants. There is no need for a sophisticated econometric model to predict what will happen. All we need to do is to ask ourselves: Would we not do the same? Would we be willing to starve in a village in Burkina Faso or Tanzania? The challenge for both the developed countries and the LCDs is to ensure that the poor countries can grow at a sufficiently rapid pace so that they can provide enough food for their populations, raise their standard of living and, most importantly, give them hope. As a background for an evaluation of the strategies that have been suggested in order to meet that challenge, the chapter evaluates the factors that have slowed down the growth process in the African subcontinent in the last two decades.

3 Obstacles to Growth in SSA

The improvement seen in the trends of economic growth in many SSA countries since the early 21st century started few years earlier. As a result, the average rate of growth of the real GDP per capita rose from −1.1% in 1990–1994 to 2.0% in 1995–1999. In 15 countries, the growth rate of the real GDP per capita exceeded 5%, and the share of countries with real per capita growth rates above 3% increased from 14% in the 1980s to 28% in the second half of the 1990s and to 23% in 2000–2003.

The percentage of countries with zero or negative rates declined from 58% in the 1980s to 12% in 1995–1999 and 16% in 2000–2003. However, these growth rates are still low considering the rates required to halve income poverty in SSA by the year 2015 and meet the poverty MDG; in order to achieve that goal, the real GDP would have to grow at an average rate of 7.5% (Fig. 6.2). The

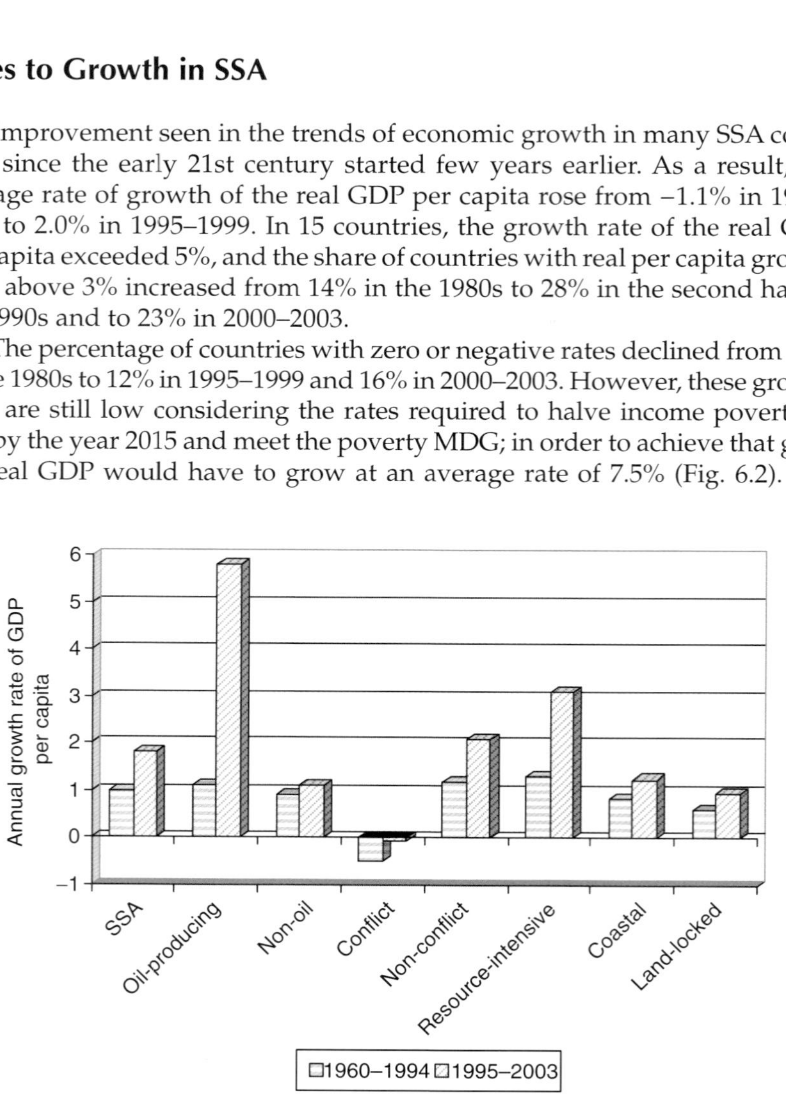

Fig. 6.2. Growth of real GDP per capita in the SSA countries (according to main country groups).

region's improving terms of trade since the late 1990s contributed to the growth recovery, but this is only part of the explanation of the region's more rapid growth. Perhaps the most important reason is the decline in military conflicts and civil wars and the establishment of democratic governments in an increasing number of countries.

Studies that focus on the economic factors that can bring about a more rapid reduction in poverty emphasize the importance of economic growth. The studies of Bruno *et al.* (1998), the World Development Report (2001), Ravallion (2001), Dollar and Kraay (2002), Bhalla (2002), and a considerable number of other studies demonstrated the impact of economic growth on poverty reduction. In the long run, growth was always pro-poor and brought about a reduction in poverty, but in the short run, there were large differences between the 'poverty elasticity of growth', namely the rate at which poverty is reduced as an effect of a given rate of economic growth.

The main reason for the differences is the impact of growth on income inequality, and in the countries of SSA, there were large changes in income inequality and considerable differences between countries in the rates at which their income inequalities changed. Other reasons were the high inequality and the low per capita income in the region already in the 1980s. Besley and Burgess (2003) estimated that global poverty elasticity to per capita GNP growth was −0.73, and the elasticities ranged from a high of −1.14 in transition countries to as low as −0.49 in SSA.

The rise in inequality in many African countries was partly due to the increase in inequality within urban areas and partly due to the widening gap between urban and rural populations as an effect of the stagnation of the agricultural sector and the pro-urban bias of government policies. Ghana and Uganda are two important examples: Ghana began the 1990s with low inequality by African standards. However, since then, both growth and poverty reduction concentrated in Accra and in the mineral-rich rural forest zone, whereas the rural savannah and rural coastal areas were lagging way behind. In Uganda, the rural–urban divide widened due to slow growth in agriculture, while in the urban areas growth accelerated, leading to an increase in the country's growth. Moreover, in the urban areas, growth was heavily skewed towards the top quintile during the 1990s, further increasing the income inequality. In the northern and the southern regions, ongoing military conflicts disconnected these regions from the centre of the country and prevented their development.

In countries that relied primarily on exports of agricultural export crops like coffee, cocoa, cotton and sugar, export earnings fluctuated widely due to the fluctuations in world prices (Table 6.1), the growing gap between the world prices and the prices that farmers received due to the growing share of the transnational corporations along the supply chain and the overvalued real exchange rates (Ng and Yeats, 2002).[2]

[2] The policy of maintaining an overvalued exchange rate was common in many African countries and reflected the urban bias of their policies.

Table 6.1. Indices of world price of selected commodities, 2001–2004. (From UNCTAD, 2004.)

	2001	2002	2003	2004
All food	100	103	107	121
Coffee (Arabicas)	72	72	74	93
Coffee (Robustas)	66	72	88	86
Cocoa	123	200	198	174
Tea	80	72	78	80
Sugar	106	84	87	88
Fish meal	118	147	148	157
Agricultural raw materials	96	94	112	123
Cotton	81	78	107	104
Non-coniferous woods	98	105	118	136
Tobacco	100	92	89	92

Index, 2000 = 100.

3.1 Growth and industrialization

Growth strategies of the SSA countries were fundamentally different from the growth strategy of the EA countries that invested their resources in the promotion of industrialization, often at the expense of the rural sector. Today, the choice between a growth strategy that promotes industrialization and a strategy that promotes agricultural production presents a difficult dilemma for the SSA countries and raises several key issues to consider:

- What are the industries in which the African countries can potentially have a comparative advantage over the industries of the Asian countries?
- How should a country choose which industry to promote and where should that industry be established? What should be the role of the public sector in making this choice and what should be the share of the public sector in the costs of the direct investments in the local industries at this initial stage? (Dollar and Shang-Jin, 2007).
- What are the overall public resources that are necessary to promote industrialization, including the direct investments and the public investments in infrastructure? What are the economic costs and benefits of these investments and what is their impact on poverty and on the income distribution?
- What are the direct and the indirect benefits from public investments in promoting a certain industry in terms of the impact on the country's growth, employment and poverty compared to investments in another industry or in the agricultural sector?

One specific industry that has been most suitable at the initial stages of industrialization and urbanization of practically all countries is textile and apparel. This is a highly labour-intensive industry that requires mostly unskilled labour and it has served as the foundation for the development of a solid industrial base in all the Asian countries from Japan, South Korea and Taiwan to China, India and Bangladesh. This industry went through very significant changes in the past decade that were not only the result of major technological developments, but also of the Multinational Agreement on Textile and Clothing. The implications of this agreement for the African countries will be discussed later on in this chapter, but even without going into details, the fundamental question is whether the African countries can follow the footsteps of all developing countries and develop a local textile industry by taking advantage of their cheap labour given the existing industries in the Asian countries.

The development of local industries in SSA can also be based on the countries' local resources, including minerals and agricultural products. These industries can increase the value-added of a country's exports and benefit from the country's comparative advantage due to the access to the local resources. Local agricultural products also offer advantages for local industries, ranging from the processing of tropical fruits and vegetables that are in high demand in the developed countries to the production of ethanol from sugarcane or maize that have lately become all the rage for industrial development. The development of local industries based on agricultural products seems to be an effective way to reach the rural population which still constitutes the majority of the population in the African countries, as well as an effective way to reduce poverty given the high concentration of the poor in the rural areas. However, there are a number of obstacles for this strategy which will be discussed in detail below.

The promotion of local industries also presents organizational and administrative challenges. To meet these challenges, the African countries would have to implement a range of measures, among them measures to strengthen their weak contract enforcement and reduce corruption. It is issues like these that are currently major obstacles in most SSA countries, where 16 of the top 20 countries in the world with the most difficult business conditions are located. It would also require large public investments to improve the poor infrastructure in rural areas and investments in education and training that are essential for industrial development.

Nevertheless, the prospects for economic development since the early 21st century are more promising than at any time since Africa's liberation from colonial rule. The change in the economic environment owes much to the change in the political climate that is reflected in the New Partnership for Africa's Development (NEPAD) adopted by the African Union in July 2002 (Hinkle and Schiff, 2004). It presents a vision of how Africa, in close partnership with international donors, can assume responsibility for its own development. In support of its governance objectives, NEPAD adopted the African Peer Review Mechanism, which measures progress in terms of political, economic and corporate governance. The contribution of peace and economic stability to economic growth is reflected in an annual per capita growth of 3–5% in more than a dozen African countries, home to more than one-fourth of the continent's population; there has been an acceleration of growth across the continent, and Africa has grown faster than LA for several years in a row.

African policy makers now face the challenge of how best to carry this positive vision forward. The progressing economic performance is encouraging, macroeconomic policy has improved very significantly, fiscal deficits are under better control, inflation has come down dramatically, foreign debts have been substantially reduced and the share of international trade in the economies of the African countries has risen. But the region still has a lot of catching up to do in order to achieve the MDGs. The current global environment offers opportunities for growth on which to build a forward-looking reform agenda; towards that end, the African countries should aim to maintain macroeconomic stability, improve the business climate, strengthen the financial sector, promote trade, strengthen the fiscal institutions and improve governance generally.

However, the previous decades of economic stagnation have taken a toll, and so far, the recent improvements in Africa's economies have contributed only little to reducing the income gap between SSA and the other regions of the world. The EA countries, which in the 1960s had lower per capita incomes than SSA, have since made the transition to middle-income status and their growth is considerably higher. The share of the SSA countries in world trade, after declining from 4% in the 1980s to less than 2% at the end of the 1990s, is only now showing signs of a slight upturn. Standard indicators of financial depth are much lower in SSA than in other developing regions and the poor infrastructure is a major obstacle to growth. To accelerate growth and improve the investment climate, these countries will have to implement comprehensive trade and financial sector reforms and build more efficient public and private institutions.

The goal of accelerating economic growth is to improve the living standards of the general population, particularly of the poor. That requires the government in each country to make difficult choices on how to allocate its resources between investments for future growth and consumption for alleviating the existing poverty. With the unprecedented debt relief and the promises to scale up aid, local populations should benefit from better education and health services. In implementing the relevant plans, governments must ensure that the increase in their spending is consistent with the economy's absorptive capacity and with the goal of maintaining macroeconomic stability and low inflation.

The significant improvements that have been achieved in recent years in debt sustainability and in the economic situation in general are attracting new lenders, both private and official. Some African governments are again tempted to contract new loans on non-concessional terms in the open financial markets if they cannot meet their needs and their plans for greater spending from the available resources that are given on concessional terms. They must, however, be very cautious in assuming new non-concessional loans and safeguard debt sustainability. They must consider very carefully the conditions of such loans in order to secure the best conditions in the highly competitive international financial market. Similarly, governments should be careful about granting tax or other concessions for FDI, whether in natural resources or in other sectors, which might be a drain on their future income.

Sustained growth requires the government to support larger enterprises that operate in the formal economy, have access to credit and are engaged in trade with connections to the global supply chain. However, so far the bureaucratic maze, irrational legislation and high corruption have driven most business activities to the informal urban sector and discouraged small industrial enterprises from increasing the scale of their activities, thus avoiding registration, legal contracts, taxes or accounting. The immediate focus must therefore be on creating a better investment climate, in which firms can be formally established and operate freely. Encouraging private investment is critical for sustaining growth in the economy. So far, public investment in Africa has been double that of private investment.

3.2 The predicament of Africa's geography

Geographical conditions are blamed by some experts for making the African countries in general and their rural populations, in particular, more vulnerable to poverty; among the factors most often cited in this context are inhospitable climatic conditions and difficult terrain that lowers yields, creates greater difficulties for efforts to diversify production and imposes great difficulties on travel as well as the transport and trade of products within and across countries. In some regions, the geographic conditions require great expenses for infrastructure due to difficult physical, environmental and climatic conditions. Landlocked countries and populations living in remote areas to which access is difficult or in areas that are frequently flooded are most vulnerable. Reducing these vulnerabilities would require large investments in solid infrastructure, and only the government can mobilize the necessary funds.

Extensive research by Radelet and Sachs (1998) found that landlocked countries have lower growth rates by around 1.5%: Using the CIF/FOB (cost + insurance + freight/free on board) margin, they found that cost of freight and insurance for landlocked developing countries was, on average, 50% higher than for costal economies; other calculations show that the total cost of crossing a border in Africa is the same as inland transport costs of over 1000 miles or sea transport costs of 7000 miles. In comparison, the cost of crossing a border in Western Europe is equivalent to only 100 miles of inland transportation. This theory certainly does not offer a ray of hope, but it is actually not very convincing for two simple reasons: First, most African countries are not landlocked and they are much better located and better endowed than the EA countries, and they still share the stagnation of the entire subcontinent. Second, the African countries had the same geographic conditions in the 1950s and 1960s, and yet most of them were growing very robustly during these years. Indeed, Uganda has been one of the fast-growing SSA countries since the late 1980s even though it is landlocked.

Limao and Venables (2000) found that 'African economies tend to trade less with the rest of the world and with their neighbors than would be predicted by a simple gravity model and the reason for that is their poor infrastructure' (p. 25). The infrastructure shortcomings include inefficient seaports and, even more importantly, the inadequate internal infrastructure of roads, rail and telecommunications. Collier and Gunning (1999, pp. 71–72) analysed the impact of Africa's poor road infrastructure which is much less developed than in most other developing countries; thus, for example, the density of the rural road network is only 55 km per 1000 km^2, compared with over 800 km in India.

Africans who live in the interior of the continent face enormous transport costs for shipping goods between coastal ports and the places where they live and work. Moreover, the Sahara effectively cuts SSA off from high-volume overland trade with its major high-income trading partner, Europe, thus further increasing transport costs. The isolation that is effectively enforced by the high transport costs increases production costs due to the small market size. Recent evidence underscores the extremely high transport costs in SSA and the severe impact on trade (Fig. 6.3).

Coastal countries tend to have considerable advantages compared to the landlocked countries of the Sahel and Central Africa. A group of African countries – including Burkina Faso, the Central African Republic, Chad, Mali, Mauritania and Niger – are so burdened by their extreme climate, related problems of health and disease, as well as poor geographic position that it is not clear that any economic model offers them a path towards development.

The quality of other components of Africa's infrastructure is also lower: Compared with Asia, there are only one-tenth the telephones per capita and the telephone system has triple the level of faults; the proportion of diesel trains in use is 40% lower; freight rates by rail in Africa are on average around double of those in Asia; port charges are far higher and air transportation is four times more costly than in EA. In addition, most of the international transport is cartelized, reflecting the regulations that many African governments are imposing in order to support the national shipping companies and airlines. As a result of these high costs, freight and insurance payments on

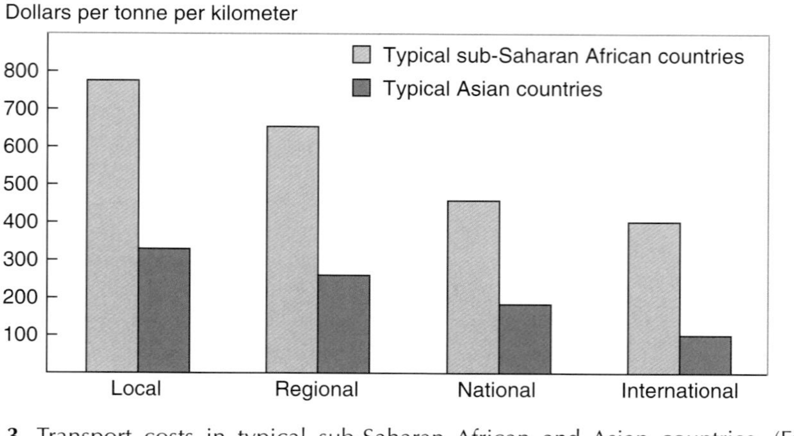

Fig. 6.3. Transport costs in typical sub-Saharan African and Asian countries. (From Starkey *et al.*, 2002.)

trade are more than a quarter of the total export earnings, whereas the average for the developing countries is only 6%.

In addition to the far higher transport costs facing Africans who live in the interior of the continent, problems of isolation are compounded in many countries and regions by the small market size: Small developing countries with little access to global trade tend to grow more slowly than countries with easy access to international trade or countries with large internal markets. A study of IFPRI (2006) shows that the majority of the rural population concentrates in areas that are relatively close to the urban centres and in practically all the four countries studied, the share of the population in the relatively more remote areas was small even when the share of the cropland in the more remote areas was much larger (Table 6.2). Bigman *et al.* (2000) showed in a study on Burkina Faso that the incidence of poverty in villages that are 6–8 km away from the main highway is 2–3 times higher than in villages that are 2–3 km away from the main road.

The poor infrastructure that prevents the development of trade even within the country increases the incidence of poverty and limits the use of the available cropland to areas that are not too far from the urban centre or the main highway; there is also a reduction in the efficient use of cropland because it is more difficult for the local farmers to use fertilizers. Since the poor infrastructure and the high transport costs are major obstacles to increasing agricultural production, using the croplands more efficiently and developing trade both with the urban centres and for exports, Africa must make large investments in an extensive road system both between the urban centres and from the villages to all-weather roads.

Calderon and Serven (2004) have shown the huge differences in investments in infrastructure between the world's main regions (Fig. 6.4). Their results highlight both the fundamental differences between the developed and the developing countries and, within the developing countries, the differences between the SSA countries and the other developing countries. In the past decade, both the LA countries and the EA countries made large investments to improve their infrastructure; the SA countries, primarily India, also made larger investments to slow down the erosion of their infrastructure. In

Table 6.2. Distribution of cropland and rural population by market access zones (%). (From IFPRI, 2006.)

Access to towns >50,000 (hours)	Burundi		Eritrea		Ethiopia		Kenya	
	Rural population	Cropland	Rural population	Cropland	Rural population	Cropland	Rural population	Cropland
<2	58	49	38	9	18	16	37	17
2–4	34	39	29	29	31	26	38	25
4–6	8	8	15	34	24	29	19	25
6–8	1	3	10	20	13	17	6	16

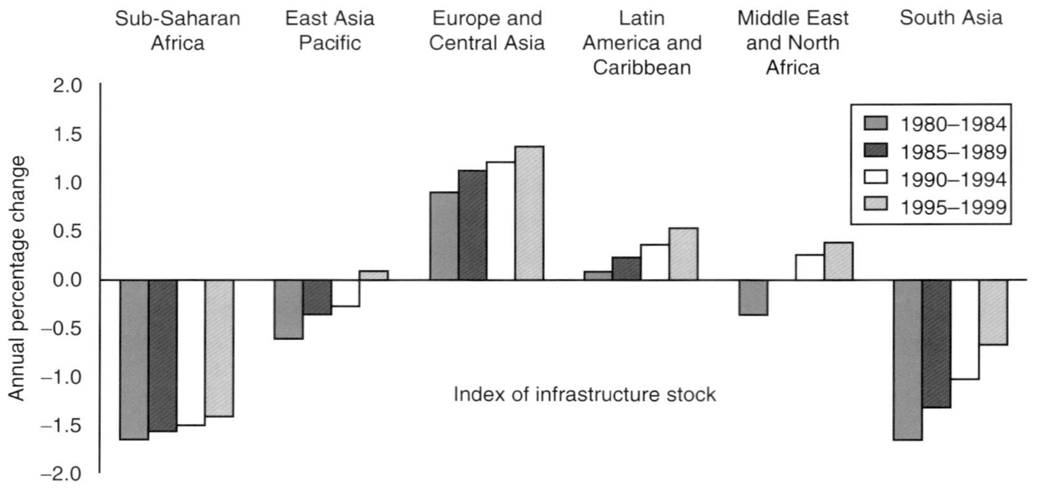

Fig. 6.4. Investments in infrastructure in developing countries. (From Calderon and Serven, 2004.)

SSA, in contrast, the road system and the other infrastructure continued to deteriorate, and state investments were very minimal.

3.3 The obstacles of governance and institutional capacity

The World Bank's report *Doing Business 2007: How to Reform* ranks 175 countries according to the ease of doing business; the average rank of an SSA country was 131. In practically all the SSA countries, there are obstacles in all private sector activities: licensing, employment, credit and transactions with the government are all tedious and time- and paper-consuming activities (Fig. 6.5). 'Under the table' payments can slightly ease or spur on the process, but the costs are high and the process is still horrendous. For instance, it takes an average of 11 procedures to start a business in a SSA country compared with an average of eight procedures in a SA country; this process takes 2 months in SSA compared with 1 in SA; running a private business costs three times as much in terms of income per capita, partly due to administrative complexities, corruption and cumbersome legal systems, and partly due to high expenses of such essential business services as telecommunications and energy.

This is one reason why despite the recent acceleration of growth, the share of investments in the GDP has not increased compared to its level in the early 1990s, and FDIs in most SSA countries, other than in oil and mineral exploration and production, are still very low. South Africa has become an important source of investment in other parts of SSA, and investment from China and India is also picking up. The World Bank's investment climate survey also found that SSA has begun to reform business regulation: two-thirds of the SSA countries made at least one positive reform in 2005/06.

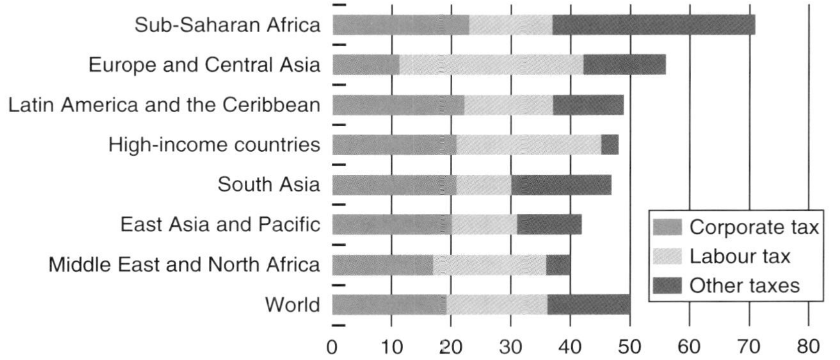

Fig. 6.5. Total average business tax rate (%). (From World Bank, 2006.)

Tanzania and Ghana ranked among the top 10 reformers in the world, and plans for reforms in other countries would reduce business costs throughout the region. The investment climate is also influenced by the very high business tax rates in Africa which are by far higher than in all other regions, driving companies to the informal sector and thus reducing the overall tax receipts. The other way to avoid these practically impossible taxes is with bribes to the various officials in charge.

Corruption filters through all levels of government: low-level bureaucrats are paid so little that their survival often depends on taking bribes. Politicians commonly take cuts from any deal they approve. Ministers often sign false contracts with government suppliers. In Kenya, a new president, Mwai Kibaki, came to power in 2002 when the public was weary of the long and corrupt regime of Daniel arap Moi. But in 2006, the new government began to unravel as evidence of high-level corruption started to spread and a report that described corruption at the highest levels forced the finance minister to resign. Liberia's president has dismissed in 2006 all of the political appointees in the country's finance ministry because of graft. Even in South Africa, President Thabo Mbeki sacked his deputy for corruption in 2005.

Foreign investors, especially in Africa's oil industry, have long paid massive bribes to politicians to win contracts or the rights to pump. Indeed, the countries reckoned to be most corrupt in Africa – Chad, Nigeria, Equatorial Guinea, Angola – are also large oil exporters and their local industries are dominated by western oil firms. African politicians are also under enormous pressure to ensure that financial rewards are passed to extended families and members of the leader's ethnic group.

The financial sector in the SSA countries is the Achilles heel of the private sector: in the 1990s, the financial sectors suffered from chronically weak banks, often publicly owned, interest rate controls, high government-directed lending rates and weak compliance with international financial standards. Many of these impediments have been corrected in recent years, but on standard indicators of financial depth, SSA countries still lag far behind other developing countries. For example, bank deposits as a share of GDP in the low-income SSA countries are half the level of other developing regions, in part because Africans do not have much trust in the banking system and prefer to be more cash intensive. Quite a few SSA banks still persistently violate international banking regulations, especially those that pertain to diversification of risk. Interest rates on loans are very high, in part because banks are reluctant to expand their domestic lending portfolios, preferring to accumulate substantial holdings of bonds, loans to large corporations and foreign assets despite their low returns, due to their much lower risks. In rural areas the coverage of the formal banking sector is extremely poor; in recent years informal financial institutes are cautiously expanding, but their interest rates are very high.

A number of administrative and policy bottlenecks still suppress the expansion of the financial sector. They include excessive reliance on regulatory monetary instruments (such as reserve requirements imposed on banks), the lack of reliable sources of information on potential bank borrowers, inadequate land titles that restrict the use of collateral and legal systems that do not yet accommodate financial instruments like leasing (Fig. 6.6).

Public institutions are also very weak: more than any other region, Africa is handicapped by its inability to implement economic policies due to its weak public institutions. Any future effort to support its development must therefore focus on strengthening institutions by providing training and technical assistance to build the local capacity; the IMF has contributed to this effort with the establishment of regional technical assistance centres (AFRITACS) in SSA. Governments need to supplement this effort by making the civil service more effective and donors need to coordinate their technical assistance with the countries (World Bank and International Finance Corporation, 2006).

According to Hernando de Soto (2000), one reason why the weak, inefficient and corrupt public institutions are particularly damaging is due to the obstacles they place on obtaining formal, officially registered rights of ownership of the assets that poor people own. As a result of the inefficient and corrupt organization of property rights, poor people lack legal ownership of their assets and, in the absence of such legal instruments, these assets cannot be purchased, exchanged, sold, bequeathed, lent or transferred in any way other than within the framework of limited networks, often subjugated to parasitic

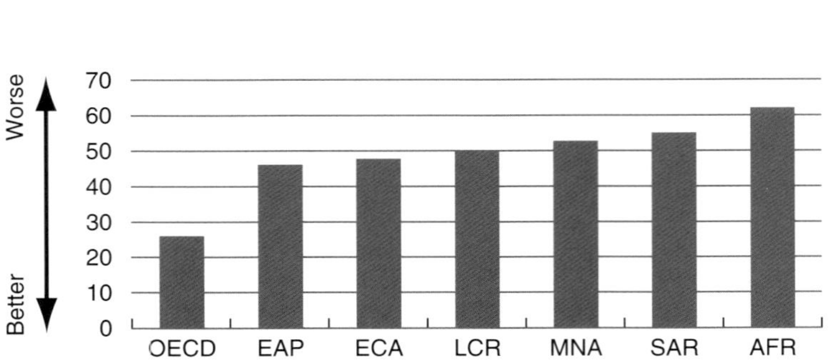

Fig. 6.6. Administrative obstacles on conducting business. (From World Bank, 2006.)

local institutions of power. As a result, these assets cannot serve as the basis for an efficient and dynamic accumulation of capital. They become, in de Soto's words, 'dead capital', and the informal sector becomes what is usually called undercapitalized and as a result cannot realize its full potential.

With the increased resource inflows, whether from exports of oil and other commodities or from scaled-up aid, good governance also requires urgent measures to secure a transparent and efficient system to supervise the flow of these resources and minimize the leakage out of the formal system. Oil exporters are now in the fourth year of an earnings spree, and low-income oil-importing countries receive higher aid inflows. In a number of countries, debt relief from the enhanced HIPC Initiative and the MDRI has also freed up domestic resources. However, the promises of scaling up aid by donor countries and multinational organizations have not yet materialized. The debt-relief programmes should prevent the African countries from accumulating new debts and all concessional lending must be transparent, carefully monitored and well understood. The African countries should also be concerned with the macroeconomic impact of resource inflows, in particular with the implications for their export competitiveness resulting from an appreciation of their real exchange rate; government supervision must carefully monitor the distribution of the resource inflows in order to secure that the poor and poverty alleviation programmes are included and receive their due share.

4 Food Insecurity: Production or Income Shortfall?

Hunger and malnutrition are the most extreme manifestations of poverty in Africa. Although SSA accounts for 13% of the population of the developing countries, it accounts for 25% of the undernourished people in these countries and that share is growing. Between 1990–1992 and 2001–2003, the number of undernourished people in SSA increased from 169 million to 206 million, and only 15 of the 39 countries for which data are reported managed to lower the number of undernourished. Food insecurity remains the major problem in SSA, and it is the region with the highest proportion (one-third) of people suffering from chronic hunger. During 2001–2003, the prevalence of food energy deficiency varied from around 40% in Tanzania, Kenya and Uganda to over 70% in Ethiopia, Burundi, Malawi, Zambia and Rwanda.[3] Even in Uganda, which experienced very robust growth in the past decade and an increase in the per capita GDP at an average annual rate of 3.8% and a more modest growth of 1% in per capita agricultural GDP, the number of undernourished people increased from 4.2 million to 4.6 million between 1990–1992 and 2001–2003, although the prevalence of malnutrition declined from 24% to 19%.

In the last three decades, food security in SSA has substantially worsened. While the proportion of malnourished individuals remained around

[3] See FAO Report: State of World Food Security, January 2007.

one-third of the population since 1970, the number of malnourished people has increased substantially with population growth, from around 88 million in 1970 to an estimated over 200 million in 1999–2001 (FAO, 2003). This record is in stark contrast to that of other developing regions such as SA and EA and LA, which made significant strides in combating malnutrition during these years. The proportion of undernourished children in the region has also increased in Africa during the last 30 years, from around 27% in the 1970s to over 33% more recently. Africa is the only region in the developing world where the number of malnourished children has been increasing (IFPRI, 2005).

Since the early 1980s, there have been minor structural changes in the LDCs as a group – although there were large differences between them. The share of agriculture in their GDP is declining very slowly; and while both industry and service activities are gradually expanding, much of the increase in the industrial value-added is concentrated in a few LDCs and the increase in their industrial activities concentrates in mining, exploitation of crude oil and the generation of hydroelectric power, but not in manufacturing.

The challenge that Africa is facing in its efforts to achieve food security and make more comprehensive structural changes in its agricultural sector is to make these changes part of a wider reform in the entire rural sector that should include large investments in infrastructure, strengthening public–private partnerships, programmes to empower women and investments to secure water availability for drinking and for irrigation. The poor infrastructure in the rural sector is one of the main obstacles to a more rapid agricultural development in Africa.[4]

Several studies have shown the links between poverty, growth and rural infrastructure in the continent as well as in other developing regions. The main factor that hinders subsistence farmers to increase their output and their income is access to the product markets. New roads would also offer more employment opportunities and enhance trade between the urban centres and the villages. The current dismal state of the infrastructure in much of SSA impedes the development of more productive agriculture and the diversification of agricultural production.

This is perhaps the main reason for the large gap in income, food security and standard of living in the rural areas between the African region and the other regions where the government made much larger investments in the rural infrastructure, including investments in roads and also in telephone lines and electricity. A growing problem in all countries is the increasing scarcity of water, and presently less than half of the population in SSA has access to safe drinking water.

The main challenge in Africa is to prevent man-made food security crises: in 2006, 39 countries, mostly in Africa, faced food crises and required emergency assistance (FAO, 2006). Over the last two decades, the number of food

[4] In 1999–2003 capital formation was only 22% of GDP in the LDCs as a group, and domestic private investment was particularly weak. Capital formation in the LDCs is far below the rate which is estimated to be required for sustainable growth (UNCTAD, 2006).

emergencies has risen from an average of 15 per year in the 1980s to more than 30 per year since 2000. Much of the increase has occurred in Africa, where the average number of annual food emergencies has tripled (FAO, 2004). Human actions have often been the cause of many food crises, primarily through wars and civil conflict. Of the 39 countries that had food crises in 2006, 25 were caused primarily by conflicts and their aftermath.

Even in countries where the food supply could meet the energy needs of the population, large income inequalities and deepening poverty led to an increase in malnutrition. In addition, problems of diet quality are becoming more widespread due to the increase in food insecurity in the urban centres with the continent's growing urbanization. However, the deepest and more severe food deficiencies are still in rural areas where the population is significantly poorer and more exposed to climate-induced transitory food shortfalls and where yields are much lower than in other countries.

Efforts to reduce food insecurity in the continent have been hampered by natural and human-induced disasters, including military conflicts and civil strife, the spread of HIV/AIDS and periodic droughts in different regions. The main reasons for the increase in food insecurity since the early 1990s were disasters in five war-torn countries: Burundi, the Democratic Republic of the Congo, Eritrea, Liberia, Sierra Leone and the Darfur region of Sudan. These countries combined account for more than 30 million or three-quarters of the region's total increase in the number of undernourished people. Particularly alarming was the extreme starvation in the Democratic Republic of the Congo, where the number of undernourished people tripled from 12 million to 36 million, and the prevalence of hunger rose from 31% to 72% of the population. The ethnic conflict in Darfur, in the southern region of Sudan, has deteriorated into outright genocide.

In order to increase their trade and produce for exports, African farmers would need an extensive road system both from the coast to the interior and in the interior from the villages to the local urban centres. Investments in the road infrastructure are obviously far beyond their means, and their capacity to trade therefore remains very limited. The high transport costs also reduce their capacity to use fertilizer or pesticides or to have access to advanced technologies, and they are forced to farm on soils that are increasingly depleted of nutrients, and without modern methods of water management. Their food production and their income from their land are therefore falling, while the population in their communities is rising, thereby further reducing their capacity to make any investments that would raise their yields.

An increase in food production is the key to the reduction of chronic undernutrition in rural areas, and a rise in productivity is the most effective way to increase the farmers' food production, raise their incomes and reduce food prices. However, in the last 20 years, Africa was the only region in the world where agricultural productivity declined (Table 6.3). While there are some success stories, yields of many important food crops, such as maize, millet, sorghum, yams and groundnuts, are no higher today than they were in 1980, and access to domestic, regional and world markets is still limited for most small farmers in Africa. In one-third of the countries, there was a rise in per capita food production, but it

Table 6.3. Agricultural commodity yields: ECA, Africa and global, 2003. (From FAO, 2004.)

Commodity	ECA	Africa	Global
Maize	1.39	1.61	4.47
Rice	1.12	1.87	3.84
Wheat	1.38	2.03	2.66
Sorghum	0.67	0.88	1.30
Millet	0.47	0.70	0.82

was too low and not sufficient to reduce the number of the poor. (On the very different Indian Experience see Rodrik and Subramanian, 2004).

The high food insecurity in nearly all SSA countries mirrors their stagnation and the region's high levels of poverty. In the last two decades, SSA was the only region in which the number of the poor continued to increase, and according to the World Bank's *Global Poverty Monitoring* database, all measures indicate that during the 1990s there was no progress in the reduction of consumption poverty (Fig. 6.7).

One problem is that Africa has the lowest share of irrigated cropland of any major region of the developing world, leaving the local farmers much more vulnerable to weather vagaries, since rainfall in much of the continent is erratic, vulnerable to high seasonal and year-to-year fluctuations, and the land is subject to high rates of evapotranspiration, which is also due to the high temperatures.

The following factors contributed to the prolonged agricultural crisis in the region:

- Dualistic development policy which is biased against agriculture;
- High population growth rate;

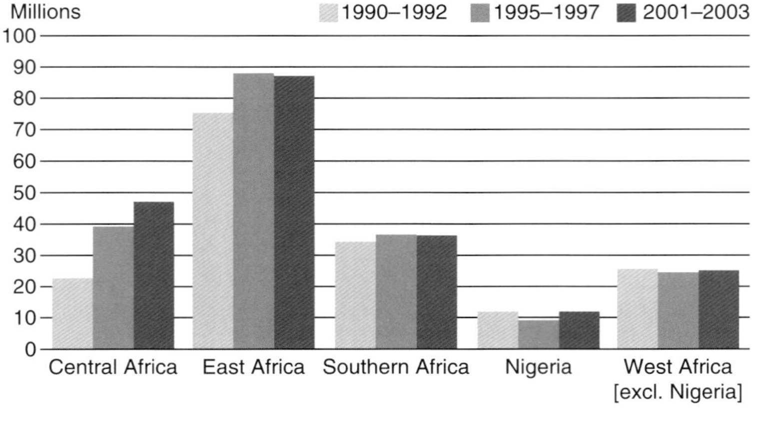

Fig. 6.7. Number of undernourished people in the main regions of SSA. (From FAO, 2007.)

- Slow pace of development of the arable land and harvested areas;
- Slow pace of technological changes and the use of rudimentary technologies in many rural areas.

The stagnation of the agricultural sector also slowed down the region's industrial progress and the contribution of the industrial sector to the GNP or to domestic employment is far smaller than in all other developing regions (Table 6.4).

Poverty remains high, food and nutrition security low, commercialization of smallholder agriculture limited and agricultural productivity low. Therefore, the prospects of achieving a significant improvement in food security are not very promising, and there are grave doubts whether the SSA countries will be able to attain the growth rates required to achieve the UN MDG to halve poverty rates by 2015 (Pattillo *et al.*, 2005).

The decline in agricultural labour productivity led to a reduction in the production per worker in 2000 to a lower level than four decades earlier. This region-wide contraction in labour productivity concentrated in countries where the entire economy was in decline, but agricultural growth in the entire region did not keep pace with population growth. Because most of the region's population resides in rural areas and depends on agriculture for income and sustenance, and given the low levels of productivity growth in the sector, poverty rates remained high and undernutrition deepened.

As a result of the decline in productivity and in total agricultural production in SSA, there was a sharp decline in the share of the region in the total output of agricultural products in the world. In 2000, the share of the region

Table 6.4. Selected indicators on agricultural production in SSA. (From Sachs *et al.*, 2004.)

Indicator	Tropical sub-Saharan Africa	Rest of developing world (average)
Population distribution		
Share of population within 100 km of coast	24.9	66.3
Share of population in tropical ecozones	62.0	34.9
Share of population in subhumid and arid ecozones	81.9	38.7
Share of population living at low density	45.2	26.5
Agriculture		
Irrigated land as share of total agricultural land	3.5	10.6
Cereal yield (kilograms per hectare)	1,102.0	2,364.6
Fertilizer consumption (100 grams per hectare of arable land)	95.0	1,606.2

declined to around one-third of its share in 1980. The rise in the share of the developed countries in world production and trade of agricultural products took place despite the very significant decline in the share of agriculture in total employment of these countries, and it was the result of the sharp rise in productivity and the introduction of much more advanced production methods.

5 Africa's Trade Policies

The share of the countries of SSA in world trade has continuously declined since the early 1970s, and for the 30 most important non-oil export products combined, their average market share declined during these years from 20% to less than 10% – an annual trade loss of well over US$10 billion. While world trade in these products grew at an annual rate of 12%, their exports from SSA grew at an annual rate of around 6%; as a result, the region's share in world exports has been continuously declining and the African countries failed to take advantage of the expanding world trade. In part, this was due to the decline in the price of many of their export products, and the decline in the profitability of their exports due to the pro-urban exchanges rate policies in many African countries that led to their appreciation. However, the main reason for the decline in their exports was the decline in the local output of these products that was primarily an outcome of the fall in productivity. The share of the region in world imports also declined, while the share of *all* the developing countries in world exports increased from 18.5% to 26.8%, and in world imports from 18.7% to 25.3%. In addition, Africa's exports continue to concentrate in primary commodities with little diversification, and there was only a meager flow of FDI into the region that concentrated in oil and minerals exploration and extraction.

Two factors explain the region's marginalization in world trade: one is the very limited openness of the African countries to trade, as measured by the export to GDP ratio or trade to GDP ratio relative to the world averages, and the other is its very slow growth, as measured by the region's GDP growth relative to the world average. There were considerable variations among countries in the relative significance of these two factors, but they both contributed to the region's marginalization. The marginalization of the region in world trade further slowed down its growth and, in some countries, increased the external debts, thus bringing about a decline in imports and adversely effecting investments.

There were also significant changes in the direction of trade during the 1990s: most notable was the reduction in the share of the EU, the largest trading partner of the region, from around 50% to nearly 40% – despite preferential access to the EU market that was given to the African countries under the Lomé Convention. The share of intraregional trade in the region's total trade increased only marginally, and that increase concentrated in two countries: South Africa and Kenya; these two countries supply most of the intraregional exports but receive only a small part of intraregional imports.

Intraregional trade remained rather limited because the countries of the region have very similar factor endowments and they trade in similar goods; another reason is the restrictive trade practices in most African countries that inhibit export expansion and diversification. This subject is further discussed below.

Given the region's small and fragmented economies and their low per capita income, their exports cannot be based solely on the domestic markets and an aggressive development strategy in the region must be based on the promotion of exports to other countries and other continents. In addition, in order to fully exploit their comparative advantage in factor inputs and in agriculture and minerals, the region needs to diversify into labour-intensive exports, including agroprocessing and manufacturing that uses the local inputs and services. This requires a stable macroeconomic environment with competitive real exchange rates, and an elimination of the anti-export bias in their trade regime.

Despite considerable improvements in recent years, Africa's openness to trade is still lagging behind. Since 1970, total trade (exports plus imports) in SSA has grown at three-fourths of the world's average rate of growth and only at about half the rate of the Asian countries. Africa's share in world trade has thus fallen from about 5.4% in 1950 to 4% in the 1970s and to 2% in 2005 (Fig. 6.8). Its trade openness (measured by the trade to GDP ratio) has also increased more slowly than that of any other major developing region, and in 2001, Africa supplanted LA as the region which is least open to trade.

The share of total trade of the SSA countries (excluding South Africa and Nigeria) in their GDP increased only marginally from 45.0% to 50.4% during the entire 20-year period from 1980 to 2001. The large drop in the terms of trade between primary commodities and manufactured goods until the early 21st century accelerated these changes and had a significant effect on both their short-term macroeconomic performance and their long-term growth owing to the importance of these commodities in their economies and exports. During the 1990s, the terms of trade of non-oil commodities deteriorated to such an extent that by 2000 they represented less than one-third of their pre-1920 levels.

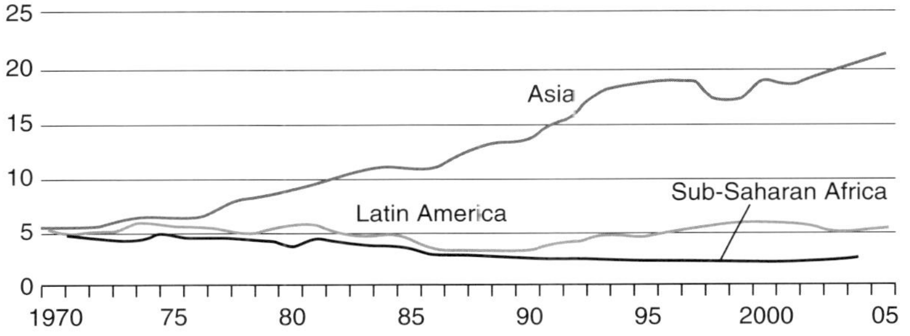

Fig. 6.8. Share of the SSA countries in world trade (%). (IMF, Direction of Trade Statistics, various years.)

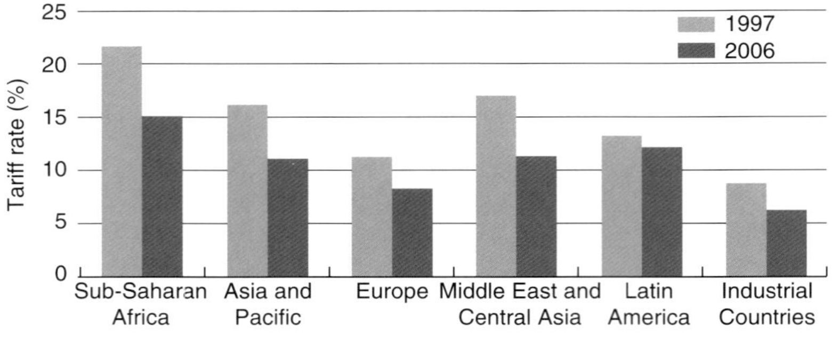

Fig. 6.9. Average tariff rates in main regions (simple average rate by region, percent). (From IMF staff estimates.)

Trade policy in many African countries has been tightly controlled by various restrictions that are imposed on trade. The protectionist trade policies were initially influenced by the perceived need to stimulate local industrial development, under the banner of import substitution and infant industry protection. In many African countries, tariffs and quantitative restrictions were the most important form of trade restrictions, but many import products were entirely prohibited. However, trade barriers in Africa were excessive, and nearly all countries used different combinations of quantitative restrictions, tariffs, licensing, import bans and foreign exchange regulations to control the flow of imports and exports. Thus, for example, the average tariffs and NTBs imposed by the African countries on imports from the OECD countries were 26.8% compared to 8.7% by the group of fastest growing exporting countries (Fig. 6.9). These controls were clearly excessive and highly detrimental: Following an extensive cross-country analysis in 117 countries of their openness to trade, Sachs and Warner (1995) concluded that the lack of openness was by far the largest contributor to the dismal economic performance of the SSA countries during these years.[5]

In many African countries, exports were subjected to similar restricting measures, with rules making it illegal to export 'strategic' items or subjecting exports to high taxes. Special marketing agencies and boards were instituted to ensure compliance. In some countries, farmers or traders needed to obtain special permits to export surplus agricultural or 'controlled' products. Tariffs on agricultural products were, on average, 23%, compared with 7.3% in the

[5] Sachs and Warner (1995). In this study, they constructed a trade openness index based on five important aspects of trade policy, and classified an economy as open if (i) the import duties averaged less than 40%; (ii) the quotas covered less than 40% of imports; (iii) the black market premium on the exchange rate was less than 20%; (iv) a state monopoly of major exports was absent; and (v) the economy was not socialist. In their cross-country regressions to explain growth between 1970 and 1989 in 117 countries, they found a strong association between openness and growth, both within the group of developing and the group of developed countries. Open economies grew annually 2.5% more rapidly than closed economies.

fast growing exporters. These restrictive measures produced a large anti-export bias in the African countries that lessened their ability to be competitive in international markets and contributed to the erosion in the share of their major export products in world trade. For example, during the 1960s, copper alloys were SSA's single largest export and the region supplied 32% of the copper alloys imported by the member countries of the OECD. By the mid-1990s, this share had dropped below 10%.

Africa's exports remain dominated by the exports of primary commodities, including oil, that account for about 40% of total exports, and agricultural products that account for more than 25%. The share of primary goods in exports has been nearly two times higher than the share in other developing countries even before the rise in their prices. The share of manufactured goods in Africa's total exports has stagnated at around 30% – well below the average share of manufactured goods in other developing regions. Exports of most African countries are concentrated in one or two products and only few countries, including Zambia and Kenya, have achieved some diversification of their exports. However, with the rise in the world prices of these primary products, the African countries that export these products are well positioned to benefit from the high global demand for these commodities. In particular, the rapidly growing Asian countries, led by China and India, already constitute significant export markets for the African products, and their importance will increase in the coming years.

Despite the current large profits of exports of primary products with the sharp rise in their price, Africa's concentration on commodity exports is disadvantageous in the longer run. The region's countries have a narrow industrial base and their exports have low value-added, primarily of semi-processed raw materials or labour-intensive products like textile and apparel that have preferential access to industrial countries. Many developing countries that are now dynamic exporters of manufactured goods initially relied on exports of primary commodities. This is particularly true of resource-rich countries such as Chile, Indonesia, Malaysia and Thailand. As recently as the mid-1980s, China, too, relied on exports of primary commodities, including oil. But as their earnings from the exports of primary commodities increased and their growth accelerated, these dynamic exporters increased their exports of manufactured products. The EA countries built their manufacturing industries by giving special incentives to their exports, by reducing their import tariffs, by creating export processing zones and by providing rebates or exemptions of duties on the materials that they imported for export production. Some EA countries attracted significant FDI with tax incentives, which helped them to upgrade their domestic technology and fill in gaps in management and marketing skills.

By contrast, SSA remains one of the world's most protectionist regions (Fig. 6.9). Although almost all African countries reduced their tariffs in the last two decades, the average (unweighted) tariff rate for SSA was still around 15% in 2006. Nigeria is often brought as an example of a country that heavily protects its market from imports, setting duty levels for agricultural products at an average rate of over 50%, and for non-agricultural products at an average rate of 25%. Coupled with the significant and costly NTBs that exist in most SSA countries, these rates keep the cost of imports artificially high and the benefits of access to a wide variety of foreign products extremely low. In addition to high tariffs and quantitative restrictions, imports of many products are prohibited by administrative rules. The controls over their imports and exports still include a mixture of quantitative restrictions, tariffs, licensing, import bans and foreign exchange regulations.

Transparent, liberal, outward-looking trade policies are critical for attracting foreign investment. Private capital flows to Africa have risen much less than those to other developing countries, and Africa has missed out on the benefits that often accompany such flows – the transfer of technologies and management and organizational skills, and the creation of jobs. Between 1980 and 1997, private capital flows to Africa rose only to US$16 billion from US$6 billion, while flows to developing countries as a group soared to US$140 billion from US$12 billion.

However, trade liberalization has considerable constraints. Trade policy reforms in general and measures to open the economy to trade have been a central part of the adjustment process in the African countries. The reforms include such policies as tariff adjustments, import liberalization, liberalization of the foreign exchange, deregulation of domestic market prices and controls and institutional reforms that particularly affected revenue generating bodies such as Customs and Excise. The reforms contributed positively to export performance in some sectors where enhanced technology transfer and exposure of local firms to international competition improved efficiency and the quality of the products. However, in quite a few sectors and many other enterprises, the sequencing of the trade reforms resulted in unnecessary damage to the local manufacturing sector. Radical import liberalization was implemented in these sectors at a time when local manufacturers faced severe resource and management constraints. When these factors are combined with the exchange rate losses suffered by many firms and the increasingly high cost of credit, firms were unable to adjust promptly to face external competition. The result was that many local firms – especially small and medium-sized enterprises, went out of business.

In the past decade there has been a considerable increase in the exports of the SSA countries to China and India. Exports of the SSA countries as a proportion of their exports to the industrialized countries rose from 1.8% in 1990 to 5.8% in 1997, 8.5% in 2001 and 10.5% in 2004 (Fig. 6.10). These exports are mostly oil, iron ore, diamonds, logs and cotton. As Asia industrializes, its demand for natural resources increases and Africa has responded to the export opportunity. China and India, which are undertaking major investments in the African continent, together account for about 10% of both SSA exports and imports, 25% more than these two countries' share in world trade. Although Africa's main trading partners remain industrial countries, the share of its exports to developing countries has more than doubled since 1990 (Subramanian and Tamirisa, 2001).

There has been a similar increase in the imports of the SSA countries from China and India, mostly of capital goods and cheap consumer goods. Manufactured goods dominate the imports from developing countries, and

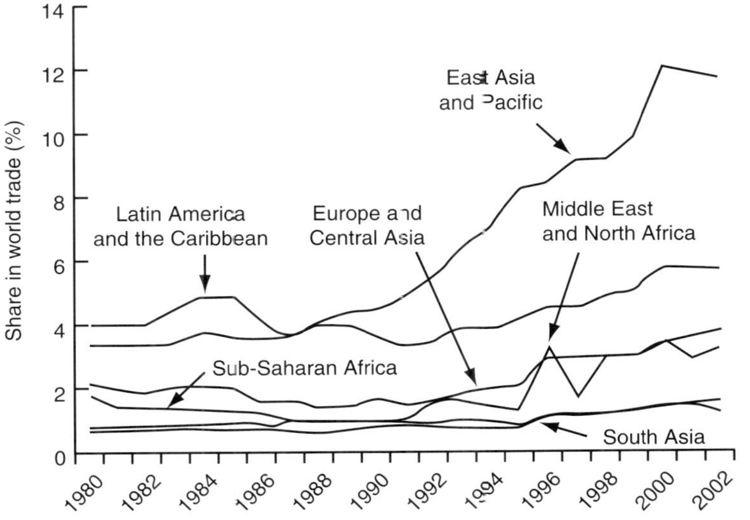

Fig. 6.10. Intraregional exports as share of world exports (excluding oil, in %). (From World Bank, 2005.)

within manufactured imports, consumer goods have the larger share, although imports of machinery and equipment from China and India are growing. Imports of consumer goods from the EA countries tend to compete with local products and bankrupt many local producers, but African consumers have benefited from their low cost. The imports of consumer goods have a negative impact on domestically produced clothing and furniture manufactures in Ghana, South Africa and other African countries that have been displaced by imports from China. In Ethiopia, competition of imported Chinese shoes bankrupted 28% of local producers and forced 32% to downsize their activity (Kaplinsky and Morris, 2006). Moreover, the Asian countries are now taking more of Africa's exports of textiles and clothing – often regarded as a spearhead of manufactured exports.

A World Bank (2004) study of trade complementarity attempted to identify areas of potential bilateral trade between China and India and the countries of SSA. The study computed a 'Trade Complementarity Index' based on SSA's revealed comparative advantage in exports, and China's revealed comparative advantage in imports, and concluded that the potential for future growth in bilateral trade with China is quite weak. The complementarity with the Asian economies in the early 21st century was weak because at that time, China and India obtained their raw materials mostly from domestic sources. Their rapid growth in recent years increased their demand for raw material, and, consequently, also their imports from SSA. These exports of primary products from SSA have very low local value-added and they contribute only marginally to local employment.

The impact of the reduction in local production due to the competition of Chinese and Indian producers on local poverty and on the livelihood of the textile workers in the African countries has been very substantial. The scale of job losses arising from the end of MFA quotas is alarming. Most of the jobs that were lost involved women, and the impact on their families was severe since most of them were left without alternative sources of employment. To compete in the more open clothing markets, African producers would have to improve management practices and increase productivity. They could also try to occupy niche markets, for example, garments with African design, but a considerable decline in production is inevitable for quite some time.

5.1 Trade preferences for exports from SSA

The marginalization of the African countries in world trade and the decline in their per capita GDP occurred despite the significant trade preferences granted to them under the OECD's Generalized System of Preferences (GSP) schemes and through the Europeans Union's Lomè Convention and its successor, the Cotonou Agreement, which extended low tariffs for African exports to the OECD area. Moreover, the region's LDCs were granted even lower tariffs and their exports were ensured by quantitative restrictions on exports from other countries. In 2001, the EC adopted the Everything but Arms (EBA) amendment to its GSP scheme, which extends duty- and quota-free access to all products originating in the LDCs, except for arms and ammunition. The programme includes all agricultural products, except fresh bananas, rice and sugar. The EU is now planning to extend the trade agreements into economic partnership agreements (EPAs) that will be discussed in Section 6.

Despite their generally open trade regimes, industrial countries tend to have restrictions on imports of agricultural products, where much of Africa's export potential is concentrated. In 1997, for example, the EU's average most favoured nation (MFN) tariff was about 15% for imported unprocessed agricultural products and 25% for processed agricultural products, compared with 4% for other goods (excluding textiles). These tariffs were the subject of the negotiations at the Doha Round, but so far, i.e. by early 2007, no agricultural trade agreement has been reached.

Moreover, these figures understate the level of protection, because tariffs are generally low or zero on goods that the EU does not produce, such as coffee and tea, and high on imports that compete with domestic products, including semi-processed and processed agricultural products, which have higher value-added. Equally important, NTBs in the form of producer price supports, export subsidies and marketing arrangements also keep out agricultural imports. These types of measures amount to agricultural subsidies estimated, on average, to be 1.5% of the GDP of the countries belonging to the OECD.

The trade preferences for developing countries are therefore of particular significance for Africa. The Lomé Convention, signed in 1975 and extended in 1980 and 1985, gives certain products originating in developing countries in Africa, Asia, the Caribbean and the Pacific duty-free access to the EU market. In 1968, Australia, Canada, the European Community, Japan and the USA accepted the GSP, which stipulated that industrial countries would grant preferential tariff treatment to manufactures and semi-manufactured products from developing countries (Krueger, 1999).

Although the Lomé Convention expired in 2000, in practice its impact, as well as that of the GSP, on the volume and pattern of trade flows has been quite small; in 1997, for example, only 17% of exports from developing countries to industrial countries benefited from the GSP. The main reasons are that these agreements typically apply to goods with low value-added for which the MFN tariff is already low or zero, and the application of preferences to exports of raw materials, combined with higher tariffs moving up the value-added chain, discouraged countries from diversifying into higher value-added exports. Moreover, access to markets in industrial countries is often restricted by complex administrative requirements with which African exporters have difficulty complying – some 40% of goods eligible for preferential treatment under the GSP do not receive it because of such requirements; as a result, the higher-income developing countries reaped greater benefits from preferential treatment than the poorest countries; for example, in 1996, Brazil, India, Indonesia, Malaysia, the Philippines and Thailand accounted for 75% of all GSP imports into the USA.

The African countries are also eligible for various trade preferences for their exports to the USA under the US African Growth and Opportunity Act (AGOA), which is contained in the US Trade and Development Act 2000.[6] The AGOA extends the product coverage of the US GSP scheme, but only to African countries. Under the AGOA, eligible countries qualify for duty-free and quota-free access to the US market (except for 'wearing apparel' products) for a range of products, including selected agricultural and textile products until 2015.[7] To be eligible, African countries must make progress in establishing a market-based economy, developing the rule of law and political pluralism, eliminating discriminatory barriers to US trade and investment, protecting intellectual property, combating corruption, protecting human and worker rights and removing certain practices of child labour.[8]

AGOA eligibility does not automatically imply eligibility for the 'wearing apparel' provisions; these are governed by a separate set of conditions and associated rules of origin.[9] To export apparel (and certain textile items) to the USA duty-free under the AGOA, countries must implement a 'visa system' that ensures compliance with the required rules of origin. Kenya, Tanzania and Uganda were declared eligible for the preferences under the apparel provisions in July 2004. Measures implemented in order to better

[6] The AGOA was revised in November 2003.

[7] Until its amendment in 2004, the AGOA was due to expire in 2008.

[8] African Growth and Opportunity Act online information; available at: http://www.agoa.gov/index.html.

[9] Under the AGOA amendments, the waiver from the normal rules of origin for wearing apparel, as applicable to 'lesser developed beneficiary countries', has been extended from September 2004 to September 2007.

utilize the opportunities under the AGOA include concessionary import duties of 2.5% on industrial equipment used by the beneficiary country to produce AGOA exports, and the establishment of AGOA-dedicated desks at the country's ports (Box 6.1).

The most significant export of manufactured products from SSA is clothing and textile, and until 2004 that sector benefited from significant preferences by the USA. Until then world trade in textiles and garments was restricted by the MFA and its predecessor, the ATC, which allowed importing countries to impose quota limits on the imports of textiles and apparel mainly from the most competitive Asian suppliers, particularly China. The African

Box 6.1. The African Growth and Opportunity Act (AGOA).

The AGOA is a policy measure implemented by the USA in order to increase trade, investment and economic cooperation between the USA and eligible SSA countries. To achieve these goals, the USA provides duty-free access to the US market for most imports from the region. The aim is to transform AGOA into a FTA by its expiration in 2015.

The purpose of AGOA is to use preferential trade access to the US market as a catalyst for economic growth in SSA by encouraging governments to open their economies and build free markets. It amends the US Generalized System of Preferences (GPS) to grant duty-free treatment to specified products from eligible countries. Congress passed AGOA as part of the Trade and Development Act of 2000, and President Bill Clinton signed it into law on 8 May 2000. In August 2002, President Bush signed amendments that expanded preferential access for eligible SSA countries, and 2 years later, President Bush signed the AGOA Acceleration Act of 2004, which extended preferential access for imports from eligible SSA countries until 30 September 2015, and extended and clarified textile-related provisions in the Act. Currently, nearly all imports from eligible countries in SSA enter the US duty-free through 2015.

AGOA's trade incentives are intended to draw African governments into improving their political and economic governance. SSA countries are not automatically eligible for AGOA benefits and the US President must designate eligible countries based on their progress towards establishing market-based economies, representative government, strengthening the rule of law, combating corruption, eliminating barriers to US trade and investment, protecting intellectual property, reducing poverty, expanding health care and educational opportunities and adopting labour standards.

The overall trade between the USA and Africa has increased sharply since AGOA was adopted, and by 2005, 37 of the 48 countries in SSA had become eligible for AGOA benefits, and over 98% of US imports from AGOA-eligible countries entered the USA duty-free. The USA was then the region's single largest export market, and USA total imports from Africa rose to US$50.3 billion, while US total exports to SSA rose to US$10.3 billion. While a large part of this increase is due to increased oil imports and higher oil prices, the legislation has also helped increase non-oil imports from AGOA-eligible countries since 2000.

The goal of AGOA is to integrate SSA nations into the global economy through trade and economic liberalization. The USA can use AGOA preferences as a lever for trade liberalization in the region through incremental steps that would have to be met in order to maintain AGOA eligibility.

textile and apparel producers were exempted from these quotas; together with tariff preferences under AGOA, which exempted the African countries from 10% to 20% tariffs that other exporting countries to the US market had to pay, textile and apparel producers in Africa were enabled by these preferences to dramatically increase their garment exports to the USA.

The end of the MFA ended these advantages of the African garment producers and raised doubts about their ability to continue to produce and compete with the Chinese producers. Altogether, an estimated US$40 billion of textile and apparel production was expected to be transferred to China from the developing world. Other beneficiaries are India and Pakistan. Scale economics not only make production by these large producing countries less expensive, but also make it much cheaper for big retailers in the USA and the EU to source their products from a limited number of countries, thus further increasing China and India's competitive advantage. With these changes, prices for garments declined by about 20%.

The outcome of removing the MFA quotas has been quite devastating for the African industries: after a rise in the share of SSA exporters of textile and apparel in US markets between 2001 and 2004, that trend was reversed in 2005 when the AGOA agreement was discontinued under the MFA. The major casualty was South Africa, whose AGOA exports collapsed to half the pervious value, but also in other African countries production had to be reduced. For the time being, the African clothing producers are not able to compete in the US market with the Chinese and Indian producers because the scale of production in most SSA plants is much smaller, the bureaucratic obstacles to production and exports are high and the physical infrastructure is very poor; in addition, productivity is low, in part due to the poor and inexperienced organization and in part due to the low skills of workers.

The global trade landscape has significantly changed in the past decade, but the African trade system is trailing behind. Traditionally, African trade policy was geared to gaining preferential access to the markets of the industrial countries, but the value of these preferences was significantly eroded as global tariffs were falling and regional trade agreements (RTAs) proliferated. Consequently, exports to industrial countries that depended on specific preferences (e.g. textiles) became less important.

6 The Impact of Regional Trade Agreements

The economies of the African countries, with the exception of South Africa, are too small, undeveloped and poor to provide local producers sufficiently large markets that will enable them to reach adequate economies of scale for efficient production and profitable investment. As many as 15 countries are landlocked and their exports are prohibitively costly. In all countries, the road infrastructure is terrible and still deteriorating and all other components of the infrastructure, from power grids to electricity generators, railway systems and telecommunication, barely function. Given the small scale of their economies, investments in the local infrastructure cannot be profitable and must therefore be made by the public sector. Regional integration will enable producers in these countries to increase the scale of production and its cost effectiveness; their governments can collaborate in infrastructure projects and share the costs, and for the 15 landlocked countries, and agreements with neighbouring countries that will give them access to the coast would be highly beneficial (Schiff and Winters, 2003; Steven, 2005; Yang and Gupta, 2005).

However, regional cooperation in infrastructure projects can work only in the context of effective and binding agreements between the countries in designing the projects, carrying them out and dividing the costs and benefits. A necessary step in that direction is to open the economies of these countries for trade among themselves so that both consumers and producers can benefit from the cooperation. More than 30 preferential and FTAs between subgroups of SSA countries have already been concluded since 1990 in order to give incentives to integrate the national markets into subregional markets, thus increasing the profitability of intraregional trade, spur producers to increase the scale of production and its efficiency and promote collaborative investments in regional infrastructure.

All SSA countries belong to at least one RTA, and many countries are members of several agreements. In principle, the PTAs can bring significant benefits to the member states, but, with only a few exceptions – such as the Southern African Customs Union and the West African Economic and Monetary Union – Africa's Preferential Trade Agreements (PTAs) have been quite ineffective. Most agreements deal only with tariffs, many agreements are subject to a wide range of exemptions, concerns about the loss of fiscal revenues led governments to delay the reduction in tariffs and prolonged the transition periods, and member states that are members in several PTAs often found it difficult to fulfil their multiple and sometimes conflicting obligations.

As a result, the PTAs between the SSA countries did not contribute much to expand the share of regional trade relative to their overall trade. In the early 21st century, the intracontinental trade was only about 12% of the countries' total exports, up from 8% a decade earlier (*Economic Report on Africa 2003*).[10] At the initial phase, the PTAs had primarily the effect of diverting trade to member states rather than generating new trade between them, largely because their MFN tariffs were still relatively high and the lower tariffs under the agreement gave incentives to trade diversion by allowing local producers and traders to lower their prices to consumers in the member states via their trade with each other rather than searching for more competitive prices in other countries or in the world market. The limited intraregional trade reflects the high barriers still maintained by these countries on the trade between member states despite the PTAs.

The reason for the limited impact on trade creation was the rather similar factor endowments and production conditions in these countries that were therefore not natural trading partners that complement each other's produc-

[10] The recorded change in imports is biased, however, because when tariffs are high, the recorded level of imports is low due to tax evasion. When tariffs come down, the increase in imports reflects in part an increase in recorded imports due to reduced tax evasion.

tion and consumption needs. Africa's heavy concentration in commodity trade further hinders intraregional trade and gives these countries fewer opportunities for product differentiation and trade. Over time, the PTAs are likely to augment trade creation by forcing producers in regional markets to produce more efficiently and to specialize in their production under the pressures of the wider competition and the economies to scale in production. Both trade creation and diversion reduce the price to consumers, but trade creation contributes to increase the economic efficiency in production, while trade diversion reduces the economic efficiency by preventing imports from more efficient producers in the world market.

One problem with Africa's RTAs is the number of overlapping agreements that are often internally inconsistent. These regional agreements (including the Common Market for Eastern and Southern Africa (COMESA), the Cross-Border Initiative for Eastern and Southern Africa, the Southern African Development Community and the Southern African Customs Union), may thus require their member states to meet incompatible and sometimes conflicting obligations, rules and administrative strategies. As a result, the complexity of many of the agreements reduces the gains from regional collaboration and undermines the incentives for joint investments. The internal inconsistencies and conflicting regulations in these agreements hinder the creation of a larger market. Under the East African Community (EAC) transition arrangements, for example, Uganda and Tanzania may charge higher tariffs on goods imported from Kenya than they can charge on these goods when they are sourced from other member countries of the COMESA. Ineffective and uncoordinated regional trade policies and agreements reduce the drive that countries need in order to make policy reforms that will open them to the regional trade; instead, they foster preferential treatment of local interests that trade liberalization and trade agreements are supposed to dissipate.

It is not only the small size of the African economies, with the exception of South Africa, that reduces the capacity to obtain significant economies to scale in production and keeps the gains from regional integration rather limited; other, perhaps more important factors are the poor road infrastructure between them, the multitude of trade barriers and the onerous regulations. The main effect of the various administrative trade barriers and high import tariffs is to raise the costs to importers that are paid as bribes in the border crossing and thus raise the price to consumers. It has been estimated that effective trade liberalization within SSA could increase the intraregional trade by over 50%. At present, though, that trade is much smaller than in other regions despite the large number of trade agreements. Figure 6.11 shows the small share of the intraregional trade in the total exports of the African countries. The large and rapidly growing share of the intraregional trade of the EA countries is largely due to the growing share of segments of production that have been outsourced to countries within that region as they increasingly specialized in production according to their comparative advantage. The segmentation of production and the outsourcing of separate segments to different countries in the region increased the complementarities and their growing specialization in production.

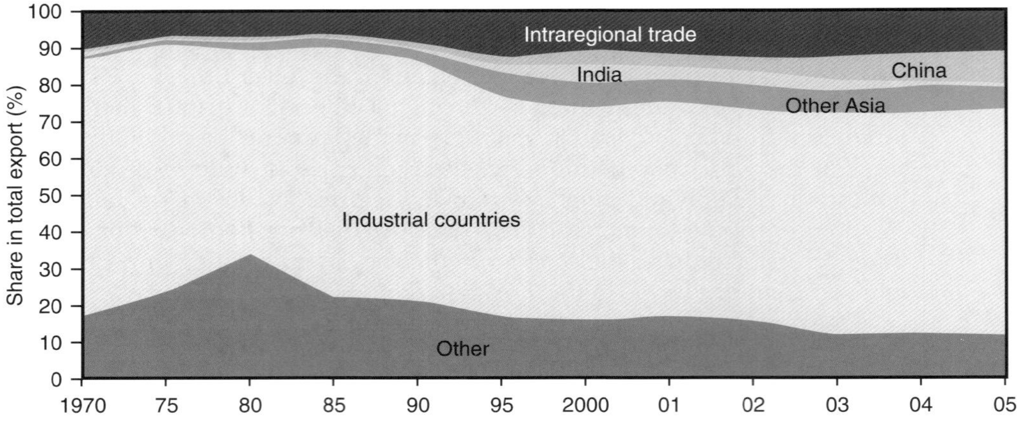

Fig. 6.11. Destination of exports from SSA: 1970–2005 (%). (From IMF, Direction of Trade Statistics.)

During the past decade, the African countries have recognized the benefits of regional collaboration and intraregional trade as an effect of consolidating subregional markets, increasing efficiency in production and coordinating their trade policies. The coordination of the regional trade policies has become more significant in the multilateral trade negotiations under the WTO and the growing significance of RTAs and inter-regional trade agreements.

The longer-term objective of the SSA countries is to establish a common market of all countries in the region that would allow traders and people to move and work freely across borders, establish unified procedures and common institutions for trade and trade policies and increase competitiveness by enabling producers to benefit from the larger scale and greater specialization in production and by promoting more diversified production across the subcontinent to increase the value-added of exports and the competitiveness of producers in the world markets. The African Economic Community (AEC) established the foundation for a common market and common procedures and institutions for trade and trade policies of all the SSA states. The current efforts of the AEC focus on strengthening the economic collaboration between the African countries in order to reach a FTA and a customs union. In parallel to these efforts, the AEC is trying to reach an agreement on a common monetary policy of a joint Central Bank that will gradually lead to a monetary union of the entire region.

However, the progress thus far has been quite limited. Even the PTAs were only very partially effective and failed to reach most of their goals. Thus, for example, the tariff rates of COMESA member countries decreased from an average rate of 11.3% in 1994 to an average rate of 5.5% in 2003; during that period, the MFN tariff rates were reduced even more than the preferential tariff rates (from 17.9% in 1994 to 10.2% in 2003), but these tariffs were on average higher than preferential tariff rates in both 1994 and 2003.

Several factors slowed down the process of reaching greater coordination in the RTAs and increasing the share of Africa's intraregional trade:

- Failure of countries in some PTAs to abide by the rules of the agreement and implement the agreed reductions in trade barriers;
- Inability to devise equitable and commonly agreed arrangements for sharing the costs of and the gains from integration;
- Limited effectiveness of agreements on common external tariffs due to a multitude of requests for exemptions;
- Failure to adjust the existing economic ties with countries that are not members of the agreement;
- Wide differences in the degree of dependence on trade taxes as a source of revenues for the governments of the member states;
- Lack of sustained political commitment due to erratic economic policies and political instability in many countries.

The trend to establish RTAs is including an increasing number of developing countries and a growing share of the South–South trade. Between 1990 and 2003, the developing countries signed 70 new RTAs (WTO, 2003). However, their contribution to increase the South–South trade has so far been quite limited because the agreements are between countries with very small markets and their benefits to the scale economies of producers in member states have not been very significant. The existing agreements in SSA are still too small, and there is not much that the countries in another trading block can gain from an agreement with, say, COMESA, either by increasing their scale economies in production or by gaining from the comparative advantage of the COMESA member countries (Mayda and Steinberg, 2007). Large countries do not have great incentives to be members of these agreements, and large trade agreements do not have great incentives to include these countries in their agreements. The dependence of the governments of these countries on the revenues from the tariffs on imports is another obstacle to a FTA.

The statistical analysis of the impact of the RTA or an FTA concentrates on their impact on the volume of trade between the member countries. However, these statistical measures may fail to take into account several indirect but often more important effects of the agreements. One effect is the political leverage that the agreement gives to member countries in their negotiations with other trading blocks or in the multilateral trade negotiations.

Another effect of a trade agreement concerns the political relations between member states even when its economic impact is limited. The mutual benefits of member states from establishing joint institutions, coordinating trade policies and creating closer relations between their producers and traders can give them a motivation to cooperate in other areas, economic and political, and thus form more stable relations and strengthen their political connections. Obviously, the stronger the economic relations between countries, the higher the costs of a conflict between them that can endanger these relations, which will therefore strengthen popular support for the collaboration between countries.

In SSA, the existing agreements will also serve as an incentive to strengthen the AEC and make progress towards the establishment of an effective African common market. All these measures may therefore contribute to stabilize the political relations between countries. Thus far, the fragile political relations between the countries of the region often deteriorated into open military confrontations, which, in addition to violent internal conflicts, have been the main obstacle to economic development.

The direct economic and the indirect political gains from the RTAs led to a proliferation of these agreements since 1990 to more than 300 world wide by the early 21st century. So long as the main multilateral trade negotiations were conducted within the framework of the WTO, the RTAs were formed primarily between neighbouring countries on a geographic basis. The African countries, with the exception of South Africa, have not been included in these agreements.

With the stalemate of the Doha Round negotiations and the ebbing influence of the WTO, there is a growing tendency to strengthen the relations between trading blocks. The small economic weight of the African PTAs has thus far reduced their leverage in negotiations with other trading blocks. The large increase in the world's demand for oil and for other natural resources available in the African countries, and the resulting sharp rise in prices, can give these countries an important leverage in strengthening their relations with other trading blocks and reaching agreements that will help them to increase their exports.

6.1 Regional Trade Agreements and the EU Economic Partnership Agreement

However, the existing and potential regional agreements between the SSA countries must now compete with the EPAs that the European community is planning between the EU and the African, Caribbean and Pacific (ACP) countries. The EPAs are planned as expanded cooperative agreements based on partnership and collaboration in trade and in the political dialogue between the EU and the ACP countries. The plan is to combine in the agreements trade, development aid and political alliance. The key principles of the agreements are reciprocity, the removal of trade barriers and the coordination of trade and aid. The longer-term plan is to develop the EPAs into a FTA that will replace the current non-reciprocal trading preferences of the ACP countries with reciprocal arrangements in compliance with the WTO rules of non-discriminatory trading arrangements.[11]

Figure 6.11 highlights the much greater significance of the trade relations of the SSA countries with the EU relative to the intraregional trade. The EPAs are therefore much more important for the African countries than the RTAs.

[11] In order not to be challenged on the grounds of discrimination under the most favored nation (MFN) clause of the WTO, ACP countries under the EPAs will be expected to grant EU originating imports of goods duty-free access to their markets.

Around 50% of the trade of the SSA countries with the industrial countries is with EU countries, and that share is much larger than the share of their interregional trade. The plan to establish the EPAs is therefore bound to weaken the incentives of the SSA countries to form RTAs.

In principle, the specifics of the EPAs will be negotiated individually with each ACP country, but the key principles listed above form the basis for all agreements. Since each agreement will be with an individual country, there is no incentive to form a regional agreement, and the EPAs will therefore weaken the motivation of the African countries to form RTAs. Moreover, since not all the SSA countries will be included in the EPAs, these agreements further reduce the motivation to form agreements or establish collaboration between the African countries and to make progress in the formation of the AEC.

The EPAs would contribute to expand the trade between the African countries and the EU. Part of that expansion would be the result of trade creation, triggered by the incentives provided by the agreement and the larger market the agreement will form. In the analysis of the trade creation, a distinction should be made between 'export creation' and 'import creation'. Export creation represents an increase in a country's exports as an effect of its comparative advantage; that increase will, however, force the producers of these products in the importing countries that do not have the comparative advantage to reduce or cease their production. Import creation represents an increase in a country's imports of products in which its producers do not have a comparative advantage. In the EPAs, under the 'reciprocity' principle, the import creation will force farmers in the SSA countries to cease the production of agricultural products that their countries will be able to import from the EU at dumping prices. The trade creation therefore does not benefit everyone.

A significant part of the additional trade with the EU and with the other ACP countries will be due to trade diversion from the rest of the world and most likely also from the intraregional trade of the African countries themselves. The reciprocity principle in the EPAs will therefore have a negative impact on the efforts to strengthen and expand the trade relations between the SSA countries, and it will also have a negative impact on many of their farmers. Thus, the EPAs will actually weaken the regional integration in SSA, reduce the intraregional trade, reduce the incentives to form a common market and weaken the incentives to solidify regional cooperation in other areas as well.

The EPAs therefore raise a number of important questions, and they will force the African countries to make several difficult choices:

- Are the African countries, individually and collectively, likely to gain or will they lose from these new bilateral trade agreements that are governed by the reciprocity principle?
- What sectors and population groups in the SSA countries are most likely to lose from these agreements and what sectors would gain?
- What are the likely effects on the income distribution in the SSA countries and what are the overall welfare effects of the EPAs, including the effects on countries that will not be included in the agreements?

- How will the formation of EPAs affect the plans of the African countries to expand their exports and what are the likely effects on trade creation and on trade diversion?
- What are the expected fiscal implications of the EPAs and the expected losses in the fiscal budget?

Given the support for regional integration, the plan to expand the economic partnership between the ACP countries and the EU raises questions whether the EPAs could be modified in a way that would strengthen the incentives of the African countries to deepen their commitment to the RTAs and to the AEC. This can be done by making more favourable agreements with subgroups of countries that establish a PTA.

A possible obstacle for an expansion of African exports to the EU of agricultural fresh and processed food products is the requirement to meet the health, sanitary and phytosanitary, technical and market standards maintained by the EU and approved by the WTO (Box 6.2). The health standards and the NTBs to trade pose a difficult challenge to African producers and exporters. Most African countries lack the expertise, the institutions and the resources to put into practice and effectively supervise the implementation of these standards. These standards therefore reduce the capacity of African

Box 6.2. Food safety standards.

Article 20 of the General Agreement on Tariffs and Trade (GATT) allows governments to act on trade in order to protect human, animal or plant life or health, provided they do not discriminate or use this provision as disguised protectionism. In addition, there are two specific WTO agreements dealing with food safety and animal and plant health and safety, and with product standards.

The main problem is how to ensure that a country's consumers are being supplied with food that is safe to eat. What standards of safety should be considered appropriate? How to ensure that strict health and safety regulations are not being used as an excuse for protecting domestic producers?

A separate agreement on food safety and animal and plant health standards (the Sanitary and Phytosanitary Measures Agreement or SPS) sets out the basic rules.

It allows countries to set their own standards. But it also states that the regulations must be based on science: They should be applied only to the extent necessary to protect human, animal or plant life or health, and they should not arbitrarily or unjustifiably discriminate between countries where identical or similar conditions prevail.

Member countries are encouraged to use international standards, guidelines and recommendations where they exist. However, members may use higher standards if there is scientific justification. They can also set higher standards based on appropriate assessments of risks so long as the approach is consistent, not arbitrary.

The agreement thus allows countries to use different standards and different methods of inspecting products. An exporting country can ensure that the practices it applies to its products are acceptable in an importing country if it can demonstrate that the measures it applies to its exports achieve the same level of health protection as in the importing country.

farmers to produce for exports to the EU even if they diversify their production and try to specialize in agricultural products in which Africa's agroclimatic conditions and cheap labour give them a clear comparative advantage.

Currently, only a small number of African agricultural producers, primarily large producers that acquired the expertise and are connected to transnational corporations, are able to meet these food safety and technical standards. Only 34 countries in SSA belong to the International Organization for Standardization (ISO), and only a small number of their farmers are actually active participants. For the African farmers, the difficulties and higher costs of maintaining these standards must be taken into account together with the fact that many of their trading partners are countries in the South that do not maintain these standards. The African farmers must therefore diversify their output not only to different crops but also to different methods of production according to the expected country of destination. Therefore, for many farmers, this system of production is likely to be much more risky and expensive.

7 Why Is Africa Falling Behind?

The 50 LCDs with a total population of 500 million people failed to integrate into the global economy and to share the benefits from the global economic growth. They include most of the countries in SSA. During the 1980s and 1990s, the SSA countries suffered from deteriorating and volatile terms of trade with the decline in the price of many of the primary commodities that were their main exports. Nevertheless, these countries remain dependent on exports of primary products and they failed to exploit their comparative advantage with their abundance of cheap labour. In *Breaking the Conflict Trap: Civil War and Development Policy* (2003), a team of World Bank experts led by Paul Collier concluded that economic forces such as entrenched poverty and heavy dependence on natural resource exports were more often the primary cause of civil wars than ethnic tensions and ancient political feuds. On these grounds, the report suggests three sets of actions to prevent civil wars: more and better-targeted aid for countries at risk, increased transparency of the revenues derived from natural resources, and better-timed post-conflict peacekeeping and aid.

7.1 Can SSA manage the next wave of globalization?

For SSA countries, the future looks rather grim: as a recent publication of the World Bank entitled *Global Economic Prospects 2007* (2007) predicts, the countries that are likely to be left behind will increasingly concentrate in SSA, thus augmenting the global geographic polarization. Accordingly, by 2030, SSA alone would be home to nearly 55% of the world's poorest decile – an 80% increase from its share in 2000.

The report emphasizes three main factors that are driving Africa's decline: high initial income inequality, rapid population growth, and the lowest per capita income growth among the developing countries. However, these are essentially technical factors that list the correlates of growth but do not explain the reasons why more and more African countries and African people are trailing behind while the rest of the world enjoys continued progress. The stagnation in Africa is not limited to the poorest 10% of the world's population but to the entire bottom of the world's income distribution. For example, in 2000, 59% of the population of SSA and 25% of the population of EA concentrated in the bottom one-third of the world income distribution. By 2030, more than three-quarters of the population in SSA, but only 16% of the population in EA would, according to the World Bank predictions, be among the world's poorest one-third. SA will continue to be the largest region in the three lowest deciles in 2030 owing to the region's very high initial poverty rates. But the entire Asian population is moving up to higher levels of the global distribution thanks to these countries' relatively rapid per capita growth, and most of SA's poor will be in the second and third deciles by 2030. In SSA, by contrast, the average African is 30% more likely to be in the three bottom deciles in 2030 than he was in 2000.

In the next 25 years, globalization is likely to spur faster growth than during 1980–2005, and the developing countries will have a central role in that growth. However, unless managed carefully, that rapid growth could be accompanied by a sharp rise in income inequality between and within the developing countries; another problem is the potential that severe environmental damage could ensue. Growth in developing countries will reach an average rate of 6% in 2007 and 2008, after the near record rate of 7% in 2006. This growth will be more than twice the rate of growth in the high-income countries, which is expected to be between 2% and 3%.

The World Bank report predicts in the 'central scenario' that the global economy would expand from US$35 trillion in 2005 to US$72 trillion in 2030. The LDCs would find it increasingly more difficult to be part of and benefit from that growth process, in part because they are latecomers: World demand for manufactures is necessarily limited by the world income and a large share of world production will be agglomerated in clusters and controlled by transnational corporations that will benefit from both the horizontal and the vertical economies to scale that this organization of production makes possible. Their demand for low-cost labour will be met by the abundant supply of cheap labour in China, India and the neighbouring countries, and the SSA countries would not have any comparative advantage to offer that could attract these producers.

Broad-based growth in developing countries that is predicted to prevail at that time period would also significantly affect global poverty, and according to the estimates, the number of people living on less than US$1 a day could be cut in half, from 1.1 billion now to 550 million in 2030. However, the highly uneven distribution of world growth would also lead to a highly uneven distribution of the world's poor across regions. In some regions, notably Africa, large shares of the population are at risk of being left behind and remaining poor. Income inequality between regions and countries is likely to rise very significantly and income inequality within countries would also continue to rise, although more countries are likely to take more active measures to slow down the rise that may become a major threat to social stability.

Global trade would be the main driving force for the rapid global growth and the share of the developing countries would continue to rise at very high rates. However, the African countries would have a very small share in that trade and, as a result, they would not have access to and would not benefit much from the spread of more advanced technologies. Thus, their labour will remain largely unskilled, education would not be widely available, and the young people who join the labour force would have very limited opportunities to acquire higher skills or higher education. In the other developing countries, the majority of the population will rise to the 'global middle-class', and benefit from unparalleled affluence, while in SSA, the majority of the population will still be undernourished. The African countries may also suffer from a more rapidly deteriorating environment, partly as an effect of the transfer of more polluting industries to the region, and partly due to their lower investments on improving their environment.

7.2 Can Africa claim the 21st century?

An earlier World Bank report entitled *Can Africa Claim the 21st Century?* was prepared in 2000 in collaboration with the African Development Bank, the UN Economic Commission for Africa, the Global Coalition for Africa and the African Economic Research Consortium. The main conclusion of this report is that Africa will be able to achieve sustained economic growth if, and only if, it invests heavily in its people, both to reduce poverty and to enhance the continent's ability to compete globally. The report put part of the blame for Africa's limited progress so far on the interventions and policies of the donor institutions themselves, through their SAPs, which have left Africans 'cash poor and project rich'. It also claims, contrary to practically all previous and subsequent bank reports that I am familiar with, that pushing the continent towards sweeping economic liberalization and reducing the role of governments was a mistake. In a frank and unabashed language, the report recognizes the 'limits of a market-driven approach', and points to the need for a 'strong and capable state'.

The 2000 report agrees that the majority of the African countries have realized in recent years the need for macroeconomic stability and the essence of private sector promotion. However, the report claims that donors focus mostly on social and state institutions, poverty reduction and enhancement of the Africans' skills and capacities. In fact, this change in focus took place only in later adjustments of the SAPs and the transition to the Poverty Reduction Strategy Programme (PRSP). The report also recognizes that many Africans, from leaders to the common people, resist and are sometimes hostile to the conditions that the financial institutions impose on governments in order to qualify for debt relief or development assistance. The report, therefore, urges the promotion of industrialization so that Africa can begin to export a broader range of processed goods and move beyond its current dependence on primary commodities. This is one of the few reports in which the resistance of the developing countries to the conditionalities imposed on them by the SAPs is recognized by the International Development Organizations.[12]

An earlier World Bank's 1994 study *Adjustment in Africa* maintained that 'adjustment is working', but the 2000 report acknowledged that many 'reforms have failed to address some difficult underlying institutional problems – and in some cases may have worsened them'. Thus, some countries accumulated large domestic debts, in part because the financial markets were hastily liberalized before fiscal deficits were brought under control.

In the agriculture sector, the 2000 study notes, policy reforms have opened up domestic and external marketing arrangements, yielding some increases in producer prices to small-scale farmers. However, without improved infrastructure, services and rural institutions, price incentives had a limited impact, especially at a time of declining world market prices for many of Africa's agricultural exports. Moreover, with the reduction in government subsidies, farmers' costs have doubled or tripled, especially the costs for inputs like fertilizers and pesticides.

The report thus agrees that in the African countries, governments and their institutions remain essential to maintain security and promote development, thus reflecting a shift away from the anti-state position that was previously expressed in the SAPs. At the most basic level, the government must function to preserve peace, security and law and order. The report notes that around a quarter of all Africans still live in countries affected by conflicts that make any progress in developing their economies virtually impossible. Even in countries that are not at war, crime and violence combined with ethnic conflicts and hostilities between rival religious groups are a serious drain on the economy.

In addition to enforcing law and order that make daily life possible and help the private sector to operate and develop, the report recognizes that the government must also play an active role in the economy, particularly in areas that are essential for development. Macroeconomic and monetary policies on the one hand, and investments in infrastructure on the other hand, are perhaps the most obvious examples or areas where the public sector must take the lead. Thus, for example, rural roads are the main obstacle for agricultural growth in many developing countries, and very low public investments to build rural roads were a major hindrance for efforts to reduce rural poverty. Better health and education are also vital public goods that only the governments can provide.

A central part of the 2000 report is the chapter entitled 'Investing in People'. This chapter emphasizes that an integral part of the efforts to accelerate growth

[12] The following story may reflect this resentment: The Nigerian government negotiated with the World Bank a project to improve the country's three railways that are falling apart. This train system is notoriously inefficient and corrupt, and the World Bank set conditions to tackle corruption. After months of negotiation, the bank and Nigeria's government agreed on a $5 million project that would allow private companies to come in, but just as the deal was about to be signed, the Chinese government offered Nigeria $9 billion to rebuild the entire rail network – no bids, no conditions and no need to reform. Similar experiences have recently been observed in other developing countries that received contributions from Iran, Saudi Arabia and Venezuela (IHT Feb, 15, 2007).

and reduce poverty is public investments in education. SSA is the only region in which elementary schools are not free and obligatory for all and a significant number of school-age children are not attending school. The number of young people who manage to get higher education at local universities is also relatively very small, and a large percentage of them are then making all efforts to leave their country and go abroad, thus draining the country of a precious resource.

Another important chapter in the 2000 report emphasizes the need to implement distributional measures in order to keep income inequalities in check and secure the share of the poor in their countries' growth. The report makes two obvious observations: first, if pervasive inequalities are not addressed 'growth will not be sustainable and will not reduce poverty'; second, '[though] the poor are willing to pay for human development investments, their resources are limited . . . user fees have deterred primary school enrolment and health centre use'. There is yet another, more fundamental, aspect of the rising income inequalities – the social rivalry that it creates between income classes and regions. In many countries, the social classes also have clear religious or ethnic identities that sharpen the hostilities and lead to open civil wars.

The subsequent report *Can Africa Claim the 21st Century?* 2007 also argues that the African countries must open up their economies more fully to international trade and that their governments should implement policies to promote exports. It claims that in most African countries an 'anti-export bias is still considerable' compared with other developing regions. Had the continent maintained the same share of world trade it held in the late 1960s, the report's analysis shows that Africa today would have earned an additional US$70 billion a year. Until the early 21st century, the world market prices did not provide Africa sufficient incentives. Between 1970 and the late 1990s, the cumulative losses in the region's terms of trade (the purchasing power of export earnings) were equivalent to a staggering 120% of the GDP of the non-oil exporting African countries.

This report strongly advocates greater diversification of the African countries' exports that currently concentrate on one or two primary products, but it also recognizes that diversification into manufacturing would be extremely difficult and would demand very high investments in transport, electricity and other basic infrastructure. Therefore, for most African countries, the possibility of diversification beyond primary commodities is presently very limited.

7.3 Mapping sub-Saharan Africa's future

Perhaps the most pessimistic perspective on SSA's future is presented in a report by the National Intelligence Council (2005) entitled *Mapping Sub-Saharan Africa's Future*. This report was prepared by the National Intelligence Council and provides a summary of a day-long discussion held in 2005 with a group of top US experts on SSA that focused on likely trends in the region in the next 15 years. The conference participants agreed that most of Africa will become increasingly marginalized as many states struggle to overcome sub-par economic performance, weak state structures and poor governance. Globalization will accentuate the differences between the African countries as some of them manage to build more orderly state institutions and a relatively well-organized economy, while the others deteriorate into complete chaos. Reform efforts in all countries will continue to be complicated by structural obstacles within the countries and by 'neighbourhood effects' such as the cross-border spillover of conflicts and flows of migrants who leave their countries either to save their lives or to escape the hopelessly poor living conditions.

The report highlights the problem that African societies characterized by poverty and malnutrition are unable to establish a sense of national unity, preserve the rule of law and maintain effective governance. These societies are more likely to be engaged in conflicts between rival groups that compete over scarce resources – land, water or any other resource needed for survival – and they would be willing to fight for them if they lose the trust that they would receive a fair share and secure their survival otherwise. They are driven by their belief that only close collaboration with the members of their tribe, clan or ethnic group would give them enough power to fight and receive their fair share and that only the combined power of the tribe could prevent their exploitation by other tribes.

The motivation for a collaboration between individuals and between subgroups of individuals – the tribe, the clan, the religious group – within the framework of a nation is the basic trust of each and everyone of them that the collaboration within this framework will be beneficial for all of them and will give them higher gains than a collaboration within any other framework. If individuals lose that trust, they also lose the motivation to maintain and preserve their national identity. If that happens to large subgroups within the nation, it inevitably weakens the central government, loosens its central control and leads to the fragmentation of the national identity into rival subgroups that seek to improve their lot by competing for the scarce resources.

The report therefore concluded that, as a result of that fragmentation, SSA is likely to become less important to the international economy in the next two decades and its growth will be significantly slower than that of most other nations. The report fully takes into account the riches that some African countries are accumulating with the sharp rise in the price of oil exports, but the supply of oil is limited and is concentrated in only a few countries. In fact, there is evidence that the booming exports of oil are actually retarding rather than promoting African development, by encouraging patronage and misrule by dictatorial leaders rather than promoting national development that will be shared by all segments of society. The dismal performance of some of the large countries affects other countries because of the 'neighbourhood effects'. As a result, the report concludes that African countries are likely to remain bogged down by their poor economic performance, weak state structures and inefficient governance.

This fate need not be shared by all African countries; some countries (Botswana, Mauritius) have already achieved high growth, and a few others have established relatively well-functioning democracies (Ghana, Benin, South Africa). Nevertheless, in large parts of West and Central Africa on the one side of the continent and in Guinea, Sierra Leone, Liberia and Cote d'Ivoire on the other side, lawlessness and anarchy are still dominating. Although in a growing number of countries autocracy and one-party rule are gradually and hesitantly yielding to more open multiparty democracy, few countries have managed to consolidate a solid national identity and many others are still scarred by past conflicts between rival groups. Quite a few countries had a rather promising economic performance, but they still have an authoritarian political system. As a result of all these developments, there are both political gaps between countries in their transition to a more democratic multiparty system, and economic gaps between the level and rate of development of different countries. These gaps increase the difficulties to conduct an open dialogue between countries with the aim of establishing a regional common market and greater political coordination.

Countering the pessimistic predictions, several positive developments give a ray of hope that future developments in the subcontinent could take a more positive direction than the quoted reports envisage: First, the advent of multiparty elections and the increasing participation of people from both urban and rural areas in these elections present a significant counterweight to dictatorial regimes, and this trend is spreading throughout the entire subcontinent. Governments are becoming more responsible and more responsive to their people, partly by putting the egregious fiscal imbalances under control, and corruption at high levels has become less acceptable. The road to the establishment of effective and non-corrupt democratic institutions is still long in most SSA countries, but the people in a growing number of countries express a strong desire to make progress in that direction and willingness to fight in order to achieve this goal. People are better informed, more open in expressing their opinions and less willing to accept the whims of their governments. While the road is long, there is a noticeable progress in that direction and future progress may be more rapid.

Reports that describe the impact of globalization on the developing countries tend to highlight the importance of cell phones and remittances, while too little attention is given to the impact of the free flow of information and of ideas. The main impact of the flow of information is the increase in the awareness of the people in all countries of their rights and their power when they see how much power people have in other countries and how much influence they can exert in order to change their living conditions and demand their rights. The more they learn about it from what they see in other countries, the less willing they are to accept a regime in their own country that arbitrarily ignores their aspirations and rights and leaves them impoverished and miserable. Corruption at high levels is also less acceptable and less tolerated when people come to realize that it is at their expense.

How long it will take for this process to bear fruits and how fast it will spread is of course difficult to predict. The answer cannot be provided by any econometric model and it is inevitably a mixture of a vision and the experience of other countries. I believe, however, that the free market of ideas cannot be sealed by any supervisory wall, and the greater people's awareness of the way other people in other countries are fighting to realize their rights, the less tolerant they will be to a regime in their own country that denies them their rights and their fair share in their country's resources and the standard of living that it could offer.

An important factor that can accelerate progress in that direction is the flow of migrants that flee the countries where conditions are abysmal, either leaving to go to other countries in the subcontinent where conditions are better or to Europe and other developed countries. These flows of migrants will force the countries to which the migrants are fleeing to take active and persistent measures to give these people enough incentives to stay in their countries. The power of tighter controls along the borders is limited, and the only effective way is to use both the stick and the carrot, by applying pressures and by giving aid in order to improve the political and the economic conditions in the countries that the migrants are leaving.

The migrants spare no risk or effort to flee their homeland with one clear goal in mind: instead of fighting over the dwindling resources in their own countries, they take whatever risk and make whatever effort necessary in order to reach the promised land of the developed countries where work is available and food is plentiful. Every year, hundreds of thousands if not millions of people are trying to do just that. Many are caught or die on their way; but those who were caught and sent back will most likely join another wave of migrants and take their chance one more time. This will continue unless they have reasons to hope that things will change for the better in their own country.

As more people in Africa and other LDCs become aware that the present system in their countries works against their interests, they will refuse to accept that fate and will join forces to demand a change. Otherwise, with Africa being left behind, people would have to passively continue drifting. But more and more people may take their fate into their own hands and do whatever it takes to escape the fate that awaits them. The resulting flood of migrants is bound to grow unless the developed countries make a joint effort to lift the African countries themselves with the tide of globalization and make sure that they are not left behind.

Section 8 offers a review of the joint effort of the international community to do just that by establishing the challenge of the MDGs. Although this effort still continues, the conclusions that can be drawn from the progress that has been made thus far are not very promising. Against the background of all the past experiences, and the risks of having yet another failure, the section seeks to make a careful midterm analysis of this programme.

8 Can SSA Achieve the Millennium Development Goals?

In September 2000, the largest ever gathering of world leaders adopted the UN Millennium Declaration that committed their nations to a global partnership to reduce poverty, improve health and promote peace, human rights, gender equality and environmental sustainability. Soon after, world leaders met again at the March 2002 International Conference on Financing for Development in Monterrey, Mexico, establishing a landmark framework for a global development partnership in which developed and developing countries agreed to take joint actions for poverty reduction. Later on that same year, UN member states gathered at the World Summit on Sustainable Development in Johannesburg, South Africa, where they reaffirmed the goals as the world's time-bound development targets.

The MDGs represent a unique commitment of all the 181 members of the UN General Assembly to contribute jointly to this collaborative effort to bring about a reduction of, and eventually end, the disgraceful poverty of such a large share of the world population still suffering from hunger. Contrary to previous declarations that proved to be just that – declarations – this UN declaration was truly historic by setting very specific quantitative and measurable targets for the following goals: eradicate extreme poverty and hunger; achieve universal primary education; promote gender equality and empower women; reduce child mortality; improve maternal health; combat HIV/AIDS, malaria and other diseases; ensure environmental sustainability; and build a global partnership for development. These commitments establish specific measures to determine the capacity of the world community to mobilize enough goodwill (Box 6.3).

Box 6.3. The millennium development goals.

The MDGs are eight goals of global development that were declared by 189 nations and signed by 147 heads of state and governments during the UN Millennium Summit in September 2000. The aim is to achieve these goals by 2015. The following is a list of these goals:

1. Halving between 1990 and 2015 the proportion of people who suffer from hunger
- Target 1: Between 1990 and 2015, halve the proportion of people whose income is less than US$1 a day
- Target 2: Between 1990 and 2015 halve the proportion of people who suffer from hunger

2. Achieve universal primary education by 2015
- Target: Ensure that by 2015, children everywhere, boys and girls alike, will be able to complete a full course of primary schooling

3. Eliminating gender disparity in primary and secondary education and empower women.
- Target: Eliminate gender disparity in primary and secondary education no later than 2015

4. Reducing by two-thirds the mortality rate of children under 5.
- Target: Reduce by two-thirds, between 1990 and 2015, the under five mortality rate

5. Halving the proportion of people without access to safe drinking water.

6. Combat HIV/AIDS, malaria and other diseases. Halting by 2015 the spread of HIV/AIDS.

7. Improving by 50% the level of adult literacy.

These goals recognize explicitly the interdependence between growth, poverty reduction and sustainable development; acknowledge that development rests on the foundations of democratic governance, the rule of law, respect for human rights and peace and security; are based on time-bound and measurable targets accompanied by indicators for monitoring progress; and bring together the responsibilities of developing countries with those of developed countries, founded on a global partnership.

Since 2001, the UN Secretary General presented the Road Map towards the implementation of the UN Millennium Declaration. These are integrated and comprehensive overviews of the situation, outlines of the potential strategies for action designed to meet the goals and commitments of the Millennium Declaration.

The analysis in this section concentrates on the first four goals of the MDGs that also came to represent its central commitments. Each goal has a specific quantitative target, except for the first goal that has two separate targets, one for poverty and the other for food security.

However, a progress report submitted in 2005 to the UN Secretary General, entitled 'A Practical Plan to Achieve the Millennium Development Goals' was not very encouraging. The most disconcerting part was the progress report on the countries of SSA. As noted above, these countries became the core of the world poverty problem, and progress in reducing poverty and increasing food security in SSA is the most important target of the entire MDG programme. The eight MDGs are aimed to promote human development and reduce poverty, hunger and disease, even in the poorest countries. While some developing regions and countries are making progress in several of these goals, including food security, the situation in Africa remains stagnant and even worsening. The progress report emphasized that not much progress was achieved in the poorest African countries since 2000, and in many of them, food insecurity even worsened. This lack of progress casts a shadow on the capacity to achieve the MDGs in SSA, and, to some degree, it also casts a shadow on hopes that the entire plan will achieve its goals.

Although the goals are undisputedly commendable and there is a common agreement about their significance, there is a dispute about the theoretical model that is underlying the plan for achieving these goals and about the programme that was prepared on the basis of this model: The theory that is used for the analysis in the UN Millennium Project (2005) is based on the assumption that the poor countries are in a 'poverty trap'; the key for escaping that trap is to raise the capital stock in these economies to the point where their downward spiral would end and a self-sustaining economic growth would take over. In order to escape the poverty trap, it is therefore necessary to give these economies a 'Big Push' of large investments between now and 2015 in public administration, human capital (nutrition, health, education) and key infrastructure (roads, electricity, ports, water and sanitation, accessible land for affordable housing, environmental management).

The concept of a low-level poverty trap is a long-standing hypothesis in the theory of economic growth and development, and it was very influential in the 1950s. The basic argument of that theory is that a Big Push of foreign aid engineered by administrative planners will work to pull a poor economy out of the poverty trap to a self-sustained growth. According to Sachs, the role of foreign aid is to increase the capital stock enough to cross that threshold level with a Big Push: 'if the foreign assistance is substantial enough, and lasts long enough, the capital stock rises sufficiently to lift households above subsistence. . . . Growth becomes self-sustaining through household savings and public investments supported by taxation of households' (2005, p. 246). That Big Push involves a combination of: (i) a big increase in foreign aid; (ii) a simultaneous increase in investment in many different sectors, as well as a package of complementary policy changes and technical interventions; and (iii) a national plan and administrative apparatus to direct the investments, the technical interventions and the policy changes. The somewhat polemic and futuristic plan of Sachs *et al.* for the Big Push in SSA is described in Box 6.4.

In a detailed study that was already published 3 years ago, Sachs, who is the head of the UN Millennium Project, and other researchers who work on this project (Sachs *et al.*, 2004) called for a major investment push in the African countries, that is essential, according to their analysis, to give these countries a chance to meet the MDGs. Although these goals were adopted by the world's leaders in September 2000, and the UN Millennium Declaration committed their nations to stronger global efforts to, among other goals, reduce poverty and hunger, not much has happened since then, and the prospects of achieving these goals are practically evaporating.

The Millennium Project recommends a list of 449 technical interventions and investments that are arguably well attuned to local needs and conditions, including planting nitrogen-fixing leguminous trees, distributing bed-nets to protect against malaria, giving fertilizer to farmers, building health clinics, constructing battery-charging stations, giving each village a truck and a truck repairman, etc. In his 2005 book, *The End of Poverty*, Sachs emphasized that these interventions must be applied systematically, diligently and jointly since they strongly reinforce one another (p. 208). Given the time left to plan all these interventions in the different countries and make all the necessary preparations and arrangements to make the investments in the different sectors, and given the lack of proper institutions in the countries themselves to administer and supervise these interventions, it is doubtful

Box 6.4. The MDG for SSA. (From Sachs *et al.*, 2004, p. 121.)

'The standard diagnosis of Sub-Saharan Africa is that it is suffering from a governance crisis. This is too simplistic. Many parts of Africa are well governed considering the income levels and extent of poverty, yet are caught in a poverty trap. The region's development challenges are much deeper than "governance." Many countries require a big push in public investments to overcome the region's high transport costs, generally small markets, low-productivity agriculture, adverse agroclimatic conditions, high disease burden, and slow diffusion of technology from abroad.

An MDG-based strategy for SSA needs to focus on rural development for a 21st century African Green Revolution and strategies to make Africa's fast-growing cities much more productive, especially for labour-intensive exports. Africa's public health systems require major investments to address the pandemics of HIV/AIDS, tuberculosis, and malaria; to tackle the unconscionably high levels of child and maternal mortality; and to provide sexual and reproductive health services that will enable better timing and spacing of births and a voluntary reduction to desired family sizes. Education strategies need to focus on increasing the supply of infrastructure and human resources and the demand-side incentives for girls and vulnerable students. The continent also requires major investments in infrastructure for water resources management and energy.

Mobilization of science and regional integration also need to be energized. In all aspects of development, Africa's strategies need to pay special attention to the situation of girls and women, who tend to face major legal, social, and political barriers and biases.'

whether, in spite of all the good intentions, these interventions can indeed be implemented systematically enough to achieve self-sustained growth. The programme was supposed to have disbursed US$10 billion from 2003 to 2005, but it has so far disbursed almost nothing.

The Big Push theory is still influential, but it is not widely accepted. The Appendix provides more details on the theoretical model and shows why it is not very relevant for today's world. Hausmann, Pritchett and Rodrik (2004) studied 'growth accelerations' episodes and found that growth accelerations were more likely in the poorest countries than in the more affluent ones. Easterly (2005) found that the poorest countries, rather than being caught in a poverty trap, did in fact better in earlier periods when they were poorer than in later periods when their economies were more developed. Evidence to support the narrative is scarce. Poverty traps in the sense of zero growth for low-income countries are rejected by the data in most time periods. There is evidence of divergence between rich and poor nations in the long run, but this does not imply zero growth for the poor countries. The idea of the take-off does not garner much support in the data.

There are also more mundane reasons to be sceptical about the magic of the Big Push. Capital by itself would not increase production. There is a need to reorganize the entire production process and the supply chain, and that requires massive investments in infrastructure including roads, deep water ports, drinking water systems, electricity, trained labour force, etc. There is also a need for effective management and, most importantly, a coherent plan. In a discussion on the report of Sachs *et al.* after its presentation, Richard Cooper disagreed sharply with the authors' proposals for greatly expanding aid to SSA at this time, unless it could be ensured that the aid would be used effectively. He argued that the aid would be wasted in many countries in the region because they lacked the elementary functions of law and order that would be needed to deliver aid effectively.

To make the required investments, there is also a need for effective organization to plan, allocate the resources between sectors and between regions and supervise the implementation of all the plans. This organization must also guide the markets in these countries as long as they are not sufficiently developed in order to prevent monopolistic controls in carrying out the investment plans and in developing the countries' main sectors. The Appendix discusses in more detail the forms of central interventions that are necessary in certain sectors to ensure efficiency and sustainable growth. Can this be done? Is this process feasible? The target date of 2015, is, in all likelihood impossible. Can these goals be reached and can this programme be realized by 2025? Here the answer is cautiously more optimistic, but success depends on the capacity to draw the right lessons from the experience of previous structural adjustments in Africa and in other developing countries.

The reason for the optimism is that some progress has to be made because the lack of progress is damaging for all. The reason for the caution is not only because of the failure of most previous aid programmes in SSA and the fact that over the last 50 years these countries received more than US$1 trillion in aid and yet today many of them are poorer than they were 50 years ago. An equally if not more important reason is that since 2002, more than one-third of the African countries, including many of the poorer ones, already received a Big Push. These countries made huge profits from the rise in the price of oil and minerals that are much larger than the disbursements envisaged by the Millennium Development programme. And yet, these profits had practically no impact on the poor. Large amounts accumulated in the accounts of the very rich elite in foreign banks, and very small amounts were spent on programmes that have an impact on wider segments of the population.

The World Bank and the IMF issued a joint report on the MDGs in April 2005 that calls for 'a Big Push in aid'. The two Bretton-Woods institutions call upon the developing countries and the foreign aid donors to implement jointly the Big Push as follows:

> For coherence and effectiveness, the scaling up of development efforts at the country level must be guided by country-owned and -led poverty reduction strategies (PRSs) or equivalent national development strategies. Framed against a long-term development vision, these strategies should set medium-term targets – tailored to country circumstances – for progress toward the MDGs and related development outcomes. And they should define clear national plans and priorities for achieving those targets, linking policy agendas to medium-term fiscal frameworks. Donors should use these strategies as the basis for aligning and harmonizing assistance. Official development assistance (ODA) must at least double in the next five years the support to the MDGs, particularly in low-income countries and Sub-Saharan Africa
>
> (IMF and World Bank, 2005a,b, p. xviii)

Countries are now required to prepare a PRSP and all the administrative, planning and supervision measures listed above must be carried out by the country's own institutions. Although under these conditions the 'capital push' will be given only to countries that have good governance, the conditions are not likely to be a guarantee that the Big Push will be effective in the poor African countries where the institutions are barely functioning. The MDG progress report suggests the following additional details for the process:

> Our core operational recommendation is that each developing country with extreme poverty should adopt and implement a national development strategy ambitious enough to achieve the Goals. The country's international development partners – including bilateral donors, UN agencies, regional development banks, and the Bretton Woods institutions – should give all the support needed to implement the country's MDG-based poverty reduction strategy.

The Big Push that is now suggested will more than double the amount of aid that is currently given to the poor countries to an annual amount of US$75 billion. Given the lack of any progress in the MDGs thus far, a revised time schedule may have to be prepared. Given the experience with past aid programmes, and the fact that countries that received large amounts of aid have not grown any faster than countries that received hardly any aid, it is also doubtful that the Big Push will work even 10 years later unless additional measures are taken. Indeed, great scepticism that this plan would work was the reason why most donor countries have thus far refused to increase the amounts they donate for aid to 0.7% of their GNI, as they promised in 2000, and so far the average amount of aid given by the 22 rich countries is 0.25% of their GNI.

Although at the present, three-quarters of Africa's poor live in rural areas, mostly working in agriculture, the focus on increasing rural incomes and reducing chronic hunger would not provide the longer-term solution. The abject rural poverty and the rapid population growth accelerate rural–urban migration, but there is no employment in the urban centres and, as a result, urban poverty is high and rapidly rising. To reduce poverty, the African countries will have to build an industrial base and internationally competitive industries. The more difficult part of their structural adjustment is to create viable export industries, and that would require far more detailed planning and much more comprehensive structural adjustments.

The longer-term solution cannot therefore focus only on the countries' food supply or their health and educations services. It must be a much more comprehensive plan of structural adjustment that will include all sectors, all regions and all population groups. The wider objectives of these structural adjustments would have to include detailed plans to help these countries to build local industries so that they can take advantage of their natural resources and increase the value-added of their exports with more elaborate processing. With the rapid growth in their urban population, this would be their only way to provide adequate employment and raise the standard of living of the urban poor.

Without careful planning and proper supervision at all levels, including supervision over the country's central authorities themselves that prepare the plan and determine the allocation of the additional resources, there are indeed reasons to doubt whether the MDG programme will work. Supervision by the donor countries or the Bretton-Woods Institutions should not be excluded. This supervision will be an opportunity to transfer know-how and advanced technologies to the recipient countries, including both the entrepreneurs and the public administrations, and thus ensure that the best possible plan and the best possible strategy would be implemented. This process is bound to take more time, and the PRSP would have to contain more details on how to implement the various investment programmes.

Moreover, in the agricultural sector, the structural changes should include not only an increase in food supply, but also a diversification of their production so that they can export some of their agricultural products and thus increase their earnings. To be able to make these changes, they need large investments that are well beyond their present capacity, and they need technical support to build their industrial base, increase the productivity in their agricultural sector and use more advanced methods of production. In other words, to make all these changes, they need another round of structural adjustments.

The World Bank study 'Assessing Aid – What Works, What Doesn't, and Why' co-authored by David Dollar and Lant Pritchett (1998) recognizes that aid delivered as investment projects in particular sectors is highly fungible. Thus, making investments in high priority sectors such as primary health or education does not ensure that money is indeed used in these sectors. It cannot therefore be guaranteed that the new aid that will be funnelled to these countries would indeed be spent on the target projects. The MDG programme envisages that the money would be funnelled through the national governments; since this aid finance is highly fungible, how can they be sure that the money would indeed be spent on the target projects and not find its way to the pockets of the corrupt elite? The question is whether countries like Nigeria or the Central African Republic, that are rated in these studies as 'average' in terms of the quality of their governance, can indeed be trusted.

The World Bank (1998) study notes, however, that in some countries foreign aid was a spectacular success. It lists as examples Botswana and the Republic of Korea in the 1960s, Indonesia in the 1970s, Bolivia and Ghana in the late 1980s and Uganda and Vietnam in the 1990s. However, the report recognizes that in many other countries aid was an unmitigated failure. The report was written in 1998, and not all countries that were included in the report as examples of success would be included in that list had the report been written a decade or so later. Moreover, in today's world, when private capital and foreign investments are far larger that the amounts of aid that donor countries and the international development organizations can offer, well-governed countries like Botswana and Korea and even borderline countries like Vietnam, Indonesia and Ghana can and do mobilize much larger amounts in the world capital market.

In an article published in the *New York Times*, Jean-Claude Shanda Tonme, a columnist for the Cameroonian daily *Le Messager*, wrote:

> Neither debt relief nor huge amounts of food aid nor an invasion of experts will change anything. Those will merely prop up the continent's dictators. It's up to each nation to liberate itself and to help itself. When there is a problem in the United States, in Britain, in France, the citizens vote to change their leaders. And in those times when it wasn't possible to freely vote to change those leaders, the people revolted. In Africa, our leaders have led us into misery, and we need to rid ourselves of these cancers.
>
> (*NYT*, 15 July 2005)

This very harsh judgement should be kept in mind when more aid is given to Uganda – one of the few countries that everybody includes in the lists of well-governed countries – even though Yoweri Musaveni eliminated term limits so that he can remain president indefinitely.

The World Bank's *Assessing Aid* emphasized the two principal attributes that make an aid programme work:

- Financial aid works in a good policy environment;
- Improvements in economic institutions and policies are the key.

Unless these necessary conditions are met, the Big Push of aid that is suggested in the MDG programme is by no means certain to achieve the goal of alleviating poverty.

The conditions today are vastly different from those that prevailed in the 1980s when the LDCs sunk into the abyss of the debt crisis. At that time, their desperate need for credit forced them to accept the heavy strings that were attached to the adjustment loans as part of the SAPs, including the very complex conditionalities. Today, the African countries are not submerged in heavy debts, simply because the world's private financial institutions are much more cautious and avoid giving them large loans or making large investments.

Moreover, in many African countries, the governments now conduct more balanced and more responsible fiscal and monetary policies and avoid huge debts. However, their current resources are very small and they would need new loans and very significant support to be able to make the necessary investments in their economies and carry out the structural changes. The challenge for the donor countries and the international development organizations is not only to help the African countries to make these structural changes, but also to ensure that the investments and the structural changes are well planned and the entire programme is carried out according to these plans.

However, these structural adjustments cannot rely only on aid because the amounts that would be required are bound to be much larger than those envisaged by the UN MDGs that focus primarily on food security and the agricultural sector. They would need large investments and these investments can only be mobilized by the IMF and the World Bank. The aid of the donor countries can be used to subsidize the interest rate on these loans and possibly also to provide insurance for the loans, so that they can also be mobilized in the international financial markets.

The African countries that would want to participate in this programme would have to meet certain conditions and criteria, but the lessons of the past SAPs and their attached conditionalities must guide the donor community in determining the new conditions. Two important lessons from the past SAPs are particularly relevant in this context: First, both the structural changes and the conditions must be tailored according to the needs, conditions, priorities and constraints of the recipient countries themselves. Second, these countries must own the programmes by taking the main responsibility for their implementation and for the debt payments on account of these loans.

The quantitative goals specified in the MDGs can guide the next round of structural adjustments, but the time schedule would have to be adjusted according to the longer-term goals. It is already clear today that the majority of the African countries will not be able to achieve the MDGs by 2015.[13] In fact, if present trends continue, then quite a few African countries are likely to be worse off in 2015 than they are today, and some of them may even be worse off than they were in 1990.

In the new round of structural adjustments, the role of the international development organizations would be to help the countries in designing the

[13] Only 6 out of the 46 LDCs for which data are available met or exceeded the target of growth of 7% per annum between 2001 and 2004. As many as 18 out of the 46 LDCs for which data are available were unable to achieve per capita growth rates of more than 1.0% per annum during the period 2001–2004.

specific programmes for each of them and structure them according to the needs and constraints of each of the countries The guiding goals of these structural programmes would remain those specified in the MDGs, but under a different time schedule. It should be a major challenge for the entire international community to help mobilize the funds, the professional expertise, the technologies and all the other necessary resources. This wider programme should mobilize the support and secure the active participation of all countries, of international NGOs, of large corporations and of private donors to join forces and make sure that these goals will be achieved.

9 Concluding Remarks

The Appendix examines in more detail the theoretical foundations of the model that has been used to justify the need and potential contribution of the Big Push to enable the poor countries in SSA to make the transition to self-sustained growth. It also examines in detail the potential of these countries to build a solid industry that will provide employment to the urban population and offer these countries a jumping board for future growth, taking into account the conditions and constraints that prevail in these countries and in today's global economy. The reader of the Appendix will see that I did not find the theoretical model and the practical plan very convincing. I have read very carefully the three main publications of the MDG programme that are listed in the References, but I have not seen the list of 449 recommended projects of technical interventions and investments prepared for the Millennium Project, so I may not have the complete information to make that judgement. Nevertheless, while these projects may be well attuned to local needs and conditions, they still have to be selected and approved by the local authorities.

Although individual countries face many constraints, partly due to the small scale of their production and partly due to the long list of their rivals, there are today new opportunities for the African countries to take a new direction in order to accelerate their growth. Practically all the SSA countries, with the exception of South Africa, are too small to build a solid industrial base or increase their agricultural production beyond the local market and diversify their production in order to be able to export part of their output. Jointly they can achieve much more, and therefore they should have strong economic incentives to join forces and collaborate in the development of both their industries and their agriculture in RTAs.

References

Baldwin, R.A. (2003) Openness and growth: what's the empirical relationship? Working Paper 9578, National Bureau of Economic Research.

Besley, T. and Burgess, R. (2003) Halving global poverty. *Journal of Economic Perspectives* 17(3), 3–22.

Bhalla, S. (2002) *Imagine There Is No Country: Poverty, Inequality, and Growth in the Era of Globalization*. Institute for International Economics, Washington, DC.
Bigman, D., Dercon, S., Guillaume, D. and Lambotte, M. (2000) Community targeting for poverty reduction in Burkina Faso. *World Bank Economic Review* 14(1), 167–194.
Bruno, M., Ravallion, M. and Squire, L. (1998) Equity and growth in developing countries: old and new perspectives on the policy issues. In Tanzi, V. and Chu, K.-Y. (eds) *Income Distribution and High-Quality Growth*. MIT Press, Cambridge, Massachusetts.
Calderon, C.A. and Serven, L. (2004) The effects of infrastructure development on growth and income distribution. World Bank Policy Research Working Paper No. 3400, Washington, DC.
Collier, P. (2003) *Breaking the Conflict Trap: Civil War and Development Policy*. World Bank/Oxford University Press, Washington, DC.
Collier, P. and Gunning, J.W. (1999) Why has Africa grown slowly. *Journal of Economic Perspectives* 13(3), 71–72.
de Soto, H. (2000) *The Mystery of Capitalism*. Basic Books, New York.
Dollar, D. and Pritchett, L. (1998) *Assessing Aid, World Bank*. Aid Effectiveness research.
Dollar, D. and Kraay, A. (2002) Growth is good for the poor. *Journal of Economic Growth* 7(3), 195–225.
Dollar, D. and Shang-Jin, W. (2007) *Das (Wasted) Kapital: Firm Ownership and Investment Efficiency in China*. IMF Working Paper, International Monetary Fund, Washington, DC.
Easterly, W. (2005) *Reliving the 50s: The Big Push, Poverty Traps, and Takeoffs in Economic Development*. Center for Global Development, Working Paper Number 65, August.
Economic Commission for Africa (2005) Economic and Welfare Impacts of the EU–Africa Economic Partnership Agreements.
Eifert, B. and Ramachandran, V. (2004) *Competitiveness and Private Sector Development in Africa: Cross-Country Evidence from the World Bank's Investment Climate Data*. World Bank, Washington, DC.
Eifert, B., Gelb, A. and Ramachandran, V. (2005) *Business Environment and Comparative Advantage in Africa*. Working Paper No. 56, Center for Global Development, World Bank, Washington, DC.
FAO (UN Food and Agriculture Organization) (2003) *The State of Food Insecurity in the World (SOFI)*. Economic and Social Department, FAO.
FAO (UN Food and Agriculture Organization) (2004) *The State of Food Insecurity in the World (SOFI)*. Economic and Social Department, FAO.
FAO (UN Food and Agriculture Organization) (2006) *The State of Food Insecurity in the World (SOFI)*. Economic and Social Department, FAO.
FAO (2007) *State of World Food Security*. Food and Agriculture Organization, Rome, Italy.
Hausmann, R., Pritchett, L. and Rodrik, D. (2004) *Growth Accelerations*. John F. Kennedy School of Government, Harvard University, Cambridge, Massachusetts (revised).
Hewitt, A. and Gillson, I. (2003): *A Review of the Trade and Poverty Content in PRSPs and Loan-Related Documents*. ODI, London.
Hinkle, L.E. and Schiff, M. (2004) Economic partnership agreements between sub-Saharan Africa and the EU: A development perspective. *The World Economy* 27, 1321–1333.
IFPRI (2005) *Food Security Outlook in Africa to 2025*.
IFPRI (2006) Strategic Priorities for Agricultural Development in Eastern and Central Africa.
IMF (2001) *World Economic Outlook*. Various years.
IMF and World Bank (2005a) *Global Monitoring Report 2005: Millennium Development Goals: from Consensus to Momentum*. IMF/World Bank, Washington, DC.
IMF and World Bank (2005b): *Concept Note. Joint World Bank and IMF Report on PRSPs – 2005 PRS Review*. IMF/World Bank, Washington, DC.
IMF *Direction of Trade Statistics* (DOTS) (2006). International Monetary Fund, various years.
Kaplinsky, R. and Morris, M. (2006) The Asian Drivers and SSA: MFA Quota Removal and the Portents for African Industrialization. Paper Prepared for a Conference on Asian and Other Drivers of Change. St Petersburg, 18–19 January.

Krueger, A.O. (1999) Are preferential trading arrangements trade-liberalizing or protectionist? *Journal of Economic Perspectives* 13(4).

Limao, N. and Venables, A.J. (2000) Infrastructure, geographical disadvantage, transport cost and trade. *World Bank Economic Review* 15, 451–479.

Mayda, A.M. and Steinberg, C. (2007) *Do South–South Trade Agreements Increase Trade?* Commodity-Level Evidence from COMESA, IMF Working Paper, February.

National Intelligence Council (2005) *Mapping Sub-Saharan Africa's Future*. Washington, DC.

Ng, F. and Yeats, A.J. (2002) What can Africa expect from its traditional exports? Africa Region Working Paper No. 26, World Bank, Washington, DC.

Pattillo, C., Gupta, S. and Carey, K. (2005) Sustaining growth accelerations and pro-poor growth in Africa. IMF Working Paper, October.

Radelet, S. and Sachs, J. (1998) The East Asian financial crisis: diagnosis, remedies, prospects. *Brookings Papers on Economic Activity* 1, 1–74.

Ravallion, M. (2001) *Growth, Inequality and Poverty: Looking Beyond Averages*. Development Research Group, World Bank, Washington, DC.

Rodrik, D. and Subramanian, A. (2004) From 'Hindu Growth' to productivity surge: the mystery of the Indian growth transition. IMF Working Paper No. 04/77, International Monetary Fund, Washington, DC.

Sachs, J.D. (2005) *The End of Poverty: Economic Possibilities for Our Time*. Penguin Press, New York.

Sachs, J.D., McArthur, J.W., Schmidt-Traub, G., Kruk, M., Bahadur, C., Faye, M. and McCord, G. (2004) Ending Africa's poverty trap. *Brookings Papers on Economic Activity* 1, 117–216.

Sachs, J. and Warner, A. (1995) Economic reform and the process of global integration. *Brookings Papers on Economic Activity* 1, 1–118.

Schiff, M. and Winters, L.A. (2003) *Regional Integration and Development* World Bank/Oxford University Press, Washington, DC.

Starkey, P., Ellis, S., Hine, J. and Ternell, A. (2002) *Improving Rural Mobility: Options for Developing Motorized and Nonmotorized Transport in Rural Areas*. World Bank, August. Washington, DC.

Stevens C. (2005) *Economic Partnership Agreements and African Integration: A Help or a Hindrance?* Institute of Development Studies, University of Sussex, Brighton, UK.

Subramanian, A. and Tamirisa, N. (2001) *Africa's Trade Revisited*. IMF Working Paper No. 01/33, International Monetary Fund, Washington, DC.

UN Millennium Project (2005) *Investing in Development: A Practical Plan to Achieve the Millennium Development Goals*. UN Millennium Project, New York.

UNCTAD (2004) *Economic Development in Africa. Debt Sustainability: Oasis or Mirage?* Geneva.

World Bank (1998) *Assessing Aid – What Works, What Doesn't, and Why*. World Bank, Washington, DC.

World Bank (2000) *Can Africa Claim the 21st Century?* A report prepared by the World Bank, in collaboration with the African Development Bank, UN Economic Commission for Africa, Global Coalition for Africa and African Economic Research Consortium, Washington, DC.

World Bank (2001) *World Development Report*, World Bank 2000.

World Bank (2004) *International Economics and Trade in East Asia and Pacific*. International Monetary Fund, Washington, DC.

World Bank (2006) *Doing Business 2007: How to Reform*. Washington, DC.

World Bank (2007) *Global Economic Prospects 2007: Managing the Next Wave of Globalization*. Washington, DC.

World Bank and International Finance Corporation (2006) *Doing Business in 2006: Creating Jobs*. Washington, DC.

World Trade Organization (2003) Annual Report 2003.

Yang, Y. and Gupta, S. (2005) *Regional Trade Arrangements in Africa*. Special Issues Paper, International Monetary Fund, Washington, DC.

Appendix: The Saving Trap and the Big Push: The Theory and Its Traps

The starting point of the model is the standard neoclassical growth model in which output per capita q is produced by a production function $Af(k)$, where A is TFP and k is the capital to labour ratio. The national saving rate is s, and d is the rate of capital depreciation, with n denoting the rate of population growth. The rate of capital accumulation $\mathrm{d}k/\mathrm{d}t$ is given by

$$\mathrm{d}k/\mathrm{d}t = sAf(k) - (n + d)k \tag{6.1}$$

A change in the capital to labour ratio is called capital deepening. The term $(n + d)k$ is called capital widening and is equal to the amount of saving per capita that is needed to hold the capital to labour ratio constant in the face of population growth and depreciation. Equation 6.1 says that capital deepening equals saving per capita minus capital widening. The economy grows in per capita terms as long as saving per capita exceeds capital widening. If saving is lower than capital widening, the economy experiences a decline in output per capita. The standard neoclassical model is typically presented as if the economy *necessarily* grows when k is very low.

Actually, when k is very low, two other things tend to be true. First, the marginal productivity of capital also tends to be very low because a minimum threshold of capital is needed before modern production processes can be started. However, once the basic infrastructure and human capital are in place, the marginal productivity of capital may indeed become very high in a low-income country. The capital threshold is denoted in Fig. 6.12 as k_T. Small increments of k below a threshold k_T might do little to raise production,

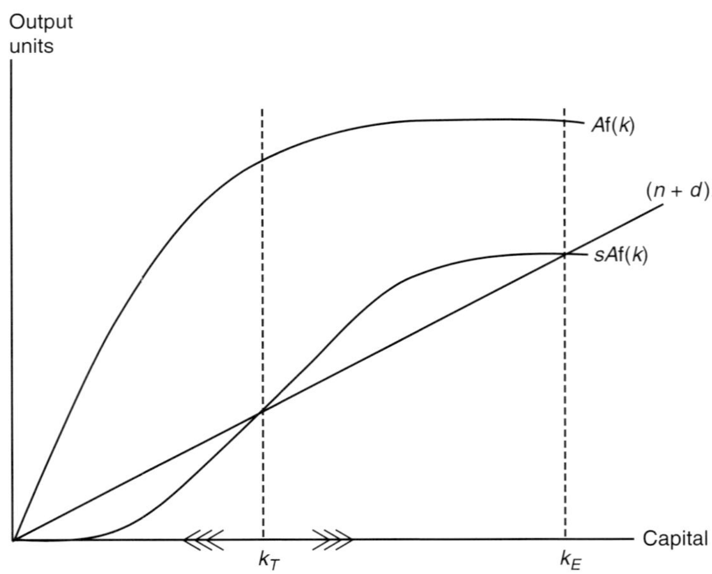

Fig. 6.12. The saving trap and the Big Push.

and the $sAf(k)$ curve might be very flat near the origin. It then becomes steep in a middle range before flattening out once again at high levels of k. In short, there is an early period of increasing returns to scale in capital before the more traditional constant or decreasing returns to scale set in.

Second, the savings rate can become very low or even negative when k is very low, because impoverished households do not save, but rather must use all (or more than all) of their current income in the struggle just to stay alive. Once basic needs are met, poor households may save quite a lot of income, but not before. The saving trap shown in Fig. 6.12. $sAf(k)$ is less steep than $(n + d)k$ when k is very low, and then turns steeper in an intermediate range. Hence, dk/dt is negative when $k < k_T$. When an economy begins with very low capital, both the capital to labour ratio and output per capita tend to *decline* over time. The very poor get poorer, pushed into more extreme poverty by the lack of capital accumulation coupled with population growth. Only when an economy has a capital to labour ratio above a minimum threshold k_T does it tend to achieve economic growth and converge to the steady-state k_E and q_E.

There are several potential traps in this model that raise questions about its relevance for the analysis of the impact of a Big Push of aid on accelerating growth in a poor country. The main objective of the remainder of this Appendix is to draw attention to these potential pitfalls in this description of the Big Push. The Appendix describes the factors that are likely to inhibit growth in the African countries despite the Big Push. The main emphasis is on the industrial sector and, as an illustration, the Appendix examines the potential to develop textile and apparel industries in the African countries and the constraints that the development of these industries may face – both because of developments in the structure of production and trade of these industries in the global economy and because of internal constraints in the African countries themselves.

The main problem of the model that is used to describe the poverty trap and the potential impact of the Big Push is that this is a one-sector model in a closed economy. There is no trade, no capital movements to and from other countries and [no] option to make a choice of the sector in which the country has a comparative advantage. A Big Push of capital gives local enterprises the resources to make investments in infrastructure and machinery in order to build an industrial base, and gives the government the resources to make the necessary public investments.

In a one-sector model, it is not possible to show the criteria for making these investments between sectors and ensuring that these investments are made in the sector(s) in which the economy has a comparative advantage. When the local industries increase their production, in an open economy they must also make a choice of what segments of production of the final product should be retained in the home country and what segments should be outsourced to other countries. A poor economy is likely to have a comparative advantage in labour-intensive industries, but when the economy does not yet have the basic infrastructure, including roads, deep water ports, electricity, drinking water, etc., its total production costs even in labour-intensive industries can be higher that in other countries where wages are higher. The purpose of the Big Push is to make these investments in the country's infrastructure, and then to build the country's industrial base.

The poor, labour-abundant country is likely to be more competitive in the world markets in labour-intensive industries that require mainly unskilled labour. The most common examples are textile and apparel that were the basis for establishing an industrial base in practically all countries. Can the Big Push help the poor countries of SSA to build textile and apparel industries that will be competitive in the world market by giving them the resources to make the necessary investments in infrastructure, train their labour force or establish efficient institutions? Are the conditions in today's global economy amenable enough so that these countries can join the global economy and provide employment to their urban populations?

Despite the very low wages and abundance of unskilled workers, and despite their very favourable agroclimatic conditions, the capacity of most SSA countries to compete in the world market is likely to remain very limited even with trade liberalization. One reason is their low productivity (Table 6.5; Fig. 6.13). This is one reason why countries in West and Central Africa that produce cotton at very low prices do not manage to proceed to more advanced stages of production and add much value and local employment to their exports due to their difficulties to develop competitive industries that can use this cheap cotton.

The textile sector requires relatively cheap inputs and the African producers have an advantage due to their access to cheap cotton that is produced in the Francophone countries. However, they must be able to compete with the leading producers in the global textile market, including China, the USA, the EU, India, Pakistan, Japan, the Republic of Korea and Turkey. In recent years, the main increase in world production and exports has been in the EA economies. The elimination of the quota import restrictions at the end of 2004 under the WTO ATC has greatly affected the textile industry, and in the coming years the share of world exports of textiles from these economies is likely to grow much faster.

The African countries that still produce textile for exports benefit from AGOA's third-country fabric sourcing allowance, which permits them to source raw material inputs from non-AGOA countries. This arrangement is slated to expire in 2007, and it will increase the need of the African producers to source inputs from local suppliers and greatly reduce their competitiveness. African textile exports under AGOA totalled US$1.4 billion in 2005, and this was already a decline of 12% from the previous year, due to increased global competition, with the elimination of the quotas under the MFA.

Even with the Big Push and the large investments in infrastructure, the African producers are not likely to be able to compete with other countries that

Table 6.5. Median value of annual wages and benefits per worker (US$).

China	India	Morocco	Bolivia	Eritrea	Ethiopia	Kenya	Mozambique	Nigeria	Uganda
1170	850	2400	790	745	310	860	675	840	415

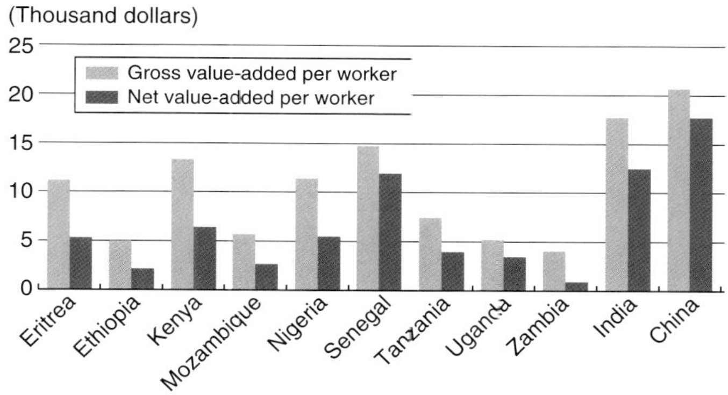

Fig. 6.13. High indirect cost in production in SSA. (From Eifert *et al.*, 2005.)

also have low labour costs, such as China, India, Bangladesh and Pakistan. Although labour costs in Africa are generally as low or even lower than those in China or India, the main factors that reduce the competitiveness of Africa in textile production are high utility costs, particularly for electricity and water, high transport costs, inefficient administration and poor financial systems and various other factors that reduce the supply reliability and prolong the delivery time of local producers. A Big Push is not enough to overcome these obstacles, and it may take years to build and improve the local institutions.

In 1995, when the ACT was reached, it was believed that the African countries would be the main beneficiaries of the agreement due to their access to cheap cotton and their close connections to the EU countries. Towards the end of the Agreement in 2004, it was clear that most textile production will move to the Asian countries and for several years, Africa had little success in attracting FDI into the textile sector; the largest share of foreign investments in the textile sector in Africa was due to the advantages it had under the preferential access to the main large markets, including the AGOA.

The World Bank survey *Doing Business* (2007) indicates that for investors and potential investors in Africa, the business environment is influenced by the nexus of policies, institutions, physical constraints and geographical conditions. Poor and erratic policies, unstable political conditions and weak private and public institutions are their main constraints (Eifert and Ramachandran, 2004). In addition, African firms incur heavy costs for transport, logistics, telecom, water, electricity, buildings, marketing, accounting, security, bribes and so forth.

In the future, investors in the African textile industries may benefit from the large and still developing regional trade zones and the possibility that regional trade agreements offer to strengthen the cotton-textile-apparel value chain, particularly in regional export processing zones (EPZs) that a number of SSA governments are trying to develop. The threat to the local industry is not only the lower input and transportation costs of textile producers in China and India that enable them to supply apparel firms worldwide with inexpensive fabric through highly developed supply chains that establish direct connection between producers and consumers, but also the lack of modern technology and the deteriorating equipment.

In the apparel industry, the potential of the African producers to develop a competitive industry for the export markets is even lower. African producers could still benefit from a number of PTAs such as AGOA and the EU's Lomé Accord, but local producers lag behind producers in other developing countries, primarily in Asia, due to poor transportation and communications infrastructure that are essential for an effective operation of apparel production, and the cotton–textile–apparel value chain that has been developed in these other countries during the last three decades and managed to acquire a monopolistic position in this sector. Apparel exporters require ready and rapid access to inputs and global markets, streamlined customs procedures, reliable transport infrastructure and supportive financial institutions that provide credit at competitive rates. Africa is lacking all that, and a Big Push cannot fill that void. A number of African countries are making an effort to develop EPZs, and local firms have already benefited by establishing themselves in these zones in Mauritius and Madagascar. These are only initial steps, and it may take some time until an efficient production system can operate.

Another direction for the development of a local industry is based on the use of the local resources, perhaps in processing and other activities that will increase the value-added of exports. These developments can go in two directions: First, processing agricultural products and, more generally, developing the production and exports of agricultural products in which Africa has a considerable comparative advantage due to very favourable agroclimatic conditions that are unique to this region and due to the low wages of rural workers.

Low public investments in scientific research and extension services in the agricultural sector in Africa, combined with poor rural infrastructure and limited trade, also had a negative impact on the use of fertilizers and pesticides; limited investments in irrigation and other advanced technologies further lowered the productivity of African farmers and most of them produced primarily for their own consumption. Only a small fraction of them had access to the high-yielding varieties (HYVs) that triggered the 'Green Revolution' in Asia and LA.

Another obstacle that farmers now face when they export their products is the need to meet high quality standards of food safety, and most small farmers and local traders in the LDCs do not have the knowledge, the necessary inputs and the facilities to be able to meet these standards or the packaging requirements. These constraints limit the ability of local farmers to diversify their production in order to increase their exports and their income. This issue is discussed in more detail in the context of the trade rules established by the WTO.

A promising strategy of diversifying agricultural production in many SSA countries is the production and processing of vegetables and fruits in which their comparative advantage is potentially the highest due to the abundance of low-wage workers, small-scale production, suitable agroclimatic conditions and the availability of these unique crops in their countries. However, the advanced methods of production and the variety and quality of the fruits and vegetables that European consumers demand require local producers and wholesalers to tailor their entire production and supply system to the current structure of the global market and to use accurate and timely market information in order to meet that demand 'on time'. The rudimentary production and delivery methods that most small African farmers and traders are using prevent them from competing in the global market, and these farmers would need extensive assistance by local and international research and extension services in order to overcome their disadvantages. It may take several years to develop the technologies that are most suitable for the local conditions and provide training to the local farmers so that they can meet these conditions.

The development of industries that use local minerals and other natural resources may be even more difficult. For one, these industries are highly capital–intensive and require skilled labour; for the other, these industries have a very low value-added and will provide employment only for a very small number of workers.

Finally, even if the Big Push is big enough to create some local industries and increase employment and income, local producers still have the option of outsourcing certain segments of production that may be relatively more expensive to produce in the home country despite the low wages. This may be due to high local transport costs, the unreliable supply of electricity, the high costs of skilled labour, etc. With this, they will put downward pressure on the local wages of unskilled workers, and the main beneficiaries from the Big Push may well be the local entrepreneurs. The skewed distribution of income will have two effects: First, it may not increase by much the demand for locally produced goods because the rich many prefer to buy relatively more imported goods. Second, although the marginal propensity to save of the rich is relatively higher, the rich have the option of saving their money in foreign banks or in foreign investment companies, where savings are less risky and/or yields are higher. The combined effect of all these factors is that in an open economy, the process of self-sustained growth that is triggered by the Big Push is likely to be much smaller than that described in Fig. 6.12 in a closed economy and it may not be sufficient to secure the reduction in poverty and malnutrition and close the gap in order to reach the MDGs (Fig. 6.14).

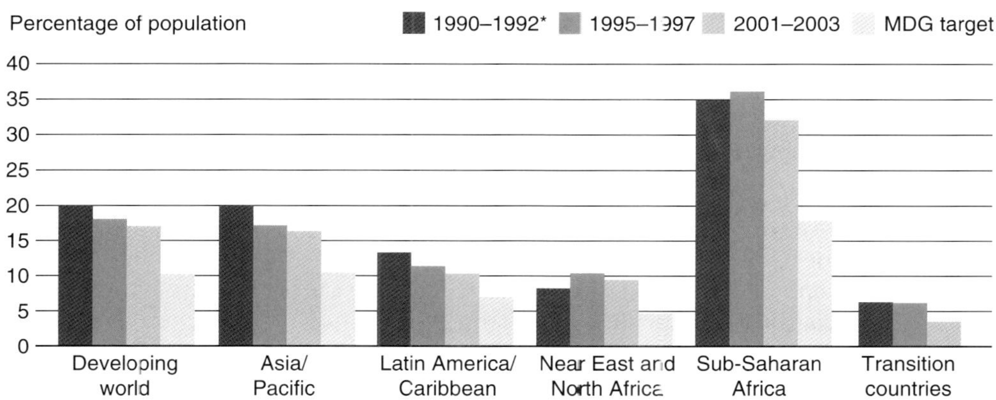

* For the transition countries: 1993–1995.

Fig. 6.14. Proportion of the undernourished people since 1990 and in the millennium development goals. (From FAO, 2007.)

Concluding Observations

At the time of writing this concluding section, the failure to reach an agreement at the Doha Round seems to be a fait accompli. Since the end of the US President's 'fast track' authority is also approaching, and since the next administration will, in all likelihood, be controlled by the Democratic Party that is much less keen to open the US economy to free trade, the failure of the Doha Round is likely to postpone the next round of negotiations on a multilateral trade agreement for quite some time. The repercussions can therefore be far-reaching. The most immediate effect will be the failure to reach an agricultural trade agreement. However, the losers will not only be the developing countries, but also the two large trading powers, the USA and the EU, even though their failure to reach an agreement on mutual reductions of agricultural subsidies was the reason for the collapse of the negotiations. This is the classical case of the 'Prisoner's Dilemma', when a failure to reach a compromise leaves both prisoners in jail. The two trading blocks are losing from this failure because the subsidies they currently give to their farmers involve huge expenses that benefit a tiny minority of their populations, and both the USA and the EU have better and far less costly alternatives to compensate their farmers if only they could reach a collaborative agreement to jointly reduce these subsidies. Moreover, on top of the direct expenses, this policy also raises the prices of agricultural products for the general population.

The wider and longer-term effect of the failure to reach a multilateral agreement at the Doha Round is the failure to reach a more comprehensive agreement to also reduce the high tariffs on a long list of manufacturing goods that are still imposed in many countries, but primarily in China and India. The main losers are producers and exporters of manufacturing goods in the EU and the USA who, with the current high tariffs, are not able to export their products to these large and rapidly growing markets. An extreme example is the high tariffs on cars in India that prohibit the exports of European cars to this expanding market and leave European producers no choice but to produce their cars in India, even though their current production lines in Europe are working below normal capacity.

The emerging economies themselves are also losing from the failure to reach an agreement, because by keeping these high tariffs, in part as a bargaining chip for any future agreement, they protect some of their inefficient industries. Perhaps more important is the fact that the failure to reach an agricultural trade agreement is highly damaging to some of the emerging economies themselves, most notably to Brazil, that stood to gain handsomely from an agricultural trade agreement that would have enabled them to export their agricultural products to the developed countries. The more damaging effect is on the LDCs that, at least nominally, were supposed to be the main beneficiaries from an agreement in what was labelled the Development Round, which had the declared goal of opening large markets for those agricultural products that are the main exports of the LDCs.

In the longer run, the most damaging effect is the failure to continue and to further extend the collaborative framework of the trade agreements that was developed during nearly six decades of arduous efforts under the General Agreement on Tariffs and Trade (GATT) and the WTO. This failure may lead to a fragmentation of the global trade system into regional and competing trade agreements, and will clearly weaken the international organizations, primarily the WTO, the IMF and the World Bank, and erode many of the achievements of the multilateral collaboration, including agreements on more controversial issues like intellectual property rights and food safety standards that were beneficial mostly to the developed countries, but required the collaboration of all countries.

The LDCs may suffer the greatest damages from the fragmentation of the multilateral agreements and the erosion of the multinational institutions that devoted great efforts and resources to help them combat poverty. The LDCs may also lose from the reduction in the preferences they received under the multilateral agreements and that reflected the commitment of the international community to help the poor countries. The following two examples illustrate these preferences:

1. The WTO's trade rules do not require duty-free trade and permit developing countries to maintain higher tariff barriers than developed nations.
2. In the multilateral trade agreement, the LDCs had the right to maintain temporary protection for selected industries or infant industries in the process of trade liberalization and to reduce their tariffs only gradually in order to ensure a soft landing for otherwise potentially successful domestic enterprises and avoid rapid and abrupt changes that can damage these countries' industrialization.

The strongest commitment to help the LDCs was made in 2000, in the Millennium Development Goals (MDGs), by all UN members who pledged to allocate more resources to help these countries to help their poor. It is now clear that the efforts to even achieve the first and most important goal of halving between 1990 and 2015 the proportion of people that suffer from hunger and chronic poverty will not succeed in the majority of the LDCs.

Perhaps a greater danger for the LDCs and for many other low-income developing countries is the potential impact of the regional trade agreements that may become the dominant structures in which world trade will be organized, because these countries may not be included in most agreements as equal members. This is because these agreements are motivated by the mutual gains that the member countries have from participating in these agreements, and quid pro quo is the main driving force for their willingness to grant tariff concessions to the other members in the agreement. The LDCs cannot offer much in exchange: their markets are too small and their purchasing power is too weak. Some developing countries have primary resources, particularly oil, that they may be able to use in these negotiations; most other LDCs have very little to offer.

As a result of the proliferation of regional trade agreements, Africa may be bypassed by the world trade system, have very limited opportunities to diversify its exports, would have to continue to concentrate on the exports of primary commodities and have very limited opportunities to develop viable industries. The flow of foreign direct investment (FDI) would therefore remain only a trickle and concentrate in the extractive oil and mineral sectors.

Moreover, since the African countries are too small and their economies too undeveloped to provide substantial economies of scale to support profitable industries, trade among the SSA countries (Africa-to-Africa or South-to-South trade) accounts for only 12% of these countries' exports, and the established regional agreements did not contribute much to increase that trade, since they share very similar structures of production. As a consequence, Africa's regional arrangements have the lowest levels of recorded intraregional trade of all such integration schemes in the world; the share of intraregional exports in the existing agreements is typically below 5%, and most exports concentrate in South Africa and Kenya. Nevertheless, most African countries have recognized the benefits of regional integration and they already participate in at least one regional organization or trade agreement, primarily in order to increase their negotiation power vis-à-vis other trade agreements.

The next wave of globalization may therefore accelerate the differentiation among African countries: the well-performing countries with relatively high governance standards and well-functioning economies will be able to access the buoyant international economy and benefit from the rapidly changing technologies. Many African countries will not share that growth. Thus far, Africa has been the continent least positively affected by globalization, and the difficulties of African countries to be integrated into and take advantage of the accelerating growth of the global economy have been exacerbated by the fierce competition in world production and trade that limit their capacity to develop competitive industries and increase the value added of their exports. The differentiation among the African countries and the deepening income gap between them may therefore be only part of a much more extreme process of rising income inequalities, growing disparities between the LDCs and most other world economies.

The World Bank study on the next wave of globalization estimates that in the next 25 years until 2030, globalization would spur faster growth in average incomes than during 1980–2005. In the next wave, the developing countries as a group will play a more central role, but the majority of the developing countries and nearly all the LDCs will not benefit from that process. The sharp rise in income inequalities between these two subgroups of developing countries is therefore likely to be the most worrying process in the next round of globalization. According to the World Bank estimates, the entire group of developing countries, that only two decades ago provided 14% of manufactured imports of the rich countries and in 2005 supplied 40%, is likely to supply over 65% by 2030. Moreover, by 2030, only 1.2 billion people in developing countries – around 15% of the world population and around a quarter of the population in the developing countries – will share that rise and join the 'global middle class', – up from 400 million today, but less than a third of the rise in the total population of the developing countries.

Amartya Sen wrote in 2001: 'Even though the world is incomparably richer than ever before, ours is also a world of extraordinary deprivation and staggering inequality.' The World Bank estimates that by 2030, inequality will reach far more extreme dimensions. These alarming inequalities and the incomprehensible depth of deprivation in the LDCs are bound to have very significant political and social implications that, unless drastic measures are promptly taken to reduce their dimensions, may well have a highly destructive impact on the global community.

Today, for many millions of people, globalization is like manna from heaven. After going through very difficult and trying times during the 1950s and 1960s, and wandering in the wilderness of the 'Great Leap Forward' with no hope of ever having a better life, the opportunities brought by the growth process made possible by globalization gave them a new horizon to look forward to and march towards. Their parents were enslaved by the great wretchedness of poverty and hunger, and they also had to overcome great difficulties, but now they have a decent life and well-rewarding work and, most important, their children are studying and gaining new skills that are necessary for the new world; for their next generation, the sky is the limit.

For many other millions of people in the developing countries, this cannot be the manna from heaven. At best, this is a genetically modified manna produced by the transnational corporations to exploit their work. Many of them manage to nourish their family, but have great worries regarding what their future holds, and they are angry about the lack of social and economic justice that deprives them of their well-deserved share in the global affluence. Many others, mostly the people in the LDCs, are left malnourished and are barely surviving amidst the global affluence.

The failure to include the needs of these people among the goods that the global economy is producing is the most malignant epidemic of globalization. Undoubtedly, the facts that nearly 1 billion people do not have enough food, cannot provide the minimum health to their families and the minimum education to their children, are the extreme symptoms of this epidemic. The goods that globalization provides should include not only consumption goods that those who can afford them are enjoying, and the public goods that the global community needs, but also the social good, social justice and basic human values that we all need.

Index

Notations

- Concepts consisting from two words or more are usually included under one of the two words, with an added notation such as (*see* and *see also*).
- With the notation (*n*) the page numbers refers to a note in the text.
- Some of the key concepts are mentioned many times in the book. The list in the index does not include the citations and this is indicated by the notation (*partial list*).